SCOPE 31

Lead, Mercury, Cadmium and Arsenic in the Environment

SCOPE 31

Lead, Mercury, Cadmium and Arsenic in the Environment

Edited by
T. C. Hutchinson

and

K. M. Meema
Institute for Environmental Studies
University of Toronto, Toronto, Ontario, Canada

Published on behalf of the
Scientific Committee on Problems of the Environment (SCOPE)
of the
International Council of Scientific Unions (ICSU)
by
JOHN WILEY & SONS
Chichester · New York · Brisbane · Toronto · Singapore

Library of Congress Cataloging-in-Publication Data:

Lead, Mercury, Cadmium and Arsenic in the Environment.
(SCOPE : 31)
Includes Index.
1. Lead—Environmental Aspects. 2. Mercury—Environmental
Aspects. 3. Cadmium—Environmental Aspects. 4. Arsenic—Environmental
Aspects. 5. Environmental Chemistry. I. Hutchinson, T. C.
(Thomas C.), 1939- . II. Meema, K. M. III. International Council of
Scientific Unions. Scientific Committee on Problems of the Environmment.
IV. Series: SCOPE (Series); 31.
TD196.L4023 363.7'384 86-9199
ISBN 0 471 91126 7

British Library Cataloguing in Publication Data:

Lead, Mercury, Cadmium and Arsenic in the Environment.
—(SCOPE : 31)
1. Pollution
I. Hutchinson, Thomas C. II. Meema, K. M.
III. International Council of Scientific Unions
Scientific Committee on Problems of the Environment
IV. Series
363.7'384 TD196.C45
ISBN 0 471 91126 7

Printed and bound in Great Britain

SCOPE 1: Global Environmental Monitoring 1971, 68 pp (out of print)

SCOPE 2: Man-Made Lakes as Modified Ecosystems, 1972, 76 pp (out of print)

SCOPE 3: Global Environmental Monitoring Systems (GEMS): Action Plan for Phase 1, 1973, 132 pp (out of print)

SCOPE 4: Environmental Sciences in Developing Countries, 1974, 72 pp (out of print)

Environment and Development, proceedings of SCOPE/UNEP Symposium on Environmental Sciences in Developing Countries, Nairobi, February 11–23, 1974, 418 pp (out of print).

SCOPE 5: Environmental Impact Assessment: Principles and Procedures, Second Edition, 1979, 208 pp

SCOPE 6: Environmental Pollutants: Selected Analytical Methods, 1975, 277 pp (available from Butterworth & Co. (Publishers) Ltd, Sevenoaks, Kent, England (out of print)

SCOPE 7: Nitrogen, Phosphorus and Sulphur: Global Cycles, 1975, 192 pp (available from Dr Thomas Rosswall, Swedish Natural Science Research Council, Stockholm, Sweden) (out of print)

SCOPE 8: Risk Assessment of Environmental Hazard, 1978, 132 pp

SCOPE 9: Simulation Modelling of Environmental Problems, 1978, 128 pp (out of print)

SCOPE 10: Environmental Issues, 1977, 242 pp

SCOPE 11: Shelter Provision in Developing Countries, 1978, 112 pp

SCOPE 12: Principles of Ecotoxicology, 1978, 372 pp

SCOPE 13: The Global Carbon Cycle, 1979, 491 pp

SCOPE 14: Saharan Dust: Mobilization, Transport, Deposition, 1979, 320 pp

SCOPE 15: Environmental Risk Assessment, 1980, 176 pp

SCOPE 16: Carbon Cycle Modelling, 1981, 404 pp

SCOPE 17: Some Perspectives of the Major Biogeochemical Cycles, 1981, 175 pp

SCOPE 18: The Role of Fire in Northern Circumpolar Ecosystems, 1983, 344 pp

SCOPE 19: The Global Biogeochemical Sulphur Cycle, 1983, 495 pp

SCOPE 20: Methods for Assessing the Effects of Chemicals on Reproductive Functions, 1983, 568 pp

SCOPE 21: The Major Biogeochemical Cycles and Their Interactions, 1983, 554 pp

SCOPE 22: Effects of Pollutants at the Ecosystem Level, 1984, 443 pp

SCOPE 23: The Role of Terrestrial Vegetation in the Global Carbon Cycle: Measurement by Remote Sensing, 1984, 272 pp

SCOPE 24: Noise Pollution, 1986

SCOPE 25: Appraisal of Tests to Predict the Environmental Behaviour of Chemicals, 1985, 400 pp

SCOPE 26: Methods for Estimating Risks of Chemical Injury: Human and Non-human Biota, 1985, 712 pp

SCOPE 27: Climate Impact Assessment: Studies of the Interaction of Climate and Society, 1985, 649 pp

SCOPE 28: Environmental Consequences of Nuclear War
Volume I Physical, 1986, 342 pp
Volume II Ecological and Agricultural Effects, 1985, 523 pp

SCOPE 29: The Greenhouse Effect, Climatic Change, and Ecosystems, 1986, 576 pp

SCOPE 30: Methods for Assessing the Effects of Mixtures of Chemicals, 1987, 928 pp

SCOPE 31: Occurrence and Pathways of Lead, Mercury, Cadmium and Arsenic in the Environment, 1987, 384 pp

Funds to meet SCOPE expenses are provided by contributions from SCOPE National Committees, an annual subvention from ICSU (and through ICSU, from UNESCO), an annual subvention from the French Ministère de l'Environment et du Cadre de Vie, contracts with UN Bodies, particularly UNEP, and grants from Foundations and industrial enterprises.

International Council of Scientific Unions (ICSU)
Scientific Committee on Problems of the Environment (SCOPE)

SCOPE is one of a number of committees established by the nongovernmental group of scientific organizations, the International Council of Scientific Unions (ICSU). The membership of ICSU includes representatives from 68 National Academies of Science, 18 International Unions and 12 other bodies called Scientific Associates. To cover multidisciplinary activities which include the interests of several unions, ICSU has established 10 Scientific Committees, of which SCOPE is one. Currently representatives of 34 member countries and 15 Unions and Scientific Committees participate in the work of SCOPE, which directs particular attention to the needs of developing countries. SCOPE was established in 1969 in response to the environmental concerns emerging at that time: ICSU recognized that many of these concerns required scientific inputs spanning several disciplines and ICSU Unions. SCOPE's first task was to prepare a report on Global Environmental Monitoring (SCOPE 1, 1971) for the UN Stockholm Conference on the Human Environment.

The mandate of SCOPE is to assemble, review, and assess the information available on man-made environmental changes and the effects of these changes on man; to assess and evaluate the methodologies of measurement of environmental parameters; to provide an intelligence service on current research; and by the recruitment of the best available scientific information and constructive thinking to establish itself as a corpus of informed advice for the benefit of centres of fundamental research and of organizations and agencies operationally engaged in studies of the environment.

SCOPE is governed by a General Assembly, which meets every three years. Between such meetings its activities are directed by the Executive Committee.

R. E. Munn
Editor-in-Chief
SCOPE Publications

Executive Secretary: V. Plocq

Secretariat: 51 Bld de Montmorency
75016 PARIS

Contents

Foreword

In his address to the sixth General Assembly of the Scientific Committee on Problems in the Environment (SCOPE) held in Washington, DC in September 1985, Dr Mostapha K. Tolba, Executive Director of UNEP, stated:

> Environmentalists and scientists addressing environmental issues have a duty to demonstrate and to keep demonstrating the advantages that accrue to all nations from environmentally sound sustainable development.

Since its initiation in 1974, SCOPE's Project on Biogeochemical Cycles has been addressing important environmental issues. Initially, the activities focused on the cycles of carbon, nitrogen, sulphur and phosphorus separately. However, realizing that these cycles are inexorably tied together in one integrated system of cycles comprising many mechanisms of mutual interaction, a study of these interactions was started and has proved to be both interesting and fruitful. In 1982, the fifth General Assembly of SCOPE recommended that a scientific advisory committee be appointed to provide for intercommunication between the various biochemical cycling projects. Shortly thereafter, the Executive Committee endorsed a project on Global and Regional Cycling of Toxic Metals. This was in response of our need to understand better the pathways and cycles of potentially toxic elements such as arsenic, cadmium, lead and mercury in the environment. We are grateful to Professor T. C. Hutchinson for agreeing to chair the Subcommittee on Metal Cycling which organized the first SCOPE Metal Cycling Workshop, held in Toronto in September 1984. The workshop concentrated on four significant environmental pollutants—arsenic, cadmium, lead and mercury—and found major gaps in our understanding of global and regional fluxes in all four elements. In particular, the scarcity of data from developing countries was noted despite the fact that all four elements have been recognized as significant environmental pollutants for at least two decades. The present report was prepared from the draft contributions prepared for and critically discussed at the Toronto workshop.

SCOPE is grateful to all those who in one form or another have contributed to the report. The financial support of the National Research Council of Canada and the Natural Sciences and Engineering Research Council of Canada in particular are gratefully acknowledged.

Obviously, this is only the start of SCOPE efforts in this important area. It is particularly important that this work be extended and that information be obtained on regional and global models of environmental fluxes of these and other potential pollutants. More emphasis will have to be placed on the tropical and subtropical climates and environments which may greatly accelerate the rates of cycling of these elements and the change in chemical form, fluxes and residence time. SCOPE will be actively pursuing funding to accomplish this task and will attempt to focus attention on this area in the future.

J. W. B. Stewart
Chairman of the SCOPE Project on Biogeochemical Cycles

Saskatchewan Institute of Pedology
University of Saskatchewan
Saskatoon, Canada S6N OWO

Acknowledgements

This volume, which includes rapporteurs' reports of the working groups and manuscripts from many of the workshop participants, is based on a SCOPE metals cycling workshop held in Toronto in September 1984. Our need to better understand the pathways and cycles of the elements arsenic, cadmium, lead and mercury and to help focus international scientific studies on these important potential toxic elements were key factors in SCOPE's decision to form a metals sub-committee.

This report forms a part of the activities of the SCOPE sub-committee on metals cycling, which is a sub-committee of the overall SCOPE Biogeochemical Cycling Committee, chaired by Professor Thomas Rosswall of Linköping, Sweden. Help in the planning of the Toronto Workshop which forms the basis of this report was received from the sub-committee members E.D. Goldberg (USA), J. O. Nriagu (Canada), P. J. Peterson (UK), P. Buat-Ménard (France), V. Subramanian (India), G. Muller (German Federal Republic), chaired by T. C. Hutchinson (Canada). Financial assistance for the Workshop and for subsequent editorial and reviewing activities were received from SCOPE (Paris), the National Research Council of Canada and from a Conference grant provided by the Natural Sciences and Engineering Research Council of Canada. Particular assistance as rapporteurs and chairmen of working groups at Toronto was provided by J. Jaworski and P. M. Stokes (Canada), S. Lindberg, A. L. Page and R. Chaney (USA), P. J. Peterson and M. Hutton (UK) and P. Buat-Ménard (France). Naomi Adamson provided the volume index, while typing help was provided by Karin Baird-Parisien. All manuscripts were externally and internally reviewed and thanks are given to the reviewers. In addition, the volume as a whole was reviewed by Dr D. Miller on behalf of the SCOPE executive. His critical and helpful comments proved most valuable.

General logistic support from SCOPE and from the Institute for Environmental Studies at the University of Toronto is gratefully acknowledged.

T. C. HUTCHINSON and K. M. MEEMA

Toronto, February, 1986

Participants in the Workshop on Metal Cycling, Toronto, 3–6 September 1984

Dr Patrick Buat-Ménard, *Centre des Faibles Radioactivities, Laboratoire mixte CNRS-CEA, P. B. No. 1, 91190 Gif-sur-Yvette, France*

Dr R. C. Chaney, *USA Dept. of Agriculture, Agricultural Environmental Quality Institute, Beltsville, Maryland 20705, USA*

Dr. A. C. Chang, *Dept. of Soil and Environmental Sciences, University of California, Riverside, CA92521*

Dr. D. C. Chilvers, *MARC, The Octagon Building, Chelsea College, 459A Fulham Road, London, SW10, OQX, England*

Dr Patrick Denny, *School of Biological Sciences, Queen Mary College, Mile End Rd., London E1 4NS, England*

Dr Robert Duce, *Graduate School of Oceanography, University of Rhode Island, Kingston Rhode Island 02881, USA*

Dr. Mohamed El-Amamy, *Dept. of Soil and Environmental Sciences, University of California, Riverside, CA92521*

Dr E. D. Goldberg, *Geological Research Division, University of California, San Diego, La Jolla, California 92093, USA*

Dr Barry T. Hart, *Chemistry and Biology Dept., Chisholm Institute of Technology, Caulfield East, Victoria 3145, Australia*

Dr Magda Havas, *Institute for Environmental Studies, University of Toronto, Toronto, Ontario, M5S 1A4, Canada*

Dr Tom C. Hutchinson, *Institute for Environmental Studies, University of Toronto, Toronto, Ontario, M5S 1A4, Canada (Chairman)*

Dr Malcolm Hutton, *Monitoring and Assessment Centre, The Octagon Building, 459A Fulham Rd., London SW10 OQX, England*

Dr John F. Jaworski, *Secretary, Subcommittee on the Heavy Metals, National Research Council, 100 Sussex Dr., Ottawa, Ontario, K1A 0R6, Canada*

Dr C. R. Krishna Murti, *Centre for Environmental Studies, Anna University, Gindy, Madras 600 025, India*

Dr. P. S. Lake, *Zoology Dept., Monash University, Clayton, Australia*

Dr Mohamed R. Lasheen, *Water Pollution Control Lab., National Research Centre, Dokki, Cairo, Egypt*

Dr S. E. Lindberg, *Oak Ridge National Laboratory, Oak Ridge, Tennessee 37830 USA*

Dr Jerome Nriagu, *Canadian Centre for Inland Waters, Box 5050, Burlington, Ontario, L7R 4A6, Canada*

Dr Moses Olade, *Geology Department, University of Ibadan, Ibadan, Nigeria*

Dr Jozef M. Pacyna, *Norwegian Institute for Air Research, PO Box 130, 2001 Lillestrom, Norway*

Dr A. L. Page, *Department of Soil and Environmental Science, University of California, Riverside, California 92521-0424, USA*

Dr Peter J. Peterson, *MARC, The Octagon Building, Chelsea College, 459A Fulham Rd., London SW10 OQX, England*

Dr Eiliv Steinnes, *Department of Chemistry, University of Trondheim, N-7055 Dragvoll, Norway*

Dr Pamela M. Stokes, *Institute for Environmental Studies, University of Toronto, Toronto, Ontario, M5S 1A4, Canada*

Dr V. Subramanian, *Jawaharlal Nehru University. School of Environmental Sciences, New Mehravli Rd., New Delhi, 110067, India*

Dr David Turner, *The Laboratory, Citadel Hill, Plymouth, PL1 2PB, England*

Dr Ming H. Wong, *Department of Biology, Chinese University of Hong Kong, Shatin, Hong Kong*

Dr Chris D. Wren, *Institute for Environmental Studies, University of Toronto, Toronto, Ontario, M5S 1A4, Canada*

Workshop Summary

T.C. Hutchinson

The Toronto SCOPE Workshop considered various aspects of the occurrence, flux, compartmentation and residence times of four elements which have been recognized as significant environmental pollutants over the past two decades. These are arsenic, cadmium, lead and mercury. Major gaps in our understanding of the global and regional fluxes for all four elements were identified, with data from developing countries in general being especially scarce. The contrasting contributions from natural versus anthropogenic sources between the elements were pointed out, with lead being now predominantly anthropogenically derived, while arsenic is estimated to be 60% from natural sources and 40% from anthropogenic. Automobile emissions, coal-fired generating plants, non-ferrous smelters and wood and dung combustion were amongst major sources of lead, arsenic and mercury, while urban, agricultural and industrial contributions overall have caused local to regional accumulation in such 'temporary' sinks as vegetation, soils, sediments and peat.

Of the four elements considered, the participants were in agreement that on a world-wide scale, elevated levels of lead represented the greatest risk to human health. The need for a fuller understanding of sources to human populations and to particular groups of high risk was emphasized, for example, the exposure of young children to lead, and the exposure of elderly females to cadmium in certain areas of Europe. The food chain sources to humans via food and drink were considered, with mercury, cadmium and arsenic being identified often as reaching people via aquatic based food-chains, including fish and shellfish. Lead sources included atmospheric particulates, especially from automobiles and from combustion processes, re-entrainment of lead-contaminated dust to children, leaded paints, and foodstuffs contaminated with solder. Lead from water pipes is still a significant problem in some places.

Even in the most widely studied environments in Europe and North America, the degree of uncertainties in elemental fluxes are high. The chemical forms of the elements are also quite poorly known, though the metal species may have a substantial influence on the behaviour and biological effect of the element in question. Methylation of mercury is one such example, with the location and microbial involvement in the methylation process requiring elucidation, especially in tropical systems. Arsenic is another element subject to microbial methylation. In contrast to mercury, this does not enhance toxicity, though it does alter solubility, pathways and sinks.

Participants presented papers on their own specialities. These involved a focus on the medium of occurrence, such as oceans, freshwater systems, soils, agriculture, atmosphere, sediments and biota. In addition, an important aspect of the Workshop was presentations and discussions dealing with regional occurrence and elemental fluxes, for example, in India, Australia, Nigeria, Hong Kong, United States, Egypt, Ivory Coast, Poland and Scandinavia. This regional aspect is to be much more systematically developed at the Second International SCOPE Metals Cycling Workshop, to be held in India in 1987, and in which particular attention will be paid to developing countries.

Significant points raised and discussed at the Toronto Workshop are described below.

Combustion temperature is a key factor in determining quantities and types of atmospheric emissions. Mercury is emitted in the form of elemental vapour, especially from the burning of fossil fuels and from mining and smelting of non-ferrous metals. Even the best maintained electrostatic precipitators are unable to remove or intercept mercury to above about 1% of levels present in the ore or coal. It is extremely volatile even at quite low combustion temperatures. Of all Hg present in coal 95 to 99% is lost to the atmosphere upon burning. Wet scrubbers intercept more Hg than electrostatic precipitators. Mercury vapour is also lost from Hg-rich soils, while chlor-alkali plants have been major sources of mercury to water bodies. The occurrence of elevated mercury levels in fish, algae and shellfish living in recent impoundments (reservoirs) and in acidified freshwater bodies in Canada, Sweden and Norway, have raised questions about atmospheric movements of mercury in polluted air masses, as well as mercury released from watershed soils and sediments under acidic conditions. The high volatility of mercury has led to its occurrence in remote regions, with a general polar directional movement suspected. Physical and chemical forms of mercury in the air need much better elucidation and the environmental and biological sites of methylation also need study. Mercury poisonings of humans in Japan and in the Middle East have emphasized (1) the dangers of eating contaminated fish by populations with a high dietary intake of fish and (2) the danger of mercury-dressed grain being eaten rather than planted. The dressing of seeds with organo-mercurials, such as phenyl mercuric acetate, acts as an excellent fungicide, markedly reducing pre-germination losses of seed. It does at the same time introduce mercury, both to the soil and to locations close to absorbing roots.

The serious concern about lead is due to the susceptibility of humans, especially young children, to significant health effects from ingestion and inhalation of lead. Lead levels in 'average' urban dwellers are considered too close for comfort to levels at which chronic lead poisoning symptoms appear in children. This concern is magnified because of the large emissions of lead into the atmosphere from automobiles in many countries and especially in rapidly developing urban centres with high human population densities. This lead is also in a form with a relatively long atmospheric residence time, that is, as fine particulates, so that dispersal to remote regions has demonstrably taken place, particularly with substantial elevations in the northern hemisphere. Southern hemisphere levels are a good deal lower overall, but not in urban centres. Environmental concerns include the low mobility of lead in soils causing accumulation in the upper litter and humus layers and in sediments. Lead accumulation has taken place in industrialized nations and is apparently gathering pace in many developing countries. Point sources of lead from smelters are also of concern, and the peeling of paint containing high lead levels has led to thousands of childhood lead poisoning cases in the United States. In humans, lead is an accumulative toxigen and brain damage to children is not reversible.

Some wildlife populations have also suffered severely from lead poisoning, most notably wildfowl populations (ducks and geese) in North America and Europe, from ingestion of lead shot used by hunters. This lead shot is present in large quantities in the mud of ponds, lakes and marshes. Many wildfowl also are hit by lead shot but are not fatally wounded. These birds accumulate lead into their blood from the wound areas and this can cause death from lead poisoning. In North America, many thousands of water fowl die this way each year.

The Lead Working Group emphasized the need to question the reliability of data, a theme echoed for arsenic and mercury, and to a lesser extent for cadmium. The lead group also estimated that anthropogenic inputs of lead to the environment were one or two orders of magnitude greater than natural sources. However, the need for much better national inventories of emissions was also emphasised, as it was for the other three elements.

One presentation from Norway dealt with the valuable regional and geographical data available on the environmental sinks of the elements, by means of carefully selected biological monitors. These include certain of the widely distributed 'feather' mosses from woodland sites, and the foliage and bark of higher plants including long-lived trees. Comparative studies of the accumulation of elements in salt-water and freshwater mussels, clams and oysters demonstrate that these are also effective monitors.

Various environmental compartments can also provide invaluable short and long-term monitors of historical changes in elemental levels, for example, herbarium and museum specimens, sediment and peat cores, annual rings in trees, annual deposition of ice in glacial cores, etc.

In determining the biological effects of lead, the chemical form was not considered as crucial as it is for mercury, for which the most toxic methylated forms are efficiently produced in aquatic systems. It was also noted that of the metals of overall environmental concern the evidence is that only mercury is definitely biomagnified, i.e., that it increases in concentration in organisms at increasingly higher (trophic) levels of the food chain. Nevertheless, several real puzzles exist regarding mercury, such as the source of mercury in tuna and especially swordfish, the occurrence of elevated mercury levels in seals from remote polar locations, etc.

Concern was expressed about the shift of mercury technologies from the northern to the southern hemisphere, and from developed to developing countries. Similar concern was expressed for this shift regarding lead. It was felt that good international communication and local education are essential to prevent the repetition in developing countries of tragic occurrences of metal poisoning already experienced in industrialized countries.

Cadmium has caused some local human health problems but the Working Group emphasized that commercial cadmium production only commenced this century, so that uncontrolled cadmium emissions could well create future problems. Cadmium inventories are very patchy on a global basis making both regional and global cycles difficult to develop at this time. It was noted, for example, that Peru and Mexico account for one quarter of the world's production of zinc, with which cadmium is always associated, so it would be important to determine cadmium fluxes in these countries. A good deal of cadmium from processing goes into municipal sewers or directly into watercourses. In addition, landfill and incinerators also are sources of cadmium to the local environment.

For cadmium, dietary intake is the biggest source to humans, and this is often via crops. Elevated cadmium levels occur in many phosphate fertilizers and can cause increases in soil cadmium concentrations. Soil concentrations also correlate well, in general, with levels in plants grown on these soils. However, as the cadmium group emphasized, the increases in Cd levels in overall dietary intake as a result of this are low relative to the Cd levels associated with human health problems. Specific diets may be a factor, as was the case with high rice diets in Japanese cadmium poisoning cases, when elevated but rather low levels of Cd in rice grain led to a major cadmium poisoning episode. The high cadmium levels in elderly women in an industrial Belgium city, correlating with increased renal dysfunction, has yet to be fully explained but food sources are suspected, perhaps via land applications of phosphate fertilizers.

For arsenic the main route to humans is by ingestion. For occupational exposure, inhalation is also significant, as it is for populations living in the vicinity of copper and gold smelters emitting arsenic oxides. Arsenic poisonings and arsenic-induced health effects do occur in areas which are naturally high in arsenic. The atmosphere is the main transport medium for this element. Anthropogenic atmospheric arsenic emissions are estimated to be 50% from copper smelting and coal combustion, since arsenic and its oxides are rather volatile and are released at quite low combustion temperatures. It was emphasized that the data for arsenic are inadequate in many ways, especially regarding natural sources and levels. Low temperature volatilization from soils, vegetation and volcanic emissions are the primary natural sources identified to date. We need better data on historical trends, on environmental sinks, on residence times in various compartments, and on the mobilization of arsenic in anoxic sediments. Inventories are needed for many developing nations on quantities of arsenic emitted from burning of wood and of animal dung. Recent work on arsenic fluxes and behaviour in the oceans has improved our understanding of this aspect of arsenic cycling.

For all elements the desirability to develop regional and global models of environmental fluxes was recognized. It was also emphasized that most of the studies to date have been made in temperate regions. Tropical and sub-tropical climates and environments may have profound effects on the cycling of these four elements and on chemical forms, fluxes and residence times. We need to understand these more completely before regional and global cycles can be established.

The presentations from India, Egypt, Nigeria, Hong Kong and Australia led to a discussion of the need for the development of standard reference materials and for reliable analytical techniques. A workshop was suggested to focus on developing country problems, including protocols for sampling, storage, analysis and interpretation of data. The need for national inventories was repeatedly emphasized.

These aspects will be a focus of the second International Workshop on Metals Cycling scheduled for 1987 in India. It is hoped that SCOPE National committees may play a lead role in these developments and that UNEP will offer strong support for development of protocols and standard reference materials for developing countries.

Part I

Group Reports

Lead, Mercury, Cadmium and Arsenic in the Environment
Edited by T. C. Hutchinson and K. M. Meema

CHAPTER 1

Group Report: Lead

J. F. JAWORSKI, *Rapporteur*
J. NRIAGU, *Chairman*
P. DENNY, B. T. HART, M. R. LASHEEN, V. SUBRAMANIAN, M. H. WONG

INTRODUCTION

The local, regional and global biogeochemical cycles of lead have been affected by man to a greater degree than those of any other toxic element. In fact, few areas on earth are now free of anthropogenic lead. The lead concentration in the most recent ice layers at the North Pole are 10- to 100-fold higher than the values in prehistoric times. Furthermore, these recent values are about 10 times greater than those of similar age at the South Pole. Even at the South Pole, the present-day rate of lead deposition is 2–5 times higher than in pre-technological times (Boutron and Patterson, 1983). In this context, urban centers can be regarded as 'hot spots' for lead where levels exceed those of pre-technological times by several thousand-fold. As a result of world-wide accumulation, especially in urban centers, lead presents a more serious environmental and health hazard than does any other element. The smelting, use and toxicity of lead were already known in antiquity; but, the greatest changes in cycling occurred after the industrial revolution and especially after the introduction of leaded gasoline in 1923 (Waldron and Stoffen, 1974) because of the widespread distribution of these sources and the small dimensions of the particulates emitted.

Lead is especially toxic to children and the young of other species (Landrigan *et al.*, 1975; Needleman *et al.*, 1979, 1982). A number of recent detailed studies aimed at assessing the health effects of chronic lead exposures in children have revealed significant effects on intelligence and on neuropsychological performance. At higher exposures in urban ghetto areas in the USA, many thousands of children suffered lead poisoning due to exposure to pealing paint, often with a 30% lead content. Also, excessive lead fallout may adversely affect some aquatic and terrestrial food chains (Jaworski, 1979). If nations are to protect their most valuable resources—their children and their environment—it is important to identify possible problems with lead and prevent them, or, failing that, identify high-risk situations and take remedial measures.

This report addresses the following questions:

(1) What is the minimum data set required to assess impacts of anthropogenic sources on local, regional and global cycles of lead?

(2) Are current data adequate to permit this assessment?

(3) What can be done to generate missing data or to improve the quality of questionable data?

MINIMUM DATA SET

To determine the impact of man's activities on the biogeochemical cycle of lead it is necessary to have some understanding of:

(1) The chemical and physical forms (e.g. particle size) of environmental lead and how they are interconverted.

(2) The critical pathways and accumulator compartments for lead in the environment, as well as lead dynamics in each compartment.

(3) The ratio of natural and anthropogenic inputs to the environment as a function of time and space as well as the order of importance within each category (i.e. a source inventory).

(4) Which organisms are relatively sensitive to lead and can thus act as indicators of incipient problems.

BIOGEOCHEMISTRY AND CYCLING OF LEAD

Lead Speciation

Lead can exist in both inorganic form as Pb(II), less often as Pb(IV), and in organic form (up to 4 Pb–C bonds). With the exception of a few sporadic measurements in urban air, marine fish, sediments, birds and in human brains (Jaworski, 1979, Forsyth *et al.*, 1984) there is relatively little information available on organometallic lead compounds (Grandjean, 1984). It seems that most, if not all, organometallic lead in the environment comes from the volatilization of gasoline additives (about 5% of these additives are emitted uncombusted to air; Morgan and Bretthauer, 1977) rather than from biotransformation of inorganic environmental lead. The flux of organometallic lead is small compared with that of inorganic lead on a global basis, but on a local basis (e.g. near gasoline stations) it might be a significant factor (Grandjean, 1984).

Inorganic lead is far more extensively studied than organometallic lead and numerous reviews have discussed speciation and cycling (Jaworski, 1979; NAS, 1980; EPA, 1983).

Lead Cycling

Figure 1.1 schematically presents the environmental pathways of lead (emphasis on inorganic lead) and highlights routes of human exposure. Depending on its nature and location, a source may deliver lead to air and/or water (see Figure 1.2 for a more detailed picture of fresh and salt water cycling of lead).

Air

The prime medium for lead transport is air because fine particulates (<1 μm diameter), generated especially by anthropogenic high temperature sources, may travel long distances before settling out via wet, dry or cloud deposition (these processes are poorly understood, see Lindberg, this volume, Ch. 8). The residence time of such fine particulates in air is from several hours to several days. With the exception of volcanoes and forest fires, natural sources are associated with lower temperature emissions which contain larger particles that tend to settle out more rapidly.

Deposition from air is greatest near a source (because of large particle precipitation) but the zone of readily detectable elevated deposition can extend to 30 km or more in the case of emissions from a tall chimney.

Land

Most of the lead particles deposited on soil are retained and eventually become mixed into the surface layer (or deeper in the case of cultivated soils). Lead accumulated at the soil surface may be taken up directly by grazing animals and by soil micro-organisms (lead deposition may be associated with decreased litter decomposition in some forests) and so enter terrestrial food chains, or, bypass food chains completely as in the case of children consuming dust and dirt during normal hand-to-mouth activity (Jaworski *et al.*, 1984).

Fresh Water

Lead in rivers comes from runoff (largely anthropogenic), erosion (mostly natural) and direct deposition from air (anthropogenic). In large lakes, most of the lead probably comes from airborne deposition. Freshwater generally contains more inorganic and organic suspended material than marine water and this material has a strong tendency to adsorb any dissolved lead.

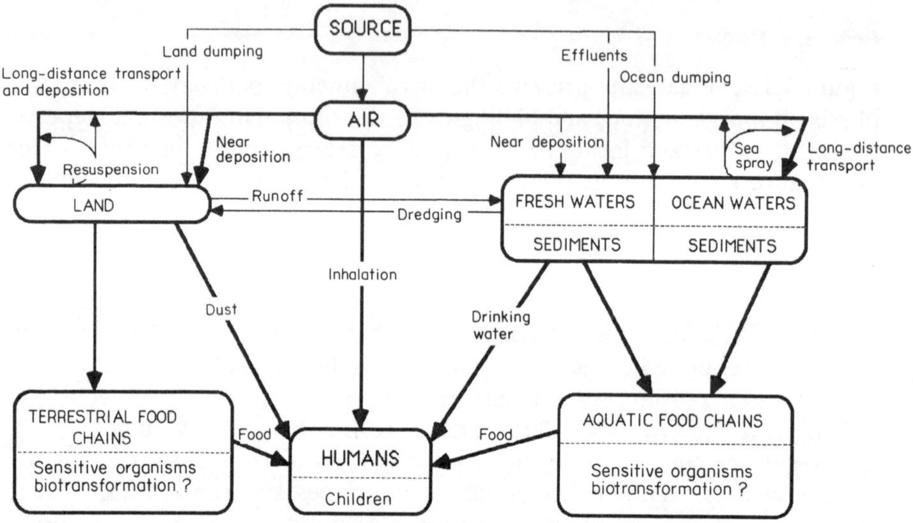

Figure 1.1 Biogeochemical cycle of lead and routes of human exposure

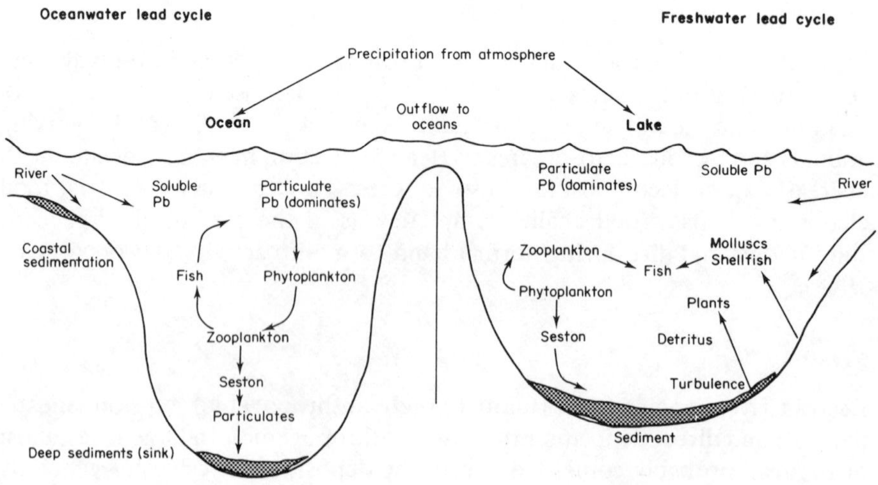

Figure 1.2 Biogeochemical cycle of lead in fresh water and in salt water ecosystems

Thus, lead movement in freshwater is closely associated with the turbulent transport of particulate matter (Everard and Denny, 1985a).

Micro-organisms and plants show a strong tendency to bioaccumulate lead from the water column (Everard and Denny, 1985a). Also, deposition of particulate matter containing adsorbed lead presents the element in a relatively available form to benthic organisms (Everard and Denny, 1984), especially filter feeders. Thus, it is observed that lead can and does enter the aquatic food chains (Everard and Denny, 1985b); however, even though individual organisms bioaccumulate lead there is no evidence for biomagnification of lead concentrations along a food chain. In fact, the reverse situation (biopurification) has been observed in some marine food webs (Patterson, 1980). Fish also lay their eggs in surface sediments, and it is known that the young (but not usually the eggs themselves) of most fish species are more susceptible to lead than the adults. Thus, contamination of spawning grounds may adversely affect fish populations and human food supplies.

A special problem is the ingestion of lead shot (from hunter's firearms) and lead fishing weights by bottom-feeding marsh birds, often with fatal results (Sileo *et al.*, 1973). Ducks, geese and swans are especially susceptible to lead since it is ground up in their gizzard and may be more available for uptake into their bodies under acidic conditions.

Ocean Water

Most of the lead entering the open oceans comes directly from atmospheric deposition rather than from rivers (see Buat-Ménard and Duce, this volume, Ch. 11). Estuaries and some coastal waters are more influenced by riverborne lead as well as direct dumping of sewage and industrial wastes. The lower concentration of particulate matter in marine waters (relative to fresh water) as well as the higher chloride concentration imply that a significant portion of lead in the water column is in dissolved form. However, on a geological time scale, lead is rapidly (residence time in the water column is several hundred years) removed to the bottom sediments by adsorption onto particulate matter or by incorporation into organisms which sink to the bottom when they die. Deep ocean sediments, except for undersea volcanic activity and other upwellings, can be considered as a sink for lead, since the ocean profiles show no evidence of significant remobilization of lead from the sediments.

ADEQUACY OF EXISTING DATA

Availability of Regional Information

In general, the availability of lead data decreases in the following order: North America, Western Europe > Australia, India > Soviet Union and

Eastern Europe, China, Middle East > South America, Southeast Asia, Africa (see regional reports included in this volume). It is not possible to derive emission data for one country using that from another country because of differences in industrial activity, climate and culture as well as the lack of statistics on production and consumption. Therefore, additional measurements are needed from many areas to fill in gaps and to increase the confidence in the data. Much of the data is not reliable and analytical and sampling methodologies need much greater attention. Data reliability would be greatly enhanced by routine use of reference materials of accepted concentrations.

Relative Magnitude of Lead Sources to the Environment

On a global basis, a number of estimates of natural and anthropogenic lead emissions to air (and to the oceans, Buat-Ménard and Duce, this volume, Ch. 11) have been made and all indicate that contribution from the anthropogenic sources are at least 1 to 2 orders of magnitude greater than natural sources (see Tables 1.1 and 1.2 for a listing and quantitative estimates). The greatest single contribution is from the use of gasoline additives and appears to remain so despite recent reductions in usage in North America. Historical records in glacial ice and aquatic sediments indicate that global emissions from anthropogenic sources have exceeded those from natural sources for several hundred years (Thornton and Abrams 1984).

Table 1.1 Estimates of global flux of lead to air from natural sources (in 10^9 gm/yr)

	Reference					
Source	Lantzy and Mackenzie (1979)	Nriagu (1979)	Weisel *et al.* (1984)	Zoller (1984)	Jaworski *et al.* (1981)	Patterson (1980)
Windblown dust	5.0	16	3.0	5.0	31	2.0
Volcanogenic	0.87	6.4	0.4	16	23	
Forest fires	6.8	0.5	—	0.5		
Vegetation	—	1.6	0.2	—	—	<0.1
Sea-salt sprays	<0.001	0.02	5.0	0.001	0.09	<1.0
Total	12.7	24.5	8.6	22	54.1	<2.0

The following data are missing or uncertain:

(1) regional estimates of source strengths for many developed and developing countries;

(2) benchmark studies to estimate as a function of time the ratio of natural and anthropogenic sources on a regional basis.

It remains to be determined:

(a) whether any significant natural or anthropogenic lead sources remain to be identified;

(b) whether lead is biomethylated to any significant extent in the environment, and if so where this occurs and what are its pathways and effects.

Table 1.2 Global estimates of anthropogenic emissions of lead to the atmosphere in 1979/1980 (in 10^9 gm/year). The figures for the USA and for Europe have been multiplied by a factor of 5 to give global estimates

Source	Nriagu (1979)	USA × 5*	Europe × 5†
Gasoline and waste oil combustion	177‡	176	176 (372)§
Waste incineration	8.9	4.2	4.0
Coal combustion	14	4.8	4.0
Primary non-ferrous metal production	85	9.8	143
Secondary non-ferrous metals	0.8	—	2.2
Iron and steel	50	4.1	73
Industrial applications	7.4	1.6	—
Wood combustion	4.5	—	2.8
Phosphate fertilizers	0.1	0.4	0.03
Miscellaneous	5.9	1.2	3.8
Total	354	202	419

* Data from EPA (1983).
† Data compiled by Pacyna (1984).
‡ This number assumes that there has been a 35% reduction in the consumption of leaded gasoline between 1975 (the base year used in the original calculation) and 1979/80.
§ The figure in brackets, which corresponds to the reported estimate for Europe, seems to be too high.

Pathways and Accumulating Compartments

The reservoirs of lead can be quantified in a straightforward manner but the rates of transfer within and between the reservoirs are only known qualitatively and semi-quantitatively. Nevertheless it should be possible, using this information as well as meteorological and physico-chemical principles, to develop models for regions where measurements have not been made.

Research is needed to confirm our concepts of long range transport of lead in air and of lead cycling in the oceans (many previous data for lead in ocean water and biota are suspect or very inaccurate due to contaminated samples).

Indicator Organisms

Many studies have shown the widely divergent responses of biota to lead in the environment. Some organisms are able to accumulate large quantities of lead without apparent harm, while others are very sensitive to lead and are easily eliminated. Suggestions have often been made as to the value of indicator species for different compartments of the environmental pathway. In the urban environment, children are sensitive to lead exposure and are the best indicator organisms. In natural ecosystems, general litter decomposing micro-organisms are sensitive to lead accumulations and measurements of their activity, such as in respiration rates, CO_2 evolution or rates of litter decomposition are often used to estimate potential ecosystem dysfunction. In agricultural soils the number of *Micrococcus luteus* is used as an indication of lead accumulation. Grazing animals readily pick up lead from herbage on which they feed, thus hair samples from cows, goats, horses, rabbits and rodents can be used for rapid survey. In freshwater systems clams, the colonial filamentous alga *Cladophora glomerata*, aquatic angiosperms and the aquatic moss *Fontinalis* sp. are useful lead indicators (Welsh and Denny, 1980). Among freshwater animals the zooplankton *Daphnia magna* is sensitive to lead in solution, and brook trout are regularly used in North America (Jaworski, 1979). In coastal and marine systems the use of mussels *Mytilus edulis* and *Cristigera* spp. have been shown to be both sensitive and reliable (Goldberg *et al.*, 1978). In terrestrial systems the horsetail species *Equisetum*, loblolly pine, red maple (Jaworski, 1979) and the feather mosses of the boreal regions world-wide, *Pleurozium schreberi*, *Hylocomnium splendens* and *Hypnum cupressiforme*, are retainers of lead and other metals (Goodman and Roberts, 1970; Ruhling and Tyler, 1971; Steinnes, this volume, Ch. 9), while in warm, temperate regions the so-called Spanish moss *Tillandsia usneoides* is a useful species for monitoring environmental lead. It is an epiphyte confined to areas with an average relative humidity of 64% or more.

Future Trends

It is not possible to anticipate all the factors which may change the mobilization, pathways, bioavailability and effects of lead on a global, regional and local basis. The following is a partial listing of some of the more important factors:

(1) changes in the world economy, which will affect the amount and geographic distribution of the production of lead and commodities containing lead;

(2) changes in national regulations concerning levels of lead allowed in gasoline;

(3) use of technology with obsolete or inefficient emission controls; this could be a future problem in developing countries;

(4) acid rain causing local or regional mobilization of lead in watersheds;

(5) deforestation;

(6) changes in the global climate causing alteration in patterns of air currents, deposition of rain and snow, changes in agricultural use, etc.;

(7) demographic changes;

(8) changes in the patterns of human and animal nutrition and diseases*.

GENERATING MISSING DATA

Organization at the National Level

Since most of the missing lead data is regional or local in character, it is obvious that there is a need for national and international or regional bodies to:

(1) gather statistics on production, consumption and disposal of commodities containing lead and other toxic substances in each country or region;

(2) organize and support monitoring of emissions and levels of lead and other toxic substances;

(3) encourage exchange of scientific information at an international level;

(4) enhance reliability of analytical procedures by use of standard techniques and protocols and use of readily available reference materials.

---

* *Editors' note*: It has been reported from many studies in Europe, Asia, Latin America and especially North America, that lead has appeared at increasing concentrations in 'sinks' located at considerable distances from industrial-urban sources of lead. These sinks include the upper organic horizons of soils. Long-distance airborne transport is implicated in this. Regions of high rain and snow deposition such as alpine areas have shown elevations to several hundred times that of background. Clean-laboratory facilities, where the normal lead contamination of glassware, collecting equipment, solvents, etc. is eliminated, have revealed that the pre-historical lead exposures in the environment were 10 to 100 times less than generally supposed. These facts suggest that vigilance is needed with respect to lead exposures, especially as childhood body burdens can rather easily exceed 'safe' levels.

Monitoring

Monitoring is required to quantify the emissions and environmental impact of anthropogenic lead sources. The following should be considered when setting up or evaluating a monitoring program:

(1) *Funding limitations* It is better to complete a program of limited scope which answers specific questions rather than have to abandon an overly ambitious or poorly conceived project.

(2) *Planning* If possible, this should be done at the national level to avoid unnecessary duplication of effort. Preliminary qualitative or semiquantitative information is required on possible natural and anthropogenic sources and pathways, and on target organisms. Sampling strategies and analytical methodology must be verified before starting and, after commencing, analytical quality should be checked by using reference samples. Emphasis should be placed on known or suspected lead sources as well as on answering specific questions about lead-sensitive environmental compartments such as areas of food production, acid soils, forest litter, aquatic sediments (especially estuaries and fish spawning grounds) and the urban microenvironment. Finally, the program should be re-evaluated at preset intervals to ensure that it is generating useful data.

(3) *Historical Record* Available sedimentary records (e.g. glacial ice, lake and ocean sediments, tree rings, herbarium samples, peat cores, etc.) should be studied to determine whether rates of lead mobilization have changed over time. Well-characterized archival samples (plant or animal tissues preserved by museums, organ banks, etc.) may provide information on temporal changes in lead exposure and bioavailability. Benchmark locations should be identified and samples taken to permit the identification of future trends in cycling of lead and other substances.

THE URBAN MICROENVIRONMENT

A 'legacy' of lead has accumulated and will remain for a long time in many cities throughout the world because of the use of lead-pigmented house paint, leaded gasoline and in some cases because of smelting and metallurgical activities. All these sources contribute to lead uptake in children via the inhalation and ingestion of dust. Each source, if sufficiently intense, may lead to toxic accumulations of lead in children or will reduce the amounts of lead required from other sources to produce such toxic accumulation. The proximity of source and receptor as well as the poor dispersion of lead in the urban corridors and especially within the home (Swedish Council for Building Research, 1984) imply that some sources (e.g. hobbies or cottage industries involving lead soldering) may have drastic effects even though their contribution to regional and global lead cycling is insignificant.

The use of lead water pipes in houses has been phased out in most industrialized countries but older housing often has a high percentage of lead piping. Water stored overnight in such pipes, especially in hot water pipes, can be a significant additional source of lead to humans.

Housing does not present an effective barrier to outside lead:

(1) the ratio of lead concentration in the air inside to outside a house is 0.6–0.85 (and is probably close to 1.0 in developing countries with hot climates);

(2) lead deposited outside of houses is continually carried inside on footwear, clothes, skin, etc.

The importance of the dust vector is emphasized by the strong influence that house cleaning, food-preparation and hand-washing practices have on blood lead levels in children.

Children are the most lead-sensitive urban dwellers; however, it is difficult to estimate their exposure if only environmental monitoring data (air, water, food, dust) are available. This is because it is impossible to sample all the places where children eat, sleep and play. Therefore, it is more reliable to monitor erythrocyte protoporphyrin fluorescence or blood lead levels (avoiding sample contamination and using an analytical laboratory of proven capability to analyze lead in blood).

COMPOUNDING PROBLEMS OF DEVELOPING COUNTRIES AND THE UNDERPRIVILEGED

The situation in developing countries differs in many ways from that in the developed countries (see Krishna Murti, this volume, Ch. 19). A greater reliance on biomass combustion for heating and cooking as well as the use of open hearth combustion (as opposed to enclosed, vented stove) leads to special health problems in the indoor and high-density urban environments (Swedish Council for Building Research, 1984; De Koning *et al.*, 1984). Economic limitations may reduce the level of emission control achieved for existing industrial operations and importation of cheap but obsolete smelters and factories can only increase pollution problems. Funding limitations hamper information gathering and environmental monitoring to assess the extent and severity of pollution. Sometimes the gathering of such information is viewed as a threat after 'more pressing' issues (e.g. industrialization) have been resolved. This attitude only creates problems for the future because pollution is more easily prevented than cleaned up. Social and cultural factors may make it difficult for a people to make the lifestyle changes necessitated by the introduction of polluting technologies. Nutritional deficiencies may increase the susceptibility of people to the effects of lead and other toxic substances.

Developed countries are not exempt from some of the problems of developing countries. Poverty, undernutrition and inadequate housing are factors which are present and are correlated with increased exposure and susceptibility to lead. The poor people in urban areas generally cannot afford to move out of contaminated ghettos into less lead-polluted suburbs nor are they found to take advantage of public education programs on lead as readily as their more affluent and more educated counterparts.

OUTLOOK

Anthropogenic sources have transferred large quantities of lead into the biosphere. In most regions this has greatly exceeded contributions from natural sources. In urban centers there have already been adverse effects on considerable numbers of children, but whether adverse effects from increased lead levels and bioavailability can be expected at present or in the future for other environmental compartments remains to be answered. Little can be done about existing lead contamination except to attempt to limit its bioavailability, for example, by removing or burying lead-contaminated topsoil from cities; however, the scope of this type of management is clearly limited by economic considerations. Natural processes of mixing and sedimentation will take decades (or more) to reduce the availability of some lead deposits, while some anthropogenic activities may act to reverse this trend, for example, production of acid rain, the movement of people from rural low-lead areas to urban high-lead areas or the building of new housing on land already used for the disposal of industrial wastes containing lead and other toxic substances.

In the short term, the present trend of removing lead from gasoline and from manufacture of food containers and other consumer goods by replacing lead with less toxic substitutes (e.g. titanium pigments in paints, barium compounds for plastics stabilizers), if continued, will significantly reduce the intake of lead by humans from present sources. Public education and effective management of existing lead contamination will reduce lead exposure for some, but not all, subgroups of the population, but it is unlikely to have much effect on the exposures of other life forms. More importantly, if governments in developing countries can be convinced of the value of preventing pollution by lead and other toxic substances through careful choice of imported or locally developed technology, then an important step will have been taken in reversing the global trend toward environmental degradation.

RECOMMENDATIONS

(1) The problem of lead pollution should be presented to governments as an important human health and environmental issue to be dealt with now by evaluating levels of environmental contamination and by acting to minimize further environmental degradation.

(2) The results of the detailed and ongoing studies of consequences of childhood exposure to lead should be widely communicated so that exposure of young children, especially, to environmental sources of lead can be minimized.

(3) Internationally accepted guidelines should be developed to assist in the selection of environmental monitoring strategies, analytical methodology and indicator organisms for lead and other toxic substances.

(4) Exchange of information among environmental scientists from different countries should be encouraged by the relevant international agencies.

(5) There is a need to establish regional inventories of natural and anthropogenic lead sources, as well as regional aspects of environmental cycling. Data should be as quantitative as possible and estimates should be made of confidence levels.

(6) Attention should be given to the replacement of those lead-based technologies which have the greatest impacts with more acceptable, less polluting technologies.

REFERENCES

Boutron, C. F., and Patterson, C. C. (1983). The occurrence of lead in Antarctic recent snow firn deposited over the last two centuries and prehistoric ice. *Geochim. Cosmochim. Acta,* **47**, 1355–1368.

De Koning, H. W., Smith, K. R., and Last, J. M. (1984). Biomass fuel combustion and health. In *WHO Int. Rept. EFP/84.64.* Geneva.

EPA (1983). *Air Quality Criteria for Lead.* US Environmental Protection Agency, Research Triangle Park, North Carolina.

Everard, M., and Denny, P. (1984). The transfer of lead by freshwater snails in Ullswater, Cumbria. *Environ. Pollut. (Series A),* **35**, 299–314.

Everard, M., and Denny, P. (1985a). Particulates and the cycling of lead in Ullswater, Cumbria. *Freshwater Biol.* **15**, 215–226

Everard, M., and Denny, P. (1985b). Flux of lead in submerged plants and its relevance to a freshwater system. *Aquatic Bot.* **21**, 181–194

Forsyth, D. S., Collette, M. C., and Marshall, W. D. (1984). Alkyllead salts in avian tissue. Paper presented at 67th Annual Conference of the Chemical Institute of Canada, June 3–6 1984, Montreal, Canada.

Goldberg, E. D., Bowen, V. T., Farrington, J. W., Harvey, G., Martin, J. H., Parker, P. L., Risebrough, R. W., Robertson, W., Schnieder, E., and Gamble, E. (1978). The mussel watch. *Environ. Conserv.,* **5**, 101–125.

Goodman, G. T., and Roberts, T. M. (1971). Plants and soils as indicators of metals in the air. *Nature (Lond.),* **231**, 287–292.

Grandjean, P. (Ed.) (1984). *Biological Effects of Organolead Compounds.* CRC Press, Boca Raton, Florida, 288 pages.
Jawororski, Z., Bysiek, M., and Kownacka, L. (1981). Flow of metals into the global atmosphere. *Geochim. Cosmochim. Acta*, **45**, 2185–2199.
Jaworski, J. F. (1979). *Effects of Lead in the Environment—1978. Quantitative Aspects.* National Research Council of Canada No. 16736.
Jaworski, J. F., Baldi, F., Bernhard, M., Brinckman, F. E., Hecht, H. P., Kloke, A., Legovic, T., McKenzie, J. M., Nriagu, J. O., Page, A. L., Sauerbeck, D. R., and Wasserman, K. (1984). Routes of exposure to humans and bioavailability. In Nriagu, J. O. (Ed.), *Changing Metal Cycles and Human Health*, Dahlem Konferenzen No. 28, Springer-Verlag, pp. 375–388.
Landrigan, P., Baloh, R., Whitworth, R., Staehling, N., and Rosenblum, B. F. (1975). Neuropsychological dysfunction in children with chronic low level lead absorbtion. *Lancet*, **1**, 708–712.
Lantzy, R. J., and Mackenzie, F. T. (1979). Atmospheric trace metals: global cycles and assessment of man's impact. *Geochim. Cosmochim. Acta*, **43**, 511–525.
Morgan, G. B., and Bretthauer, E. W. (1977). Metals in bioenvironmental systems. *Anal. Chem.*, **49**, 1210A–1214A.
NAS (1980). *Lead in the Human Environment.* US National Academy of Sciences, Washington, D. C.
Needleman, H. L., Gunnoe, C., Leviton, A., Reed, R. R., Peresie, H., Maher, C. and Barrett, P. (1979). Deficits in psychologic and classroom performance in children with elevated dentine lead levels. *New England J. Med.*, **300**, 689–695.
Needleman, H. L., Leviton, A., and Bellinger, D. (1982). Lead-associated intellectual deficit. *New England J. Med.*, **306**, 367.
Nriagu, J. O. (1979). Global inventory of natural and anthropogenic emissions of trace metals to the atmosphere. *Nature*, **279**, 409–411.
Pacyna, J. M. (1986). Atmospheric trace elements from natural and anthropogenic sources. In Nriagu, J. O., and Davidson, C. I. (Eds), *Toxic Metals in the Atmosphere*, pp. 33–52, John Wiley and Sons Inc., New York.
Patterson, C. C. (1980). An alternative perspective. In US National Academy of Sciences, *Lead in the Human Environment*, pp. 265–343.
Ruhling, A., and Tyler, G. (1971). Regional differences in the deposition of heavy metals over Scandinavia. *J. Appl. Eco.*, **8**, 497–507.
Sileo, L., Jones, R. N., and Hatch, R. C. (1973). The effect of ingested lead shot on the electrocardiogram of Canada Geese. *Avian Dis.*, **17**, 308–313.
Swedish Council for Building Research (1984). *Proceedings of the 3rd Int. Conf. on Indoor Air Quality and Climate. Stockholm, Aug. 20–24, 1984.*
Thornton, I., and Abrahams, P. (1984). Historical records of metal pollution in the environment. In Nriagu, J. O. (Ed.), *Changing Metal Cycles and Human Health.* Springer-Verlag, Berlin, pp. 7–25.
Waldron, H. A., and Stoffen, D. (1974). *Subclinical Lead Poisoning.* Academic Press, London, New York, 224 pages.
Weisel, C. P., Duce, R. A., Fasching, J. R., and Heaton, R. W. (1984). Estimates of the transport of trace metals from the ocean to the atmosphere. *J. Geophys. Res.* **89**, 11 607–11 618
Welsh, R. P. H., and Denny, P. (1980). The uptake of lead and copper by submerged aquatic macrophytes in two English Lakes. *J. Ecol.*, **68**, 443–455.
Zoller, W. H. (1984). Anthropogenic perturbations of metal fluxes into the atmosphere. In Nriagu, J. O. (Ed.), *Changing Metal Cycles and Human Health.* Springer-Verlag, Berlin, pp. 27–41.

Lead, Mercury, Cadmium and Arsenic in the Environment
Edited by T. C. Hutchinson and K. M. Meema

CHAPTER 2

Group Report: Mercury

S. LINDBERG, *Rapporteur*
P. M. STOKES, *Chairman*
E. GOLDBERG, C. WREN

INTRODUCTION

Among the metals and metalloids of concern for their potentially harmful effects in the environment, mercury is unique for a number of reasons. It is chemically distinctive (Carty and Malone, 1979) and because of its tendency to form strong covalent bonds its behavior in biological systems is also distinctive. The biogeochemical properties of mercury mean that elevated exposures tend to be quite natural phenomena. This suggests that relatively minor perturbations of key portions of the cycle (i.e. deposition, accumulation by fish) could result in major changes in the exposure to or uptake by sensitive human populations. The subtlety of mercury's early toxic effects suggests the need for careful monitoring of these key parts of the cycle and of target populations.

The major pathway to man is commonly accepted to be ingestion of aquatic organisms, particularly fish. Mercury is the only metal which indisputably biomagnifies through the food chain. In addition, unlike most other metals, mercury is efficiently transformed into its most toxic form (methyl mercury) in the aquatic environment. These facts make mercury exposure to target populations potentially serious even in remote areas. All of the above observations suggest that the effects of environmental exposure to mercury may be insidious, difficult to detect, and potentially widespread.

In the context of global metal cycles, the primary form of mercury emissions to the environment further exemplify its uniqueness. Mercury emissions, both natural and anthropogenic, are dominated by losses of vapour forms to the atmosphere (Kothny, 1973). These forms have a relatively long residence, conducive to long range transport (Lindqvist *et al.*, 1984). As a result, the bulk of the global atmospheric pool of mercury is in vapor form (Brosset, 1981). All of these facts indicate the truly global dispersion of mercury, regardless of its source. Although emission estimates have a high degree of uncertainty, several published calculations support the idea that natural emissions are quantitatively comparable with those of man (Pacyna, this volume, Ch. 7). However, some estimates suggest that natural sources dominate (Fitzgerald *et al.*, 1984). Even though anthropogenic emissions of total mercury have increased by two or three times between 1900 and 1970 (Andren and Nriagu, 1979), with several documented incidences of severe local effects, we feel that the total global atmospheric load of mercury is relatively insensitive to man's direct emissions. Therefore, we should focus on changes in the general chemistry of the atmosphere which could somehow enhance mercury deposition and thus have a significant effect on the global cycle. These changes could be reflected in major perturbations in both urban and remote areas.

The objective of our study group was to identify several major research questions which deserve particular attention at this time. These are addressed below. For recent reviews on the state of knowledge regarding mercury cycling the reader should consult Nriagu (1979) and Lindqvist *et al.* (1984).

QUESTION 1

Is there a human health problem related to mercury? Have any health effects been documented recently? If so, who are the populations at risk? Have any direct environmental effects ever been documented, or are the indirect 'effects' of food chain transfer and bioaccumulation the major sources for concern?

Food chain transfer with or without biomagnification seems to be confined to certain well-defined food chains with major aquatic links. For example, direct effects on fish-eating birds and mammals have been reported in isolated cases in Canada and Sweden. In addition to documented acute effects of human exposure to mercury we must remember that we have no data on the potential effects of long-term chronic exposure of man to levels of atmospheric Hg vapour which are measurably elevated above background levels, but not high enough to result in immediate toxic effects. For a review of the general human health effects of mercury the reader is directed to Piotrowski and Inskip (1981).

Mercury and Human Health

The primary route of exposure of methylmercury to man is through the consumption of food, especially fish. Several human populations around the world have been identified on the basis of their high consumption of fish as being in a high-risk group with respect to methylmercury poisoning.

Elevated mercury levels in many of the fish have been linked to both anthropogenic (Canada, Italy) and natural (New Guinea, New Zealand) sources (Hutton, this volume, Ch. 6). Many attempts to diagnose methylmercury intoxication have been inconclusive studies. Future clinical, epidemiological or neurological testing of humans must be statistically valid and rigorously designed.

In contrast to other toxic elements of concern (e.g. Cd, Pb) little is known about nutritional or other dietary factors (e.g. Ca interactions) which can influence methylmercury uptake and toxicity in humans or other mammals. Routine tests of humans in high-risk groups should be developed using the least intrusive but valid techniques (e.g. hair analysis). Also, surveillance of humans in areas subject to direct mercury discharge into the environment should be undertaken, regardless of fish consumption in these areas. While the use of mercury for agricultural purposes and direct industrial discharges of mercury into the environment have been largely curtailed in North America and much of Europe, this is not the case in developing countries around the world. Precautions must be taken to prevent future episodes of human poisoning by mercury.

Mercury and Environmental Health

Mercury is naturally present in low quantities in all living organisms and, under any degree of environmental pollution, may be efficiently accumulated by the biota. Mercury is unique among other toxic elements as it is the only metal which is consistently biomagnified within the aquatic food chain. However, evidence for effects in the environment have been limited to the upper trophic levels, though experimental demonstrations of effects on phytoplanktonic photosynthesis at ppb of mercury in marine waters have been reported.

Biological impacts associated with mercury in the environment have generally been associated with substantial local discharges of mercury into air or water from point sources. For example, elevated levels of mercury have been observed in a variety of aquatic benthic organisms, plants and fish living downstream of a chloralkali plant (Hildebrand *et al.*, 1980b). While toxic effects have not been documented for the aquatic biota, organisms consuming contaminated aquatic species have suffered mercury poisoning. Species affected in Canada under these conditions include a variety of fish-eating birds (Fimreite, 1979), wild mink, and otter (Wobeser and Swift, 1976; Wren, 1984). Serious contamination of terrestrial wild birds has been documented in Canada, Sweden and other countries as a result of the use of organo-mercurial compounds as seed dressings (Borg *et al.*, 1969; Fimreite, 1979). In Sweden, the species most affected were birds of prey and predatory animals such as fox and marten (Borg *et al.*, 1969).

In summary, biological effects attributable to mercury have almost always been associated with recognized point sources of mercury discharge into the environment. While mercury levels in biota, especially in aquatic species, may be elevated in certain areas resulting from more diffuse sources of mercury in the environment (e.g. fossil fuel combustion, see Question 7) no direct widespread biological effects have been unequivocally linked to these elevated tissue levels. However, many animal and plant species are sensitive to mercury poisoning, and this fact warrants continued surveillance of mercury in biota from a wide range of conditions and geographical locations. In addition, because of its capacity to become biomagnified upward in natural food chains, mercury monitoring must also be directed toward predatory species at the highest trophic levels which will be exposed to the most elevated mercury levels within a particular ecosystem.

QUESTION 2

What are the natural versus the anthropogenic fluxes of mercury to the global atmosphere? Are the same species of mercury released from each source? If natural fluxes actually do exceed anthropogenic fluxes by 3–5 times, as some have suggested, does it matter where the sources are, or merely that dispersion and deposition are occurring on a global scale?

Estimates of global emissions of mercury to the atmosphere are highly variable and necessarily imprecise because of the uncertainty about natural emission rates. Even more uncertain is the actual speciation of the emitted mercury, generally assumed to be elemental mercury vapor (Hg^{o}) (Lindqvist *et al.*, 1984). Recent estimates of global emissions from natural sources are on the order of 2700 (Pacyna, this volume, Ch. 7*) to 6000 (Fitzgerald *et al.*, 1984) tons/year while those for anthropogenic emissions are 630 to 2000 tons/year (same authors, respectively). Despite this uncertainty, both estimates result in a similar ratio of natural/man-made losses to the atmosphere (about 3–4 : 1). Earlier estimates (Andren and Nriagu, 1979) were higher for both pathways, but resulted in a ratio of the same order (2 : 1). Thus, there is general agreement that natural sources are somewhat larger than those due to man's activities. Given this and the global uniformity in atmospheric concentrations (Slemr *et al.*, 1984), it could be argued that the location of the specific sources is not particularly important on a global scale, although it can be critical on a local scale.

Because of the importance of the natural emission and cycling of mercury on a global scale, it is possible that man's major impact on the biogeochem-

* See Pacyna, Figure 7.1 for estimates of Hg fluxes for natural versus anthropogenic sources for the USA, Europe and the USSR (*Editors*).

ical cycle is of an indirect nature. That is, barring any major changes in anthropogenic emission trends, it is not the direct release of mercury which is critical, but the way in which man's other emissions may alter the global cycle. For example, our greatest influence on the cycle could be the increased emissions of oxidants and oxidant precursors which are so important in the cycle of atmospheric S and N (NAS, 1983). The existence of various theorized soluble vapor phase species is dependent on oxidation of Hg^{o} in the atmosphere, possibly by ozone or peroxides (Lindqvist *et al.*, 1984). One recent report suggests that the lifetime of mercury in the presence of elevated levels of O_3 is on the order of hours (Stevens *et al.*, 1982), instead of the commonly accepted period of years (Lindqvist *et al.*, 1984).

QUESTION 3

What are the current trends in mercury use on a global scale?

We need to document changes in both utilization and emission resulting from reductions in the use of mercury slimicides, Hg cell chloralkali processes, and Hg seed dressings. We may be biased in our view of the current global 'situation' by looking only at the technologically developed nations and not by also considering mercury use in developing countries. It is possible that there has been a shift in its industrial and agricultural use from the northern to the southern hemisphere. The technologies that caused problems in the developed world, such as the chloralkali and polyvinyl chloride industries, and the use of organic mercurials in biocides and slimicides are apparently being transferred to the developing countries in the southern hemisphere and tropics. However, there is only anecdotal evidence of such changes which would suggest a shift in environmental releases from the northern to the southern hemisphere. Global data to confirm these shifts are needed.

Since we suspect that man's releases of mercury to the atmosphere actually dwarf his releases directly to aquatic environments, we need to determine if a recent emission inventory exists. It appears that the newest data available is about 1975 for the USA and much of the rest of the world, although Swedish data exists for 1982. It is entirely possible that future trends will result in new problems of marine mercury pollution in the southern hemisphere. Perhaps a monitoring activity or other surveillance program of the mussel watch type (Goldberg *et al.*, 1978) might provide answers to these questions.

Research Proposal: The UNEP should be requested to initiate a program to obtain and publish statistics on mercury production, utilization and environmental disposal for the world scientific community.

QUESTION 4

What is the speciation of mercury in the atmosphere and how critical is it to the geochemical cycle?

Unlike the other metals considered in this SCOPE report, mercury exists in both industrial and natural emissions and in ambient air primarily as a vapor species, which is conducive to long range transport and deposition on a global scale. Common wisdom suggests that the major vapor species in air is elemental mercury (Hg^o). However, most measurements rely on various techniques which result in an operational definition of the real nature of the speciation of this vapor (Schroeder, 1982).

Knowledge of this speciation is critical to our understanding of removal processes, and perhaps more importantly, to any predictions of the influence of changing atmospheric chemistry on mercury deposition (see Question 2 above). Wet deposition is considered to be the major removal process of atmospheric mercury on a global scale (Lindqvist *et al.*, 1984), although the global rates are highly speculative. However, since Hg^o is sparingly soluble in water, typical atmospheric levels can in no way account for measured concentrations in rainwater (Fogg and Fitzgerald, 1979). Some have theorized the existence of highly soluble vapor phase compounds of mercury which are formed in ambient air but which are rapidly scavenged by precipitation (Lindqvist *et al.*, 1984). However, none of these compounds has yet been positively identified and it has been suggested that there is sufficient particle-borne soluble mercury to account for mercury in precipitation.

Analytical sampling trains need to be developed to accurately determine the true speciation of mercury vapor not only in ambient air and precipitation but in actively precipitating systems (in-cloud air) to determine the forms actually being scavenged by the raindrops. These speciation studies need to concern both chemical and physical speciation using methods validated under controlled laboratory conditions. For a review of such methods see Schroeder and Jackson (1983) and Brosset (1981, 1983, 1984).

As uncertain are our estimates of wet deposition, those for dry deposition are far worse, ranging over orders of magnitude (Lindberg, this volume, Ch. 8). Particle dry deposition can be theoretically estimated and perhaps measured in some cases; however, very little data exist with which to assess dry deposition of the largest atmospheric mercury pool, the vapor species. Deposition velocities with which to estimate global fluxes do not exist. However, some crop plants are known to absorb Hg^o directly from the air in a stomatal process (Browne and Fang, 1978), and even the limited solubility of Hg^o in seawater suggests some vapor dry deposition may occur to the sea surface. Plant physiologists have long had the ability to measure uptake of various vapors by foliage using chamber experiments (e.g. Hill, 1971). This expertise needs to be applied to mercury vapor uptake studies to determine the potential importance of this pathway. These data should then be used to estimate the extent to which direct vapour uptake by forests could be increasing the mercury burden in remote watersheds (see Question 7). If trees scavenge mercury vapor to the same extent as SO_2, HNO_3 or other major vapor species, it is possible that the commonly used bulk deposition methods of estimating mercury deposition to forests could significantly underestimate the true total input to potentially sensitive environments. One could effectively argue that atmospheric speciation is unimportant to the at-risk population which, in every case of which we are aware, is exposed directly to methylmercury in fish. However, because of the potential for increased deposition if atmospheric mercury were transformed to more soluble species, we recommend that major physical and chemical speciation studies of atmospheric Hg be instituted as soon as possible.

QUESTION 5

Can we trust the published data on mercury in the environment? If so, how far back in time?

The scientific community needs to develop and implement quality assurance plans and protocols for sampling, preservation, and analysis of mercury samples to assure accurate data for establishment of spatial and temporal trends and for input to new models.

The assemblage of the data base about mercury in the environment, about the natural and anthropogenic source terms, and about the reaction with the living system must be carried out with quality control procedures. The protocols for sampling, preservation, analysis and interpretation should be developed to satisfy totally the needs of the designated problem. Primary standards and interlaboratory comparison exercises are an essential part of the analytical procedures. Appropriate statistical analysis should be used to validate the data.

We recognize that recent intercalibration exercises of mercury analyses in sea water have demonstrated inadequacy in sampling and/or analytical techniques (Olaffson, 1982). We propose the formulation of primary reference standards for water, sediments and biological materials and extended intercalibration exercises involving these media.

Research Proposal: A workshop should be convened to develop and publish protocols for the sampling, storage, analysis and interpretation strategies to assess mercury contamination in the environment. Further, means to produce and distribute primary reference materials for mercury should be developed at the workshop.

QUESTION 6

Has the question of where methylation occurs ever been satisfactorily answered? It has been reported to occur in sediments, soils (by biotic and abiotic processes), and in fish (gut and skin). Demethylation has been reported for marine and terrestrial mammals. Has the puzzle of the lack of methylmercury in surface water been laid to rest? While these are interesting academic questions, do they matter to the ultimate receptor or next host organism, or is the only important point that the predator is ingesting methylmercury?

The first demonstrations of biological methylation of mercury were for sediments (Jensen and Jernelov, 1969; Olson and Cooper, 1976). The methylation process was shown by Wood *et al.* (1968) to be bacteriologically mediated, and the process of methylation is dependent upon not only the metabolic activity of bacteria and the concentration of inorganic mercury, but also on pH, redox potential, organic substrate and temperature (Beijer and Jernelov, 1979). Some experiments showed that anaerobic conditions promoted methylmercury production (Olson and Cooper, 1976) while others, for example, Bisogni and Lawrence, (1975), found that aerobic conditions were more favorable. Chemical methylation is also possible (Jewett and Brinkman, 1974; Carty and Malone, 1979), and it is probable that both chemical and biological processes are involved in the methylation of mercury.

More recent studies have indicated that methylation can also occur in the water column in freshwater (Furutani and Rudd, 1980) as well as marine systems (Topping and Davis, 1981). Methylation has also been shown to occur in fish intestines (Rudd *et al.*, 1980) and on fish skin (Rudd, personal communication, 1984). In general the source of the mercury is the sediment or eroding material entering the water body. The relative importance of methylation in the abiotic and biotic components, respectively, is difficult to determine. Levels of total mercury in water are extremely low (see Question 7), and the percentage of methylmercury to total mercury is about 30 in freshwaters (Kudo *et al.*, 1982; Brosset, 1981).

Microbial demethylation has also been demonstrated in mercury polluted sediments (Jernelov *et al.*, 1975) and is considered to be of fairly widespread and frequent occurrence. In fact, mercury methylation rates measured in complex systems are *net* methylation rates, resulting from a combination of both processes. Indirect evidence also exists for the occurrence of a demethylation in animals, as shown by a lower ratio of methyl to total mercury in liver and kidney compared with muscle. This type of evidence has been shown for marine mammals (Jones *et al.*, 1979; Gaskin *et al.*, 1976), for birds (Norheim and Froslie, 1978; Stoneborner and Harrison, 1981), mink (Jernelov *et al.*, 1976), otter (Wren *et al.*, 1980) and fish (Burrows and Krenkel, 1973).

In the opinion of the Working Group it is unlikely that a satisfactory elucidation of all the sites and mechanisms of methylation and demethylation, particularly the quantitative aspect, will be made for natural systems for a considerable period of time. Mercury speciation in water is particularly difficult to determine. At this time, it seems more pertinent to concentrate efforts on factors which predispose the carnivorous fish to accumulate methylmercury. These factors may include deposition from the atmosphere, and physiochemical variables which are known to be related to fish uptake and loss rate of mercury, both organic and inorganic. A mathematical modelling approach is likely to be promising, following a rigorously controlled collection of information in fish mercury by a standardized protocol for collection and measurement.

QUESTION 7

What is the mercury/rain/fish link? Is there direct evidence to support any of the hypotheses put forward to explain the trend of increased mercury levels in fish from remote lakes with low alkalinities? This relationship has been reported in several studies in Scandinavia and North America, but is it a recent phenomenon, or merely a recently discovered natural relationship? Do we have the historical data to test these hypotheses? If not, how can we design experiments or surveys to test them?

Within a given geographic area, and for surface waters which have no known point sources of mercury, the concentrations of the metal in fish muscle (of specified species and standard age) often show a statistically significant relationship with chemical and physical properties of the lake or its watershed. Most frequently repeated is the negative correlation between surface water pH and mercury in muscle (Suns *et al.*, 1980; Håkansson, 1980; Wiener, 1983; Wren and McCrimmon, 1983). Fish mercury has also been shown to be negatively correlated with alkalinity (Suns *et al.*, 1980) and negatively correlated with calcium concentration (Wren and McCrimmon, 1983). A positive correlation was shown by Suns *et al.* (1980) between mercury in perch and the watershed area/lake volume ratio.

Fish muscle containing >1.0 μg Hg/gm (wet weight) is considered unfit for human consumption in the USA and Sweden (Piotrowski and Inskip, 1981). In Canada the recommended limit is 0.5 μg/gm. Mercury in fish, although measured as total Hg for routine determinations, is generally $>80\%$ methyl mercury (e.g. Huckabee *et al.*, 1979), whose concentrations increase with body size (Phillips and Buhler, 1978; Huckabee *et al.*, 1979). Although the site of methylation is unclear, the process can be bacterially or chemically mediated. While fish cannot methylate *in vivo*, gut and skin bacteria can methylate mercury.

Total (and inferred methyl) mercury concentrations in edible flesh of piscivorous fish inhabiting waters of low alkalinity and pH < 6.0 frequently exceeds 1.0 μg/gm wet weight, thus creating a potential health risk. Furthermore, in Sweden and Canada a number of lakes have been 'blacklisted' with the issue of a warning not to consume fish. This has identified a problem in areas which have tourism and recreational fishing as an economic base.

Without historical data, we have no means to determine whether these elevations of mercury in fish from remote areas represent a recent phenomenon, or whether this is an entirely natural relationship. This means that we cannot determine from existing data whether the depression of lake pH from acidic deposition, with or without increased deposition of mercury, has increased the occurrence of mercury contamination in the areas of concern.

The major hypotheses concerning the fish mercury–lake pH relationship include the following:

(i) Bacterial methylation is enhanced by low pH. Recent studies on methylation in sediments by Ramlal *et al.* (1985) and by Steffan and Winfrey (1984) show clearly that the reverse is true, that is that low pH significantly decreases bacterial methylation rates in sediments.

(ii) Given a specific concentration of methyl mercury, the mono-methyl mercury–dimethyl mercury equilibrium is pH dependent, with mono-methyl mercury formation being favored at low pH (Jensen and Jernelov, 1969; Atmosphere–Biosphere Report 1981).

This has not been verified experimentally, but still remains as a viable hypothesis. Determination of mercury species in waters where total Hg does not exceed 10 ng/litre cannot be made directly. Miller and Akagi (1979) offer an alternative suggestion: the partitioning of mercury between sediment and water is pH dependent. They show that a decrease of 1 pH unit (for sandy sediment systems) or of 2 pH units (for organic sediments) doubled the amount of methyl mercury in the water column while the total amount of methyl mercury in the system remained unchanged.

(iii) Increased deposition of mercury from long range transport has occurred in conjunction with increased acid deposition, resulting in increased mercury in fish.

Total atmospheric deposition of mercury to a lake and its watershed is difficult to estimate for technical reasons (see Question 4). Dry deposition is particularly hard to measure, while the proportion of mercury re-emitted to the atmosphere is unknown. Mercury deposited onto land is likely to be bound in the organic layers of forest soils, and a considerable amount of this organic material will eventually find its way to the lake via runoff and leaching. Acidic precipitation could enhance the transport from watershed to lake, and possibly from atmosphere to watershed/lake system as well. Lindqvist *et al.* (1984) discuss this source of mercury to lakes, and give the range of 20–25 gm/year/km for mercury in runoff. Lindberg (1980) has also hypothesized that rain acidification could directly enhance the wet deposition of mercury to such lakes. This is because solubility of atmospheric mercury may be influenced by the acidity of the scavenging raindrop (see also Lindberg, this volume, Ch. 8).

It is possible to make rough calculations of a mercury budget for a lake, as done by Lindqvist *et al.* (1984) for southern Sweden. These authors suggest that increased mercury loading is at least partially responsible for increased fish mercury observed in remote lakes in south and central Sweden. Total mercury in water is consistently low (0.010–0.040 μg/litre) for the types of lakes under discussion; sediment mercury, while quite low (100–500 μg/kg), has increased in south and central Sweden over the last century.

Methylation of mercury (Question 6) can occur in water, sediment or in the watershed. Recent investigations (Hultberg, personal communication) suggest that non-biological methylation of mercury in the watershed stimulated by metals such as iron and manganese may result in a major contribution of the mercury into the lake in a bioavailable form. Acidic deposition would enhance this process.

(iv) Nutrients, organic content or general lake trophic status influence fish mercury indirectly.

Other factors related to fish mercury include size of fish, which may be related to lake pH in that stressed fish may grow more slowly, and trophic status of the water, the effect of which is probably indirect and is still controversial. Studies by Beijer and Jernelov (1979) suggested that fish from lakes of higher trophic status had less mercury, while recent evidence (discussed in Wren, 1983) tends to show the reverse effect. Although lake trophic status, alkalinity and pH tend to be correlated, this variable may nevertheless introduce a confounding effect in some data sets.

QUESTION 8

Do we have a significant mercury legacy in stored solid wastes? If all active emissions were stopped today, how much anthropogenic mercury would continue to be released from orphan wastes by volatilization or leaching?

As a result of legislated regulations the industrial emissions of mercury both to surface waters and to the atmosphere have been identified and considerably reduced. However, the processes of reduction in most cases resulted in a shift from direct emissions of Hg vapor or divalent Hg into air and water to emissions into landfills in the form of solid or slurried wastes containing mercury in various forms. For example, nearly 90% of the current mercury use by the chloralkali industry in the United States is deposited in waste ponds (>10^5kg/yr) (Harriss and Hohenemser, 1978). Conversion of existing Hg-cell plants to non-Hg processes suggests that existing waste ponds will be the major source of mercury release from this industry in the future unless these wastes are stored such that volatilization and leaching losses do not occur. The loss of mercury from one such orphan waste site in the United States was estimated to be 75 kg/yr, 50% to water and 50% to air, which is comparable with losses from active Swedish plants (Lindberg and Turner, 1977). In addition, air concentrations of mercury vapor near this site were close to the USA guidelines for ambient air, and concentrations of methyl mercury in downstream fish exceeded WHO guidelines (Hildebrand *et al.*, 1980a).

Because of the massive amounts of these wastes produced (>10^9 kg/yr in the USA alone), the volatility and solubility of mercury in such wastes, and the potential residence time of mercury in large waste deposits (hundreds of years), these areas must be located and monitored. An additional but similar problem involves often extensive areas of sediment in lakes and rivers which were at one time contaminated by local industrial activities. Examples exist in many countries (Nriagu, 1979). Although direct emissions to these water bodies have ceased, the sediments represent a potential long term source of mercury to the aquatic system, particularly if the sediments are dredged or otherwise resuspended (e.g., Lindberg and Harriss, 1977).

Mercury bioconcentration and methylation in aquatic food chains (see Questions 1 and 6) indicates that fish consumption is a major mercury pathway to man. This suggests that any potential large scale or long term releases of mercury to surface waters from contaminated waste storage areas or sediments must be closely monitored. As a first step we recommend that all nations inventory the locations of major mercury users, both past and present, to prevent the loss of such critical information upon cessation of active production at these sites. All too often hazardous waste sites are forgotten until years later when it is too late or too costly to remedy the problem. Once these sites have been identified and located, simple but routine monitoring programs should be instituted, concentrating on key indicator organisms such as top carnivore fish or mammals.

QUESTION 9

What is the true nature of the Hg–Se relationship recently described? Selenium is said to decrease uptake and toxicity of mercury to certain organisms.

What are the mechanisms and are we simply trading one toxic metal problem for another?

Several investigators have demonstrated that the administration of selenium together with methylmercury reduces the toxicity of the mercury (NRCC 1979). These effects have been demonstrated experimentally both *in vitro* and *in vivo* for a number of laboratory animals. There is also some suggestion that selenium may provide protection against methylmercury toxicity in marine mammals which accumulate high levels of mercury. A 1 : 1 Hg : Se molar ratio in tissues has previously been reported (Koeman *et al.*, 1975). However, other studies show that the ratio of Hg : Se is not always 1 : 1 (molar) in these organisms (Koeman *et al.*, 1975) and that an efficient demethylation process occurs in the target organs of many marine mammals and birds. This also affords protection against MeHg toxicity (Norheim and Froslie, 1978; Gaskin *et al.*, 1979).

Despite the amount of research and discussion of the subject, the mechanisms through which selenium provides protection against methylmercury poisoning are far from clear (Skerfving, 1978). The mechanism of action of selenium does not appear to involve a decrease in tissue Hg levels, but may involve free radicals formed by the breakdown of MeHg molecules (Ganther, 1978).

No recent advances have contributed to unraveling the interactions of selenium and mercury. The status quo can best be expressed from the review of Piotrowski and Inskip (1981): 'the present-day recognition of selenium's influence on the metabolism of MeHg is of limited use in the understanding of mechanisms involved in the Se : MeHg interactions'.

Other factors have also been shown experimentally to modify the toxic effects of Hg. For example, Vitamin E delayed the onset of methylmercury poisoning in animals (Welsh and Soares, 1976), while Gilbert *et al.* (1983) demonstrated *in vitro* that Vitamin E has protective potential against the genotoxicity of methylmercury. Yamane *et al.* (1976) observed that Zn acetate modified the toxicity of methylmercury to rats, although other researchers have not observed similar effects (Johnson and Pond, 1974). Despite these studies using culture tissues or laboratory animals, the interactive effects of diet, nutrition, disease and natural stresses on the uptake and toxicity of methylmercury to populations of wild organisms and to humans remains unknown.

REFERENCES

Andren, A. W., and Nriagu, J. O. (1979). The global cycle of mercury. In Nriagu, J. O. (Ed.), *The Biogeochemistry of Mercury in the Environment*, pp. 1–21, Elsevier/North Holland Biomedical Press, Amsterdam.

Atmospheric-Biosphere Interactions: Towards a Better Understanding of the Ecological

Consequences of Fossil Fuel Combustion. (1981). National Academy Press Wash. DC, 263 pages.

Beijer, K., and Jernelov, A. (1979). Methylation of mercury in aquatic environments. In Nriagu, J. O. (Ed.), *The Biogeochemistry of Mercury in the Environment*, pp. 203–211, Elsevier/North-Holland Biomedical Press, Amsterdam.

Bisogni, J. J. Jr., and Lawrence A. W. (1975). Kinetics of mercury methylation in aerobic and anaerobic aquatic environments. *J. Water Pollut. Control Fed.*, **47**, 135–152.

Borg, K., Wanntrop, M., Erne, M., and Hanko, E. (1969). Alkylmercury poisoning in terrestrial Swedish wildlife. *Viltrevy*, **6**, 301–379.

Brosset, C. (1981). The mercury cycle (preliminary communication). *Water, Air and Soil Poll.*, **16**, 253-255.

Brosset, C. (1983). *KHM 76. Emissions of Mercury Compounds with Flue Gas.* Project Coal, Health and Environment. Final Report April 1983. Swedish State Power Board. S-16287, Vallingby, Sweden.

Brosset, C. (1984). *Investigations of Mercury in Air and Natural Waters during 1983.* Report EM 1327 Swedish Water and Air Research Laboratory. Box 5207 S-40220, Gotenborg, Sweden.

Browne, C. L., and Fang, S. C. (1978). Uptake of mercury vapor by wheat: an assimilation model. *Plant Physiol.*, **61**, 430–433.

Burrows, W. D., and Krenkel, P. A. (1973). Studies on uptake and loss of CH^3–203 by bluegills. *Environ. Sci. Technol.*, **7**, 1127–1130.

Carty, A. J., and Malone, S. F. (1979). The chemistry of mercury in biological systems. In Nriagu, J. O. (Ed.), *The Biogeochemistry of Mercury in the Environment*, pp. 433–480, Elsevier/North-Holland Biomedical Press, Amsterdam.

Fimreite, N. (1979). Accumulation and effects of mercury on birds. In Nriagu, J. O. (Ed.), *The Biogeochemisty of Mercury in the Environment*, pp. 601–627, Elsevier/North-Holland Biomedical Press, Amsterdam.

Fitzgerald, W. F., Gill, G. A., and Kim, J. P. (1984). An equatorial Pacific Ocean source of atmospheric mercury. *Science*, **224**, 597–599.

Fogg, T. R., and Fitzgerald, W. F. (1979). Mercury in southern New England coastal rains. *J. Geophys. Res.*, **84**, 6987–6989.

Ganther, H. E. (1978). Modification of methylmercury toxicity and metabolism by selenium and vitamin E: possible mechanisms. *Environ. Health Persp.*, **25**, 71–76.

Gaskin, D., Stonefield, K. I., Suda, P., and Frank, R. (1979). Changes in mercury levels in harbour porpoises from the Bay of Fundy, Canada, and adjacent waters during 1969–1977. *Arch. Environ. Contamin. Toxicol.*, **8**, 733–762.

Gilbert, M. M., Sprecher, J., Chang, L. W., and Meisner, L. F. (1983). Protective effects of vitamin E on genotoxicity of methylmercury. *J. Toxicol. Environ. Health.*, **12**, 767–773.

Goldberg, E. D., Bowen, V. T., Farrington, J. W., Harvey, G., Martin, J. H., Parker, P. L., Risebrough, R. W., Robertson, W., Schnieder, E., and Gamble, E. (1978). The mussel watch. *Environ. Conserv.*, **5**, 101–125.

Hakanson, L. (1980). The quantitative impact of pH, bioproduction and Hg-contamination on the Hg-content of fish (pike). *Environ. Pollut., (ser. B)* **1**, 235–304.

Harriss, R. C., and Hohenemser, C. (1978). Mercury: measuring and managing the risk. *Environ.*, **20**, 25–36.

Hildebrand, S. G., Lindberg, S. E., Turner, R. R., Huckabee, J. W., Lund, J. R., and Andren, A. W. (1980a). *Biogeochemistry of Mercury in a River–Reservoir Sys-*

tem: Impact of an Inactive Chloralkali Plant in the Holston River–Cherokee Lake. ORNL/Tm-6141. Oak Ridge National Laboratory, Oak Ridge, Tennessee.

Hildebrand, S. G., Strand, R. H., and Huckabee, J. W. (1980b). Mercury accumulation in fish and invertebrates of the North Fork Holston River, Virginia and Tennessee. *J. Environ. Qual.*, **9**(3), 393–400.

Hill, A. C. (1971). Vegetation: a sink for atmospheric pollutants. *J. Air Poll. Cont. Assoc.*, **21**, 341–346.

Huckabee, J. W., Elwood, J. W., and Hildebrand, S. G. (1979). Accumulation of mercury in freshwater biota. In Nriagu, J. O. (Ed.), *The Biogeochemistry of Mercury in the Environment.*, pp. 277–302, Elsevier/North-Holland Biomedical Press, Amsterdam.

Hultberg, H. Personal communication. Swedish Environmental Research Institute, Goteberg, Sweden.

Jensen, S., and Jernelov, A. (1969). Biological methylation of mercury in aquatic organisms. *Nature,* **223**, 753–754.

Jernelov, A., Lander, L., and Larsson, T. (1975). Swedish perspectives on mercury pollution. *J. Water Pollut. Control Fed.*, **47**, 810–822.

Jernelov, A., Johanson, A., Sorenson, L., and Svenson (1976). Methylmercury degradation in mink. *Toxicol.*, **6**, 315–321.

Jewett, K. L., and Brinkman, F. E. (1974). Transmethylation of heavy metal ions in water. *Div. Environ. Che,. Am. Chem. Soc.*, **14**, 218–225.

Johnson, S. L., and Pond, W. G. (1974). Inorganic vs. organic mercury toxicity in growing rats: protection by dietary Se but not Zn. *Nat Rep. Int.*, **9**, 135–147.

Jones, D., Ronald, K., Lavigne, D. M., Frank, R., Holrinet, M., and Uthe, J. F. (1976). Organichlorine and mercury residues in the harp seal. *Sci. Total Environ.*, **5**, 181–195.

Koeman, J. H., van de Ven, W. S., de Goeij, J. J., Tjioe, P. S., and van Haaften, J. L. (1975). Mercury and selenium in marine mammals and birds. *Sci. Total Environ.*, **3**, 279–287.

Kothny, E. L. (1973). The three-phase equilibrium of mercury in nature. In Kothny, E. L. (Ed.), *Trace Elements in the Environment.* Adv. Chem. Ser. 123, 48–80, American Chemical Society, Reinhold, New York.

Kudo, A, Nagase, H., and Ose, Y. (1982). Proportional methylmercury to total amount of mercury in river waters in Canada and Japan. *Water Res.*, **16**, 1011–1015.

Lindberg, S. E. (1980). Mercury partitioning in a power plant plume and its influence on atmospheric removal mechanisms. *Atmos. Environ.*, **14**, pp. 227–231, and **15**, 631–634.

Lindberg, S. E., and Harriss, R. C. (1977). Release of mercury and organics from resuspended near-shore sediments. *J. Water Poll. Cont. Fed.*, **7**, 2479–2487.

Lindberg, S. E., and Turner, R. R. (1977). Mercury emission from chlorine-production solid waste deposits. *Nature, Lond.*, **286**, 133–136.

Lindqvist, O., Jernelov, A., Johansson, K., and Rodhe, H. (1984). *Mercury in the Swedish Environment, Global and Local Sources.* Report of the Workshop held at Lerum, Sweden, November, 1983. SNV PM 1 National Swedish Environ. Protect. Board, Solna, Sweden.

Miller, D. R., and Akagi H. (1979). PH affects mercury distribution not methylation. *Ecotox. and Environ. Safety.*, **3**, 36–38.

National Academy of Sciences (NAS) (1983). *Acid Deposition–Atmospheric Processes in Eastern North America.* National Academy Press, Washington, DC.

Norheim, G., and Froslie, S. (1978). The degree of methylation and organ distribution of mercury in some birds of prey in Norway. *Acta pharmacol et toxicol.*, **43**, 196–204.

Nriagu, J. O. (1979). *The Biogeochemistry of Mercury in the Environment.* Elsevier/North-Holland Biomedical Press, Amsterdam, 696 pages.

NRCC (1979). *Effects of Mercury in the Canadian Environment.* Nat. Res. Council Canada. No. 16739, 290 pages.

Olaffson, J. (1982). An international intercalibration for mercury in seawater. *Marine Chem.*, **2**, 129–144.

Olson, B. H., and Cooper, R. C. (1976). Comparison of aerobic and anaerobic methylation of mercuric chloride by San Francisco Bay sediments. *Water Res.*, **10**, 113–116.

Phillips, G. R., and Buhler, D. R. (1978). The relative contributions of methylmercury from food or water to rainbow trout (*Salmo gairdneri*) in a controlled laboratory experiment. *Trans. Am. Fish. Soc.*, **107**, 853–861.

Piotrowski, J. K., and Inskip, M. J. (1981). *Health Effects of Methylmercury.* MARC Report No. 24, Chelsea College, London. 82 pages.

Ramlal, P. S., Rudd, J. W., Furutani, A., and Xun, L. (1985). The effect of pH on methyl mercury production and decomposition in lake sediments. *Can. J. Fish. Aqu. Sci.*, **42**, 685–692.

Rudd, J. W. Personal communication (1984). Freshwater Institute, Winnipeg, Manitoba.

Rudd, J. W., Furutani, A., and Turner, M. A. (1980). Mercury methylation by fish intestinal contents. *Appl. Environ. Microbiol.*, **40**(4), 777–782.

Schroeder, W. H. (1982). Sampling and analysis of Hg and its compounds in the atmosphere. *Environ. Sci. Technol.*, **16**, 394A–400A.

Schroeder, W. H., and Jackson, R. A. (1983). A method for the separation of several volatile mercury species in the atmosphere. Presented at the 6th World Congress on Air Quality (IUAPPA), Paris, France, May, 1983.

Skerfving, S. (1978). Interaction between selenium and methylmercury. *Environ. Health Perspec.*, **25**, 57–65.

Slemr, F., Schuster, G., and Seiler, W. (1984). Distribution, speciation and budget of atmospheric mercury. (in prep.) Cited in Lindqvist *et al.* 1984.

Steffan, R. M., and Winfrey, M. R. (1984). Effect of experimental acidification on mercury methylation and volatilization in a northern Wisconsin lake. (abstract). *Abstr. Annu. Am. Soc. Microbiol.*, **84**, 207.

Stevens, R. D. S., Reid, N. W., Schroeder, W. H., and McLean, R. A. N. (1982). The chemical forms and lifetimes of mercury in the atmosphere. Presentation to the 65th Canadian Chemical Conference, Toronto, Canada, May, 1982.

Stoneborner, D. L., and Harrison, C. S. (1981). Heavy metal residues in sooty terns from the Gulf of Mexico and north-central Pacific Ocean. *Sci. Total Environ.*, **19**, 51–58.

Suns, K., Curry, C., and Russel, D. (1980). *The Effects of Water Quality and Morphometric Parameters on Mercury Uptake by Yearling Yellow Perch.* Ontario Ministry of the Environment, Technical Report LTS80-1, Rexdale, Ontario, Canada.

Welsh, S. O., and Soares, J. H. (1976). The protective effect of vitamin E and selenium against methylmercury toxicity in Japanese quail. *Nat. Rep. Int.*, **13**, 43–51.

Wiener, J. (1983). *Comparative Analysis of Fish Populations in Naturally Acidic and Circumneutral Lakes in Northern Wisconsin.* US Fish and Wildlife Service Report FWS/OBS/-80/40.16. Eastern Energy and Land Use Team, Kearneysville, West Virginia USA. 107 pages.

Wobeser, G., and Swift, N. (1976). Mercury poisoning in a wild mink. *J. Wildl. Dis.*, **12**, 335–340.

Wood, J. M., Kennedy, S., and Rosen, G. G. (1968). Synthesis of methyl mercury compounds by extracts of a methanorganic bacterium. *Nature (Lond.)* **220**, 173–174.

Wren, C. D. (1983). Examination of environmental factors affecting the distribution of mercury and other metals in Precambrian Shield lake ecosystems. PhD Thesis, University of Guelph, 153 pages.

Wren, C. D. (1984). Probable case of mercury poisoning in a wild otter (*Lutra canadensis*) in northwestern Ontario. *Can. Field Nat.*, **99**(1), 112–114

Wren, C. D., and McCrimmon, H. R. (1983). Mercury levels in the sunfish, *Lepomis gibbosus*, relative to pH and other environmental variables of Precambrian Shield lakes. *Can. J. Fish Aquat. Sci.*, **40**, 1737–1744.

Wren, C. D., McCrimmon, H. R., Suda, P., and Frank, R. (1980). Total and methylmercury levels in wild mammals from the Precambrian Shield of south central Ontario. *Bull. Environ. Contam. Toxicol.*, **25**, 100–105.

Yamane, Y., Fukino, H., and Imagawa, M. (1976). Suppressive effect of zinc on the toxicity of mercury. *Chem. Pharm. Bull.*, **24**, 836–837.

Lead, Mercury, Cadmium and Arsenic in the Environment
Edited by T. C. Hutchinson and K. M. Meema

CHAPTER 3

Group Report: Cadmium

M. HUTTON, *Rapporteur*
R. L. CHANEY, *Chairman*
C. R. KRISHNA MURTI, M. A. OLADE, A. L. PAGE

INTRODUCTION

Cadmium is a relatively rare metal which occurs in nature as a minor component of other non-ferrous metal ores. It may be considered as an element of concern peculiar to the twentieth century since commercial extraction began only at the beginning of this century. Over half the cadmium ever produced has been refined in the last 20 years (Nriagu, 1979).

Given the striking increase in cadmium consumption over recent years, it is pertinent to ask whether there has been a corresponding change in the global cycling of the metal. This question can be partially answered by examination of data from studies into the historical trends of metal levels in dated ice cores from remote areas of the world. Such studies reveal that there is no evidence for any increase in the cadmium content of recent ice cores from either Greenland or the Antarctic (Zoller, 1984). Thus, human activities have not, at present, had a measurable impact on the global, or even hemispheric cycling of the metal. This is in marked contrast to lead, where human activities have enhanced lead concentrations in recent ice cores from Greenland and the Antarctic, as well as in surface waters from mid-ocean sampling locations (Zoller, 1984).

The human health significance of cadmium can be placed in perspective by comparison with the situation for lead. In the case of lead, there is now a consensus that children, particularly in urban localities, are at risk from the neurobehavioral effects of the metal. In contrast, reports of health effects of cadmium in environmentally-exposed populations have been restricted to certain communities in Japan and, recently, elderly populations from an industrialized city in Europe (Roels *et al.*, 1981).

SOURCES OF CADMIUM

Source Inventories

To understand the biogeochemical cycle of cadmium, it is essential to identify and quantify the sources of the metal in the environment. Many of the major human sources have been known for some time, but only in recent years have efforts been made to quantify these discharges. Such studies have produced source inventories for areas ranging from extensive details of an industrial sector in one country to a much more poorly known full global inventory. Most global inventories have only been concerned with atmospheric emissions of cadmium. Little attention has been paid to discharge to land and water. In only a few instances have natural sources of cadmium release been taken into account. Table 3.1 provides a list of published source

Table 3.1 Published source inventories of cadmium for country, regional and global areas

Study area	Time period	Natural sources	Environmental compartment			Reference
			Atmosphere	Water	Land	
Sweden	1977/78	—	×	×	—	SNV, 1982
USA	—	—	×	—	—	EPA, 1979
USA	10-year period	—	×	×	×	Yost, 1979
Denmark	1977/78	—	×	×	×	Miljoministeriet, 1980
Switzerland	—	—	×	×	×	Brunner and Baccini, 1981
EEC	1979/80	×	×	×	×	Hutton, 1982
Europe	1979	×	×	—	—	Pacyna, Semb and Hanssen, 1984
North Sea	—	×	—	×	—	Norton, 1982
World	1975	×	×	—	—	Nriagu, 1979
World	—	×	×	—	—	Lantzy and Mackenzie, 1979
World	—	×	×	—	—	Jaworowski, Bysiek and Kownacka, 1981

inventories for cadmium which have been prepared for countries, regions or at the global level. This list is not intended to be exhaustive but rather to illustrate the range of inventories which have been carried out in the last five years. Examination of the data in these studies would reveal large disparities in the source strengths and emission factors (gm cadmium emitted/tonne material produced or consumed) assigned to a particular process. It will therefore be necessary to carry out a 'quality control' study in order to determine the validity of the values employed in these inventories. In addition, it is suggested that special attention should be paid to the construction of accurate inventories for those countries or regions where human activities may be mobilizing large amounts of the metal. Of particular interest will be those countries which are large producers and consumers of cadmium. SCOPE can play a significant role in this, together with such international organizations as UNEP.

Natural Sources

Volcanic action is considered to be a major natural source of atmospheric cadmium. This situation results from the vast amount of matter discharged, together with the enrichment of cadmium in the volcanic aerosol. However, considerable variation exists in the degree of enrichment reported for different volcanoes and this presents problems when attempting to estimate worldwide discharges from this source. The periodicity of volcanic events as well as their often remote and poorly monitored locations also present considerable difficulties.

Other natural sources of atmospheric cadmium include ocean sprays, forest fires and the release of metal-enriched particles from terrestrial vegetation. These sources are difficult to quantify and only in the case of release from vegetation are the quantities likely to be significant on a global scale.

The weathering of crustal materials plays a major role in the natural cycle of cadmium but rarely results in any marked enrichment of the metal in the environment. A notable exception is that of certain carboniferous shales found in various parts of the world which are enriched with cadmium. The soils derived from these deposits and the vegetation growing in the soils also contain elevated cadmium levels.

Anthropogenic Sources

The sources of environmental cadmium associated with human activities can be conveniently divided into two broad categories; (a) those which produce a marked cadmium impact in the vicinity of the discharge source and (b) those which do not. The former category includes non-ferrous metal mines, smelters and refineries together with those industries involved in the manufacture of cadmium-containing articles. The second source category contains many 'inadvertent' sources in which cadmium is a natural constituent of the material being consumed or processed, such as coal combustion, refuse incineration and the iron and steel industries. Studies around the above facilities have, in general, been unable to detect any enhancement of soil or atmospheric cadmium levels. However, the cadmium emissions from such sources may still make a significant contribution to the overall atmospheric burden of cadmium at the local or regional level. The size of the particles involved in cadmium transport is just one key factor in determining local accumulations.

Non-ferrous metal mines can be a significant source of local cadmium contamination, particularly those mines which exploit zinc and lead ores. Zinc ores invariably contain the most cadmium but the actual content varies widely between different ore bodies. This factor together with mining practices, local topography and drainage all influence the intensity of contamination. Airborne contamination can arise from the wind-blown dispersal of metal-rich spoil heaps,while water-borne transport of cadmium-rich material may result in contamination of agricultural land some distance downstream of the mine. This form of dispersal can be particularly marked during flood and storm conditions. This problem is not confined to active mines and many disused zinc-lead mines act as persistent sources of cadmium. Mine spoil heaps in Wales still discharge into rivers a century after the mines closed.

It is suggested that attention should be drawn to this long-term and persistent source of cadmium contamination in those developing countries with a large or rapidly growing mine production of zinc. Two countries of particular concern are Peru and Mexico which together account for about a quarter of the world's total mine production of zinc (International Lead and Zinc Study Group, 1983).

Cadmium is extracted as a by-product during the refining of zinc ores. The two basic processes of zinc production, thermal smelting and electrolytic refining, produce different kinds of cadmium-rich discharge. Thermal smelters are responsible for much larger atmospheric emissions of cadmium and also produce liquid wastes rich in cadmium which require safe disposal. In contrast, the only major wastes associated with electrolytic plants are solid leach residues.

In the last two decades, electrolytic refining has assumed an increasingly important share of the world's production of zinc and cadmium and at present accounts for over 75% of the market. Over the same period, the Imperial Smelting Furnace process has become the major thermal route of production. The once important vertical and horizontal retort processes, which emit large quantities of cadmium to the atmosphere, have been phased out in many countries. However, smelters of these types are still in operation in Japan, Mexico, China and the USSR (International Lead and Zinc Study Group, 1980).

Several published inventories have assigned extremely large atmospheric emissions of cadmium to zinc production. It is considered that such studies may have erroneously assumed that all zinc production occurred by thermal processes.

The primary production of lead and copper also releases cadmium to the environment, the amount mobilized being dependent on the cadmium content of the ore concentrate being processed. Generally, copper ores are only modestly enriched with cadmium, while larger discharges of cadmium can be expected from the primary production of lead. This situation may be reversed in the secondary production of the two metals as a large proportion of the feed material for lead refineries is relatively pure scrap lead while copper refineries may handle copper-cadmium alloy scrap.

Cadmium has five principal uses: (1) as a protective plating on steel, (2) in various alloys, (3) in pigments, (4) in stabilizers, and (5) in nickel-cadmium batteries. Cadmium discharges to the environment arise both during the manufacture of cadmium-containing materials and when these products are discarded. Currently, only about 5% of the cadmium consumed is recovered after disposal compared with about 40% recovery for lead. The reason for this poor recovery is that, unlike other metals of commercial importance, most cadmium is consumed in the form of compounds. These are present in materials at relatively low concentration and pose economic and technical problems for recovery.

The liquid and solid cadmium wastes arising from the manufacture of the above items are, in many countries, discharged to the sewage system. Those countries without a sewage network presumably dispose of these wastes to adjacent watercourses or to specific disposal areas. In the case of product disposal, all cadmium-containing articles can enter the municipal waste disposal pathway where their fate is either to be landfilled or incinerated. A considerable proportion of the cadmium-plated products will ultimately be recycled but these will enter the scrap steel cycle and a proportion will ultimately be emitted to the atmosphere.

In terms of identifying those areas of the world where large amounts of cadmium may be mobilized, the situation in the European Community is of particular interest. Currently, more cadmium is produced and consumed in this region than in either the USSR or USA. Several developing countries in south-east Asia exhibit large growth rates in the consumption of cadmium but at present the quantities consumed are relatively small.

Pathways for Human Exposure

Dietary intake represents the largest source of cadmium exposure in the non-smoking general population. Tobacco is an additional source in smokers, and can be as important as cadmium in food to overall body burdens. In general, cadmium levels in drinking water are low and intake from beverages is trivial compared with dietary intake. Plant-based foodstuffs represent the largest source of dietary cadmium in most populations and thus changes in crop cadmium levels may have a marked impact on cadmium exposure.

Studies of dietary cadmium exposures from several countries indicate that average daily intakes in populations from Europe and the USA are about 20–40 μg. These values are lower than the tolerable intake for dietary cadmium of 1 μg/kg body weight/day, or for Europeans, about 70 μg/day, proposed by the FAO/WHO. Certain groups are exposed to higher than average quantities of cadmium, either because their food is contaminated or because of unusual dietary habits. Examples include individuals who consume large quantities of shellfish and those who regularly eat food plants grown on contaminated soils. Such soils are those which have received heavy applications of sewage sludge and phosphate fertilizers, those in areas of mineralization and those close to non-ferrous metal smelters and other point sources of airborne cadmium.

Generally, cadmium uptake from air represents only a minor contribution to total intake, despite the relatively high uptake efficiency of the lungs. However, certain populations living close to point sources of airborne cadmium may inhale significant quantities of the metal.

Cadmium is present as a finely dispersed aerosol in cigarette smoke, resulting in an efficient uptake of the metal. It has been calculated that the daily consumption of 20 cigarettes results in the absorption of an amount of cadmium equivalent to an additional dietary intake of 25–30 μg cadmium. Individuals who smoke two or more packets of 20 will, of course, absorb correspondingly more cadmium.

There is no doubt that there has been a marked increase in the consumption and consequent environmental dissipation of cadmium in this century. There is also some evidence that there has been a corresponding increase in the exposure levels of the general population in some regions. The findings of a study of preserved wheat samples from Sweden, collected between 1916 and 1972, are equivocal but there is some indication of an increase in cadmium content of wheat in recent years. More convincing evidence exists for an increase in the body burden of cadmium during this century in populations from Europe and Japan. Comparison of the cadmium content of historic specimens of kidney with present-day values revealed that present values exceeded those found in historic specimens.

INFORMATION NEEDS

(1) Improvements are needed in the accuracy of existing source inventories for cadmium, paying attention to regional differences in industrial production technology and efficiency of control devices. Further studies are required into cadmium discharges to aquatic systems, particularly estuaries and coastal waters.

(2) Studies into the impact of human activities on the biogeochemical cycling of cadmium should pay particular attention to those areas of the

world which are either important mine and smelter producers of non-ferrous metals or are major consumers of cadmium.

(3) More information is needed on the sources of cadmium in crop plants together with the relative significance of cadmium inputs to agricultural land from phosphate fertilizers, sewage sludge and atmospheric deposition. The long-term behaviour of cadmium in arable soils requires investigation in different regions of the world.

(4) Further research is required into the significance of crops grown in areas of elevated cadmium for human dietary intake in different parts of the world. Such studies should take into account local marketing and distribution patterns of food crops.

REFERENCES

Brunner, P. H., and Baccini, P. (1981). Die Schwermetalle, Sorgenkinder der Entsorgung? *Sep. Neuen Zürcher Z. Beil. Forsch Tech. Nr 70.*

EPA (1979). *Sources of Atmospheric Cadmium.* EPA-450/5-79-006, US Environmental Protection Agency, Research Triangle Park, North Carolina.

Hutton, M. (1982). *Cadmium in the European Community: A Prospective Assessment of Sources, Human Exposure and Environmental Impact.* MARC Report 26, Monitoring and Assessment Research Centre, Chelsea College, University of London. 100 pages.

International Lead and Zinc Study Group (1980). *World Directory: Lead and Zinc Mines and Metallurgical Works.* Metro House, London.

International Lead and Zinc Study Group (1983). *Trends in Production of Lead and Zinc.* Metro House, London.

Jaworowski, Z., Bysiek, M., and Kownacka, L. (1981). Flow of metals into the global atmosphere. *Geochim. Cosmochim. Acta,* **45**, 2185–2199.

Lantzy, R. J., and Mackenzie, F. T. (1979). Atmospheric trace metals: global cycles and assessment of man's impact. *Geochim. Cosmochim. Acta,* **43**, 511–523.

Miljøministeriet (1980). *Cadmiumforurening.* Miljøministeriet, Copenhagen.

Norton, R. L. (1982). *Assessment of Pollutant Loads to the North Sea.* Water Research Centre Technical Report TR182, Medmenham, UK.

Nriagu, J. O. (1979). Global inventory of natural and anthropogenic emissions of trace metals to the atmosphere. *Nature,* **279**, 409–411.

Pacyna, J. M., Semb, A., and Hanssen, J. E. (1984). Emission and long-range transport of trace elements in Europe. *Tellus,* **36B**, 163–178.

Roels, H. A., Lauwerys, R. R., Buchet, J-P., and Bernard, A. (1981). Environmental exposure to cadmium and renal function of aged women in three areas of Belgium. *Environ. Res.,* **24**, 117–130.

SNV (1982). *Monitor 1982.* Tungmetaller och Organiska miljögifter i Svensk Natur. Statens Naturvardsverk, Solna, Sweden.

Yost, K. J. (1979). Some aspects of cadmium flow in the U.S. *Environ. Health Perspect.,* **28**, 5–16.

Zoller, W. H. (1984). Anthropogenic perturbation of metal fluxes into the atmosphere. In Nriagu, J. O. (Ed.), *Changing Metal Cycles and Human Health,* Dahlem Konferenzen, Springer-Verlag, Berlin, pp. 27–41.

Lead, Mercury, Cadmium and Arsenic in the Environment
Edited by T. C. Hutchinson and K. M. Meema

CHAPTER 4

Group Report: Arsenic

P. BUAT-MÉNARD, *Rapporteur*
P. J. PETERSON, *Chairman*
M. HAVAS, E. STEINNES, D. TURNER

INTRODUCTION

Arsenic is well recognized as an element of public concern. Although the main route of entry to man is ingestion, inhalation can also be significant for occupational exposure. In contrast to several other potentially toxic elements, local health effects are apparent in areas naturally enriched with arsenic as well as in areas contaminated by industrial activities. This element has been quite extensively studied in terrestrial environments, fresh and marine waters and air in some geographic regions. With the development of new analytical methods much greater insight into the occurrence of the various inorganic and organic forms of arsenic is being obtained. Our knowledge of arsenic cycling is, however, still limited primarily because some natural sources which may be dominant are poorly quantified. However, there are estimates of the extent to which methylated arsenic compounds contribute to ambient atmospheric arsenic levels and estimates of low temperature volatilization.

SOURCES

Arsenic is ubiquitous in our environment and has both natural and anthropogenic sources, the atmosphere being the major transport pathway for this element (Buat-Ménard, 1984). Chilvers and Peterson (this volume, Ch. 17) have reviewed current information on the various source strength estimates. Total arsenic emissions into the atmosphere from anthropogenic sources are of the order of 30 000 T/yr. About 60% results from two major sources, copper smelting and coal combustion. Uncertainties on these estimates depend mostly on the adopted value for emission factors. Since emission factors have been taken from developed countries, estimated global source strengths could be underestimated. Data are needed from countries outside Europe, the USA and Canada in order to refine present global estimates. Also, there are still gaps in the details of anthropogenic sources of arsenic such as herbicide use, agricultural and forest burning, gold mining and smelting (especially from the USSR and South Africa), and wood and cow dung burning (Africa and India being the major sources). Chilvers and Peterson have calculated herbicide use of arsenic at 3440 tons per year, coal combustion at 6240 tons As/year and combined smelter and steel production at 14 350 tons As/year.

Uncertainties in our knowledge of natural arsenic sources are much larger than for anthropogenic sources. This is probably a priority research area since source strength estimates are estimated at 60 : 40 for natural versus anthropogenic sources. According to Chilvers and Peterson (this volume, Ch. 17) volcanic activity and low temperature volatilization (biological methylation) are the two dominant natural sources. They calculate low temperature volatilization to be 26 200 tons As/year with volcanoes, on average, contributing 17 150 tons As/year to arsenic emissions from natural sources of 45 480 tons As.

For the volcanic source (Zoller, 1984) we have to distinguish between large but sporadic emissions due to explosive volcanic activity and continuous emission, magma degassing, fumarolic activity, and geothermal activity. The latter emission may be assumed through estimates of sulphur flux to the atmosphere from volcanoes and As/S ratios.

Although biological methylation of arsenic was postulated ten years ago by Wood (1974), its importance as a source for atmospheric arsenic has yet to be carefully evaluated through direct flux measurements.

While global emissions to the atmosphere from anthropogenic sources, especially copper smelters, contribute indirectly to contamination of the land and terrestrial waters, considerable amounts of arsenic are added to the land directly as landfill from the dumping of slag and sludges, and in the waste water from smelting and refinery activities (Chilvers and Peterson, this volume, Ch. 17). Such direct anthropogenic emissions to land and water are of comparable magnitude with the atmospheric emissions although their initial impact is more likely to be of local concern, rather than of regional or global significance. However, where these local imports involve high human population densities or large populations *per se*, then their significance is not to be minimized.

PATHWAYS

The general view of the Working Group is that we need more studies of the various biogeochemical and biochemical pathways of arsenic. Such studies should consider not only total arsenic but its occurrence in different chemical forms, inorganic and organic, which have different levels of toxicity and different lifetimes in the various environments of concern. While a great deal of information is now available, the completeness of the data-base is a good deal less than lead, for example.

Biogeochemical Pathways

Removal rates of arsenic from the atmosphere to terrestrial ecosystems should be investigated both on local and regional scales. There is a dearth of data on arsenic speciation in aerosols, especially as a function of particle size. This dearth of data is also evident for arsenic in rainwater, cloudwater and fog. Also, if volatile forms of arsenic are emitted into the atmosphere (either from the marine or from the terrestrial biosphere) we need to know the lifetimes of these volatile species (which might be very short, $\ll$ 1 day) as well as their end-products.

A good data-base is needed for aquatic systems such as lakes, estuaries and rivers especially outside Western Europe and North America. Of particular concern is the understanding of the local arsenic cycle in such environments: the role of redox cycling and microbial activity on arsenic speciation deserves special attention especially for assessing As(III) levels. Mining areas should be given first priority in future studies. Studies on the tolerance of various sediment bacterial species to arsenic have focused on both the occurrence of the phenomenon and on the mechanisms.

With respect to soils, there is a need to assess the relative importance of external sources (atmospheric deposition) and internal sources (rock weathering) which may be highly variable geographically (Steinnes, this volume, Ch. 9). A comprehensive picture of arsenic cycling (speciation and transport) between the biotic and abiotic soil compartments has yet to be established, as well as an assessment of transfer rates in the relevant ecosystems (plants, trees, animals). However, the data-base has improved substantially over the past 10-15 years so that the need is now for a better geographical coverage, for more details on specific ecosystems and for studies of pathways, feedback loops and mechanisms of retention.

Pathways to Man

It was suggested that because of the so-called soil/plant barrier effect, elevated arsenic concentrations in soils may well reduce crop production substantially before enhanced food chain accumulation occurred (Chaney, 1984). Moreover, the food chain can be affected directly through ingested soil which may also contain significant levels of arsenic contaminants. Again, most of our knowledge is based on a limited number of data from temperate soils. Much of this has been recently reviewed (e.g. NRC Canada, 1979; Lederer and Fensterheim, 1983).

Certain fish species are a potential source of As(III). A data-base for sea water and freshwater fishes is needed together with toxicological studies.

Perhaps the most critical pathway to man recognized at present is through drinking water. Most studies undertaken up to now relate to arsenic in tap water. However, surface waters as well as groundwaters are significant sources of drinking water for many populations and may show locally high levels above threshold range either in anomalous geochemical areas, or close to mining areas or fly ash deposits. Such areas deserve particular attention. Poisonings from arsenic-contaminated well water have occurred in many countries on a small scale.

Though it is generally thought that inhalation is a minor pathway of arsenic to man, occupational exposure to high arsenic atmospheric concentrations and arsenic-rich particulates is of concern especially in some mining and smelting industries. The Working Group recommends that arsenic monitoring should be undertaken in such areas.

SINKS

Relatively little is known about the fate of arsenic in lacustrine, coastal and estuarine sediments. Important issues are (i) the assessment of arsenic sulphide as an ultimate arsenic sink in anoxic sediments, and (ii) the rate of remobilization of arsenic as a function of redox conditions in sediments and hypolimnion. Of special concern is the release of As(III) from anoxic to oxic environments (Turner, this volume, Ch. 12).

HUMAN HEALTH CONCERNS

Because of the absence of data as indicated above, especially outside North America and Western Europe, it is difficult at present to assess what fraction of the population of the Earth is potentially subject to arsenic health effects. Research is needed to identify potential early warning biological indicators especially for freshwater ecosystems (Andreae and Klumpp, 1979). Monitoring of foodstuffs and beverages is an ongoing necessity, as is regulation of drinking water quality.

The health effects of arsenic are now much better understood. Epidemiological studies have implicated airborne arsenic, as well as arsenic in pesticide manufacturing plants, in increased respiratory cancers. The three main groups/industries at risk are: (a) copper smelter workers, (b) arsenical pesticide manufacturing workers, and (c) arsenical pesticide applicators. In each case, exposure is not only to arsenic but to other chemicals also, so the correlations are not absolute (Harding-Barlow, 1983). In all cases studied, a synergistic interaction with tobacco smoking has been found.

Though the biological half-lives of some arsenic compounds are relatively short and thus suggest no age dependent accumulation in man (Hutton, this volume, Ch. 6) they should also be determined with better accuracy for all the relevant arsenic species. It is also recommended that studies on 'hand-mouth' exposure of children should be undertaken where high arsenic levels in soils are observed, as in the vicinity of arsenic-emitting smelters or gold mines, and around arsenical pesticide manufacturing plants.

ANALYTICAL CONCERNS

It is important that studies addressing the problems noted above take full advantage of recent analytical developments allowing identification of both organic and inorganic forms of arsenic in aquatic systems (Andreae and Klumpp, 1979). The development of analogous techniques for arsenic speciation in the atmosphere, soils and biological matrices is strongly recommended, together with the development of appropriate reference materials relevant to local and regional issues (see Group Report on Standards, this volume, Ch. 5).

GENERAL RECOMMENDATIONS

Systematic studies where high levels of arsenic can be anticipated, such as mining and anomalous geochemical areas, should be undertaken.

The high levels of arsenic encountered in present and potential drinking water supplies indicate that further studies be initiated before new resources are developed. Such studies should be encouraged by national organizations.

Our knowledge of anthropogenic source strengths outside Western Europe and North America should be improved.

More accurately measured emission rates from natural sources (volcanoes, low-temperature volatilization) for each of the hemispheres, as well as the role of the oceans as a net source or sink for arsenic should be established.

Documentation at regional and global scales of the past and future trends of anthropogenic arsenic emissions is required. Historical records from ice cores or lake sediments and peat bogs are certainly worthy of study. Projection of future trends should take into account changes in technology as they are likely to occur in each country.

The level of interaction of the arsenic cycle with other cycles (P, Fe, Mn) in both marine and terrestrial ecosystems should be established.

Biogeochemical and biochemical investigations on arsenic should conform with appropriate sampling and analytical methodology. The development of more relevant reference materials should also be strongly encouraged.

REFERENCES

Andreae, M. O. (1979). Arsenic speciation in seawater and interstitial waters: the biological, chemical interactions on the chemistry of a trace element. *Limnol. Oceanogr.*, **24**, 440–452.

Andreae, M. O. and Klumpp, D. W. (1979). The biosynthesis and release of organoarsenic compounds by marine algae. *Environ. Sci. Tech.*, **13**, 738–741.

Buat-Ménard, P. (1984). Fluxes of metals through the atmosphere and oceans. In Nriagu, J. O. (Ed.), *Changing Metal Cycles and Human Health*, Dahlem Konferenzen, pp. 43–69, Springer-Verlag, Berlin.

Chaney, R. L. (1984). Potential effects of sludge-borne heavy metals and toxic organics on soils, plants, and animals, and related regulatory guidelines. *In Proc. Pan American Health Organization Workshop on the International Transportation, Utilization or Disposal of Sewage Sludge.*

Harding-Barlow, I. (1983). What is the status of arsenic as a human carcinogen? In Lederer W. H. and Fensterheim, R. J. (Eds), *Arsenic: Industrial, Biomedical, Environmental Perspectives. Proc. Arsenic Symp., Gaithersburg, MD.*, pp. 203–209, Van Nostrand Reinhold Co., New York.

Lederer W. H. and Fensterheim, R. J. (Eds), (1983) *Arsenic: Industrial, Biomedical, Environmental Perspectives. Proc. Arsenic Symp., Gaithersburg, MD.*, Van Nostrand Reinhold Co., New York, 443 pages.

NRC (National Research Council of Canada) (1979). *Effects of Arsenic in the Canadian Environment.* NRC Associate Committee on Scientific Criteria for Environmental Quality. NRCC No. 15391.

Wood, J. (1974). Biological cycles for toxic elements in the environment. *Science*, **183**, 1049–1052.

Zoller, W. H. (1984). Anthropogenic perturbations of metal fluxes into the atmosphere. In Nriagu, J. O. (Ed.), *Changing Metal Cycles and Human Health*, Dahlem Konferenzen, pp. 27–14, Springer-Verlag, Berlin.

Lead, Mercury, Cadmium and Arsenic in the Environment
Edited by T. C. Hutchinson and K. M. Meema

CHAPTER 5

Group Report: The Need for Standard Reference Material for Metal Analyses

C. R. KRISHNA MURTI, *Rapporteur*
P. BUAT-MÉNARD, M. OLADE, P. J. PETERSON, E. STEINNES

There is increasing interest globally in measuring levels of toxic elements in the environment and evaluating their long-term effects on human health and ecosystems. Since high quality analytical data is an essential prerequisite for arriving at estimates of exposure levels and quantification of global and regional cycles, quality control of the analytical data generated must be ensured, particularly in view of the matrix problems posed by the diversity of non-living and living materials to be analysed. The need to constantly update analytical skills and the ability to conform to rigorous standards of performance is appreciated even in some of the highly accredited laboratories.

Environmental pollution by trace metals has, of late, begun to cause concern in many countries outside the highly industrialized societies and some of them have already initiated limited monitoring programmes or participated in the recently concluded GEMS (Global Environmental Monitoring Systems) programme on Pb and Cd. Recognizing the urgent need to build data-bases on a regional basis, it is recommended that monitoring be encouraged and supported on a regional and sub-regional basis. In this context one can visualize a regional programme using rice, for example, as the monitor, with samples derived from more than one country in a given geographical region where rice is grown under comparable conditions of animal husbandry (Philippines, Viet Nam, Thailand, Burma, India, etc.). The use of shellfish, such as mussels, as global monitors was well illustrated by the international co-operative study known as Mussel Watch. Another example is pollution by arsenic arising out of burning fire wood as domestic fuel (many countries in Africa, Latin America, and in India). Yet another example is the co-operative study on mercury in fish and sediments organized by the UNEP regional seas programme. Besides the above specific reference standards, general standards need to be produced, so there can be a wide exchange of samples for inter-laboratory comparison on a regional basis.

Many reference materials already exist to ensure quality control in industrialized countries, such as those of the National Bureau of Standards in the USA, and the reference materials available for analysis through the International Atomic Energy Agency in Vienna. However, these materials are often beyond the budgets of scientists in developing countries and/or are of matrices and materials inappropriate, for example, for the tropics.

All countries entering into this new sphere of activity need to be provided with ready access to reference standards. The International Atomic Energy Agency in Vienna has been supplying limited numbers of biological samples as standards. It is understood that IAEA has a special responsibility to less developed countries in this connection.

Part II

Contributed Papers

Lead, Mercury, Cadmium and Arsenic in the Environment
Edited by T. C. Hutchinson and K. M. Meema

CHAPTER 6

Human Health Concerns of Lead, Mercury, Cadmium and Arsenic

M. HUTTON

*Monitoring and Assessment Research Centre,
The Octagon Building, King's College London,
University of London, 459A Fulham Road,
London SW10 0QX, United Kingdom*

ABSTRACT

The trace elements lead, mercury, cadmium and arsenic have caused major human health problems in several parts of the world. Concern over such incidents has prompted numerous investigations into the metabolism and toxic effects of these four elements. This chapter outlines their contrasting metabolism and describes the major health effects and the relative scale of such incidents. Attention is paid to the environmentally important chemical species of mercury and arsenic, the overall health significance of early biochemical effects and the limitations of certain epidemiological studies. Comparisons are made between the exposure threshold for the initial effects of each element and the exposure levels seen in the general population.

INTRODUCTION

The three metals, lead, mercury and cadmium, and the metalloid arsenic have all caused major human health problems in various parts of the world. The overt toxicity of these elements has been recognized for many years; indeed, the harmful effects of lead were known as far back as the second century BC in ancient Greece (Waldron, 1973). Over the years, physicians became increasingly familiar with the symptoms of metal poisoning arising in occupationally exposed workers and in individual cases of poisoning. In more recent times, toxicologists concerned with metal poisoning attempted to elucidate the metabolism and range of effects induced by metals in these two populations. Such studies revealed that certain effects only become apparent at particular stages of the exposure scale. In cases of high exposure, clinical signs and symptoms can be observed. At lower exposure levels clinical manifestations may be absent but effects may be observed at the physiological or biochemical level.

From the mid-1950s to the early 1960s it was recognized that poisoning by these four elements was not restricted to occupationally exposed workers or to the occasional individual. In the USA, detailed studies by health officials revealed that many hundreds of cases of childhood lead poisoning occurred every year in the dilapidated housing areas of the older cities (Lin-Fu, 1980). At about the same time in Japan, major environmental poisoning incidents caused by methylmercury and by cadmium were the subject of intense scientific investigation and attracted public attention and concern. Possibly less well known but also occurring in Japan over the same time period were two mass incidents of arsenic poisoning (WHO, 1981). These major episodes have prompted numerous studies of other populations considered to be subjected to elevated exposure levels from a variety of sources. Such studies, employing increasingly sophisticated techniques, revealed that effects could be detected at exposure levels which had previously been considered safe.

This chapter reviews the current state of knowledge regarding the health effects induced by the four elements in environmentally exposed populations. Attention is paid to the significance of the observed effects, the scale of such incidents and the limitations of recent epidemiological studies. By reference to dose response relationships, a comparison is made between the exposure threshold for the initial effects of each element and the exposure levels seen in the general population.

METABOLIC FACTORS

Before discussing the health effects caused by the four elements, it is worthwhile to consider the major metabolic factors associated with environmental exposure, as this will allow a greater understanding of the basis for the observed effects. The relevant factors are summarized in Table 6.1. In the cases of mercury and arsenic, the chemical species of importance in environmental exposure need to be distinguished. For mercury these are the short chain alkyl compounds, particularly methylmercury. In the case of arsenic, both inorganic compounds, as arsenates (As^{5+}) or as arsenites (As^{3+}), and organic compounds, such as arsenobetaine and arsenocholine, are of importance.

Uptake

Generally, the predominant route of exposure for all four elements is ingestion, although inhalation is an important source of lead, particularly in

Table 6.1 Major metabolic factors associated with environmental exposure to lead, mercury, cadmium and arsenic

	Lead	Mercury	Cadmium	Arsenic
Major routes of entry	Ingestion and inhalation	ingestion	Ingestion and inhalation (tobacco)	Ingestion
Gastrointestinal absorption (%)	~10	~95	~5	>80
Organs of accumulation	Bone, kidney and liver	Brain, liver and kidney	Kidney and liver	Keratinous tissue
Major routes of excretion	Urine	Faeces	Urine	Urine
Biological half-life	~20 years	~70 days	>10 years	10–30 hours

urban-dwelling populations. In addition, the smoking of tobacco can be an important source of cadmium. Great differences exist between the four elements in the fractional absorption by the gastrointestinal tract.

Methylmercury and the various arsenic compounds are absorbed efficiently. Gastrointestinal absorption of cadmium is enhanced considerably in individuals with iron deficiency (Flanagan *et al.*, 1978) while lead absorption may be much higher than 10% in children (Alexander *et al.*, 1973).

Behaviour in the Body

Both inorganic arsenic and methylmercury exhibit a relatively even tissue distribution, but arsenic does accumulate in hair, skin and nails, while methylmercury displays considerable affinity for the brain. About 90% of the body burden of lead is stored in the skeleton, while in soft tissues the liver and kidneys have high concentrations. Cadmium is selectively accumulated in the kidney and liver, particularly in the renal cortex. In these organs cadmium is bound specifically to a low molecular weight protein, metallothionein. Both lead and methylmercury show significant transplacental transfer. In the latter case this transfer is particularly efficient.

With the exception of methylmercury, the major route of excretion of the elements is via urine. In humans inorganic arsenic is methylated to methylarsonic acid and dimethylarsinic acid and these are rapidly excreted. The organic arsenic compounds present in seafood are excreted unchanged.

The principal route of excretion for methylmercury occurs by biliary excretion into the gastrointestinal tract. Most of the methylmercury excreted in this manner is subsequently reabsorbed, resulting in an enterohepatic circulation. In the case of cadmium, the binding to metallothionein is thought to be responsible for both the very low excretion rates of this metal and the observed accumulation with age, particularly in the renal cortex.

HEALTH EFFECTS AND POPULATIONS AFFECTED

Lead

The major health effects of lead are manifest in three organ systems; the haematological system, the central nervous system (CNS) and the renal system. Table 6.2 summarizes the various effects in these systems which have been attributed to lead in environmentally exposed populations.

Table 6.2 Major health effects attributed to lead in environmentally exposed populations

Organ affected	Range of effects reported
Haematological system	Inhibition of ALA-D (δ-aminolevulinic acid dehydratase) and haem synthetase and corresponding accumulation of ALA and FEP (free erythrocyte protoporphyrin). At higher levels of exposure, reduced haem synthesis and anaemia.
Nervous system	CNS impairment at moderate exposure in children, reflected by inattention, cognitive difficulties, fine motor dysfunction and altered EEG patterns. Under heavy exposure, encephalopathy may arise.
	Effects on the peripheral nervous system (PNS) indicated by reduced nerve conduction velocity.
Renal system	Functional impairment of the tubular region characterized by mild aminoaciduria, glucosuria and hyperphosphaturia. Morphological effects include mitochondrial damage and intranuclear inclusion bodies. Long-term heavy exposure may result in irreversible nephropathy.

The haematological effects of lead have been recognized for many years and the biochemical basis for such changes are now reasonably well established. Two of the most sensitive effects on this system arise at the first and final steps of the haem synthesis pathway. The enzyme catalysing the first step, δ-aminolevulinic acid dehydratase (ALA-D) is uniquely sensitive to lead, and erythrocyte ALA-D is often reported to be inhibited in the general population (Hernberg and Nikkanen, 1970). This partial inhibition is not considered to be a deleterious effect because the enzyme exhibits a large reserve capacity (Zielhuis, 1975). In addition, erythrocyte ALA-D in mammals has no function as such because mature erythrocytes do not participate in haem synthesis. The inhibition of ALA-D in other, haem-forming tissues will, if exposure is sufficiently high, result in an increased concentration of the substrate ALA in the body and urine (Selander and Cramer, 1970).

Lead also interferes with the last stage of haem synthesis, the incorporation of iron into protoporphyrin, catalysed by haem synthetase (ferrochelatase). This step takes place in the inner matrix of the erythroid cell mitochondria in the bone marrow. It is now thought that lead does not interfere with the enzyme but rather inhibits the transmitochondrial transfer of iron (Piomelli 1980). This effect results in the accumulation of protoporphyrin in the mitochondria, its incorporation into the globin molecule in place of haem and the presence of elevated levels of free erythrocyte protoporphyrin (FEP) in the peripheral blood.

When devising an ambient air quality standard for lead in the late 1970s, the US EPA identified the elevation of FEP as the earliest adverse effect of lead (EPA, 1977). This was based on the fact that, unlike the inhibition of ALA-D, the insertion of protoporphyrin occurs within the mitochondrial matrix and elevated levels of FEP represent an impairment of a specific mitochondrial function. In addition, there is a possibility that other mitochondrial functions will also be affected.

Some investigators now consider that the systemic accumulation of ALA is not without significance and may be involved in the neurotoxic effects associated with lead exposure (Moore, 1980). This is based on the findings that ALA can pass the blood-brain barrier (McGillion *et al.*, 1974), while administration of ALA causes neuromuscular and neurophysiological effects (Moore and Meredith, 1976; Cutler *et al.*, 1979). Lead may also induce the development of anaemia but this effect only arises at high levels of exposure (Zielhuis, 1975). Indeed, Piomelli (1980) considers that children may die from the neurological effects of lead without even developing anaemia.

Acute effects of lead on the central nervous system (CNS) are generally seen in children heavily exposed from pica and are manifest by severe encephalopathy which can culminate in coma and death. More controversial is the reported association between certain subtle neuropsychological and neurobehavioural effects and lead exposure in otherwise asymptomatic children. It is this issue which is of most concern in the environmental toxicity of lead because these effects have been reported to occur in children from the general population subjected to commonly occurring exposure regimes. Further concern relates to the greater susceptibility of children to lead resulting from the higher intake and uptake together with the greater sensitivity of the developing CNS. The major study relevant to this issue was carried out on children in the USA by Needleman *et al.* (1979) using dentine lead as the index of exposure. The study involved numerous tests of performance and intelligence in the children, classified according to their dentine lead levels. The classroom behaviour of the children was rated by the teacher. Needleman *et al.* (1979) reported that children from the high lead group showed significantly lower verbal IQ and performed less well on other behavioural tasks, after controlling for five covariates in an analysis of covariance. In addition, the teacher's assessments of classroom behaviour were reported to show a dose-dependent increase in poor ratings (Needleman, 1980). Since the Needleman study, other investigations in children from the Federal Republic of Germany (Winneke, 1983) and the United Kingdom (Yule and Landsdown, 1983) have reported an association between increased lead exposure and decreases in measurements of intelligence and behaviour. The evidence from these and other investigations may reflect a causal relationship but could also be indicative of the effects of confounding factors. Indeed, the only area where the two divergent opinions on this issue appear to agree is that the study design of retrospective epidemiological investigations will never fully be able to control for confounding factors.

Despite this, an influential paper by Rutter (1983) argued that these and other studies provide sufficient evidence to infer a cause and effect relationship. In contrast, an expert Committee concluded that the results of the study by Needleman *et al.* (1979) 'neither confirm nor refute the hypothesis that low lead exposure in children leads to neuropsychologic deficits' (EPA, 1983). The Committee concerned not only criticized the study of Needleman *et al.* (1979) for failure to control confounding variables but, moreover, noted deficiencies in data handling. In addition, a question of possible bias was raised, as large numbers of eligible subjects were rejected from analysis. However, reanalysis of the data in response to the Committee's recommendations still produced a significant association between dentine lead and signs of intelligence deficit (Needleman, 1984).

The functional morphological effects of lead on the kidneys listed in Table 6.2 occur at exposure levels rarely encountered outside of the occupational setting. Acute lead intoxication in children may elicit renal dysfunction but little is known of the extent and reversibility of this effect.

From the above, it is clear that children represent the critical group with respect to environmental lead exposure. The particular populations considered to be at most risk, based on either observed effects or elevated exposure levels, are the following:

(i) certain pre-school children from the general population living in urban areas, particularly in those countries where lead is added to petrol;

(ii) children living in proximity to point sources of lead such as smelters and scrap yards;

(iii) children living in areas with lead plumbing and served by plumbosolvent water. In this population, exposure takes place before and after birth;

(iv) children living in housing with high leaded paintwork in poor condition, particularly those individuals which exhibit pica.

Mercury

Methylmercury intoxication is characterized by effects on the CNS and the areas mainly affected are those associated with the sensory, visual and auditory functions and those concerned with co-ordination (WHO, 1976). Early effects are paraesthesia in the tongue, lips and distal extremities, while in more severe cases, blurred and constricted vision and ataxia may appear (Piotrowski and Inskip, 1981). The developing nervous system of the fetus is more sensitive to methylmercury than the adult and pre-natal exposure can result in neurotoxic effects in the infant in the absence of effects in the mother (WHO, 1976). Pregnant women are apparently more sensitive to methylmercury than are other adults (Marsh *et al.*, 1979).

In the last thirty years, there have been several outbreaks of methylmercury intoxication of two distinct types. The first, termed Minamata disease, took place in Japan in the 1950s and 1960s and was caused by long-term ingestion of contaminated fish. There is disagreement over the number of poisonings caused by Minamata disease but one study estimates about 1000 cases, 3300 suspected cases and about 100 deaths (Tsubaki *et al.*, 1978).

Several studies have examined the health of various communities of Canadian Indians because of their elevated methylmercury intakes from fish. The most recent investigation (Methylmercury Study Group, 1980) reported an association between several neurological parameters and methylmercury exposure. However, Piotrowski and Inskip (1981) question this conclusion and note the important influence of confounding factors on the putative association. Piotrowski and Inskip (1981) conclude that the evidence for a neurological effect of methylmercury in the study population is equivocal but that the exposure regimes in question may be at the threshold for effects.

Other populations with heavy fish consumption have also been identified in the Mediterranean, particularly in Italy, in Papua New Guinea, Peru, New Zealand and the Seychelles (WHO, 1976; T. Kjellström, personal communication; M. Berlin, personal communication). In some of these areas no health-related studies have been undertaken, or they are currently still under way. Those studies which have been completed have failed to provide clinical evidence of methylmercury intoxication.

The other type of intoxication is characterized by the Iraqi epidemic in 1971–72 where exposure resulted from consumption of bread prepared from grain dressed with alkylmercury fungicides. The incident in question resulted in the poisoning of about 6000 individuals (Clarkson, 1977) and the deaths of over 500 in hospital (Bakir *et al.*, 1973). Previous outbreaks of this type together with the number of poisonings are as follows: Iraq (1956), about 100; Iraq (1960), about 1000; Pakistan (1969), about 100; Guatemala (1963-65), about 45; and Ghana (1967), about 150 (Clarkson, 1977). Clearly the inadvertent ingestion of seed treated with organic mercury as a fungicide is a hazard, despite a coloured dye being used and seed bags being marked in several languages.

Cadmium

The kidney is the critical organ of intoxication after long-term exposure to cadmium. One of the initial signs of renal dysfunction is an increased urinary excretion of proteins. Cadmium-induced proteinuria is generally considered to be characterized by the excretion of low molecular weight proteins, particularly α_2,β_2 and γ-globulins. This form of proteinuria, caused by an impaired reabsorption function of the proximal tubules, is not specific for the metal and may be found in hereditary forms of tubular dysfunction.

Some recent studies (Bernard *et al.*, 1979) indicate that a glomerular pattern of dysfunction may also be an early effect of cadmium exposure, as evidenced by an increased excretion of high molecular weight proteins. Later effects on renal function are manifested by aminoaciduria, phosphaturia and glucosuria (Friberg *et al.*, 1974). Recent investigations of cadmium and renal tubular injury have used sensitive radio-immunoassay techniques to measure urinary concentrations of the low molecular weight proteins, β_2-microglobulin (β_2-m), and retinol-binding protein (RBP) as indicators of early effect. Of these two proteins, most use has been made of β_2-m but some studies have also measured RBP.

The precise significance of an increased excretion of β_2-m is still unclear, especially as markedly elevated β_2-m levels correspond to relatively small reductions in the tubular reabsorption efficiency. Furthermore, there is no significant transport of low molecular weight proteins back to the blood after tubular uptake; instead these proteins are degraded *in situ* (Cooper and Plesner, 1980). Apparently, the significance of this tubular lesion is the loss of those amino acids which are salvaged by a normal individual. However, increased urinary excretion of β_2-m is generally considered to be irreversible and is indicative of an impairment of tubular function and, as such, is considered by some workers to constitute a health effect.

Until recently, reports of health effects of cadmium in populations not occupationally exposed to cadmium were confined to Japan. Many areas of Japan are contaminated with cadmium, as a result of discharges from numerous non-ferrous metal mines and smelters. Long-term consumption of rice grown in these areas has resulted in elevated cadmium exposure levels and signs of renal dysfunction in several localities (Friberg *et al.*, 1974). A syndrome termed 'itai-itai disease' has also been identified in one area characterized by severe renal dysfunction and damage to bone structure. The disease predominantly affected elderly multiparous women with poor nutritional status, who had lived in the contaminated area for many years.

A recent investigation also examined elderly females (>60 years) from a city in Belgium subjected to long-term cadmium contamination (Roels *et al.*, 1981). These women were reported to have larger cadmium burdens and a higher prevalence of signs of renal dysfunction compared with controls from a city with lower levels of cadmium contamination. The cadmium exposure levels in this population were much lower than those encountered in contaminated areas of Japan and the corresponding signs of renal dysfunction were also less pronounced. This study suggests that cadmium may exacerbate the age-related decline in renal function at moderate exposure levels to the metal.

The mortality statistics of the two Belgian cities have also been examined (Lauwerys and De Wals, 1981). Overall mortality rates were similar in the two populations but mortality from nephritis and nephrosis was higher in the cadmium 'polluted' city. However, mortality rates from uremia were similar in the two populations, a surprising finding given the apparent disparity in mortality rates from nephritis and nephrosis.

Mortality studies carried out in cadmium-contaminated areas of Japan have produced conflicting results. One investigation (Nogawa *et al.*, 1981) of a small population reported a significant increase in overall mortality for males with proteinuria. As in the Belgian study, this population was reported to show a large excess of deaths due to nephritis and nephrosis.

Arsenic

Long-term exposure to inorganic arsenic can give rise to health effects in a large number of organs. Those effects reported to occur in populations environmentally exposed to arsenic are shown in Table 6.3 The information depicted in Table 6.3 has been extracted from WHO (1981) and Pershagen (1983).

The most characteristic effects following chronic arsenic exposure are hyperkeratosis of the palms and soles of the feet together with hyperpigmentation, particularly in areas not exposed to the sun. Skin tumours have also been commonly reported and these are often located on the hands and feet.

Haemangioendothelioma of the liver, a very rare form of cancer, has been associated with long-term arsenic exposure in several instances. Most of the reported cases refer to individuals which had been prescribed Fowler's solution, a trivalent arsenic medication.

Table 6.3 Health effects in environmentally exposed populations attributed to arsenic

Organ affected	Effects
Skin	Hyperpigmentation Hyperkeratosis Skin tumours
Lungs	Lung cancer*
Liver	Liver dysfunction Haemangioendothelioma
Cardiovascular system	Peripheral vascular disturbances leading to gangrene
Nervous system	Peripheral neuropathy Hearing defects
Haematopoietic system	Disturbed erythropoiesis with anaemia
Reproductive system	Increased frequency of spontaneous abortions*

* The role of arsenic in these effects is equivocal.

Other effects of arsenic include peripheral vascular disturbances resulting in gangrene and a disease termed 'Blackfoot disease' (Tseng, 1977).

Some studies have reported an increased prevalence of lung cancer and increased rate of spontaneous abortion in communities close to point sources of atmospheric arsenic (Newman *et al.*, 1976; Pershagen *et al.*, 1977; Nordström *et al.*, 1978). However, several detailed evaluations of these and other studies concluded that without further evidence it is not possible to say whether arsenic was the causative agent in those incidents (IARC, 1980; WHO, 1981; Pershagen, 1983).

Those populations affected by arsenic, outside the occupational setting, may be divided into four categories, as shown in Table 6.4. The most serious of the two food poisoning incidents in Japan resulted from the accidental use of arsenic-contaminated sodium phosphate in the preparation of dried milk for infants (Nakagawa and Ibuchi, 1970). A follow-up study of the survivors of this incident revealed severe hearing loss in about 20% of the sample (Yamashita *et al.*, 1972).

The most important cause of environmental arsenic impact relates to the consumption of arsenic-contaminated drinking water. In most cases, the arsenic is of natural origin but in some instances mining activities are responsible. Studies of several affected populations, particularly those in Argentina and Chile, have revealed a close association between skin cancer and arsenic exposure. Similarly, the incidence of Blackfoot disease in Taiwan, manifest by gangrene of the extremities, was found to increase in a dose-dependent manner with arsenic (Tseng, 1977).

Table 6.4 Large scale cases of poisoning associated with exposure to inorganic arsenic

Type of incident	Episodes
Food inadvertently contaminated during preparation (acute and subacute effects)	Morinaga milk incident, Japan (1955). >12 000 cases, 130 deaths Soy sauce incident, Japan (1956). >400 cases
Drinking water exposure (chronic effects)	Numerous incidents, with reports from Argentina, Chile, Taiwan, U.S.A., Canada and Japan. In excess of 0.25 million people exposed for several decades in Chile.
Communities close to point sources of airborne arsenic (chronic effects)	Evidence for effects on morbidity and mortality around copper smelters and pesticide plants. Single report of hearing effects in children living near a power plant in Czechoslovakia where a high As (~1000 μg/gm) lignite was burnt.
Exposure via medications (chronic effects)	Sodium arsenite (Fowler's solution) previously used in the treatment of psoriasis and leukaemia, resulting in severe effects in hundreds of patients.

Several epidemiological investigations of populations living close to point sources have reported an increased mortality from lung cancer but it is not possible to assign this effect to arsenic exposure. Minor hearing loss was reported in children residing close to a power plant in Czechoslovakia burning lignite with an arsenic content of about 1000 μg/gm (Bencko *et al.*, 1977). A similar study found no change in hearing in children subjected to elevated arsenic exposure levels from a large copper smelter (Milham, 1977).

The long-term consumption of Fowler's solution has resulted in the severe poisoning of large numbers of individuals. Examination of the victims has provided valuable information on the toxicity of arsenic. However, this form of contamination is not considered to be an example of environmental contamination.

EXPOSURE THRESHOLDS

One way of assessing the risk to the general population from exposure to a particular chemical is by making a comparison between the actual exposure and the exposure threshold for initial effects. This has been done for the four elements in question using data derived from dose-response analysis;

the values obtained are shown in Table 6.5. In the case of lead, no formal dose-response analyses appear to have been carried out for neurobehavioural and neuropsychological effects. The blood lead level which represents the threshold values for elevated FEP in children is similar to the actual values encountered in many children from the general population particularly those living in urban areas.

This finding and the recent evidence for the neurotoxic effects of lead in young children both show that exposure regimes experienced by the general population commonly exceed the threshold for effects. It is for this reason that of the four elements considered in this workshop, lead is the greatest cause of public health concern.

Table 6.5 Estimated threshold values for chronic exposure to lead, methylmercury, cadmium and inorganic arsenic

Metal	Effect	Estimated threshold	Reported values in the general population or background	Notes and authorities
Lead	ALA-D FEP	11 μg/dl 15–20 μg/dl Blood lead	10–20 μg/dl	Data refer to children Value for ALA-D is for a 10% prevalence rate (Piotrowski and O'Brien, 1980; Piomelli *et al.*, 1982).
Methyl-mercury	Earliest signs and symptoms	20 μg/gm hair 80 ng/ml blood	2 μg/gm 10 ng/ml	Thresholds refer to non-pregnant adults. The threshold for the fetus is considered to be lower. Background values strongly influenced by fish consumption (Piotrowski and Inskip, 1981).
Cadmium	β_2-m cancer	150 μg/day dietary intake	20–70 μg/day	Threshold refers to 50-year daily intake needed to cause elevated urinary β_2-m in 10% of the population with average body weight of 70 kg (Kjellström, 1980; Hutton, 1983).
Arsenic (inorganic compounds)	Skin cancer	200 μg/litre in drinking water	2 μg/litre	Threshold refers to 5% prevalence after lifetime (70 years) exposure (WHO, 1981).

Exposure levels of methylmercury in the overwhelming majority of the general population are significantly lower than those considered to be at the threshold for early effects. It is only certain sub-groups within the general population, particularly fishing communities, which are likely to be exposed to elevated levels.

The dose response analyses for cadmium were based on data from elderly females whose diet may have been nutritionally inadequate. This should be borne in mind when extrapolating the findings of these analyses to other populations. The 10% response value for cadmium exposure, at 150 μg/day is considerably higher than the average intakes reported for most countries around the world. Japan is one notable exception and average values for 'uncontaminated' areas can approach this threshold. It should also be borne in mind that dietary intake of cadmium displays a log-normal distribution and individuals at the upper end of the distribution will be exposed to considerably larger amounts than average.

Exposure information for early effects of arsenic are considered to be inadequate for dose-response analysis (WHO, 1981). Table 6.5 depicts the threshold value for skin cancer, derived from populations exposed to arsenic in drinking water. It is clear that the arsenic levels commonly found in drinking water pose no realistic cancer hazard to the general population.

REFERENCES

Alexander, F. W., Delves, H. T., and Clayton, B. E. (1973). The uptake and excretion by children of lead and other contaminants. *Proc. Int. Symp. Environmental Health Aspects of Lead.* pp. 319–330, Commission of the European Communities, Luxembourg.

Bakir, F., Damluji, S. F., Amin-Zaki, L., Murtadha, M., Khalidi, A., Al-Rawi, N. Y., Tidriti, S., Dhahir, H. I., Clarkson, T. W., Smith, J. C., and Doherty, R. A. (1973). Methylmercury poisoning in Iraq. *Science*, **181**, 230–241.

Bencko, V., Symon, K., Chladek, V., and Pihrt, J. (1977). Health aspects of burning coal with a high arsenic content. II. Hearing changes in exposed children. *Environ. Res.*, **13**, 386–395.

Bernard, A., Buchet, J. P., Roels, H., Masson, P., and Lauwerys, R. (1979). Renal excretion of proteins and enzymes in workers exposed to cadmium. *Eur. J. Clin. Invest.*, **9**, 11–22.

Clarkson, T. W. (1977). Mercury poisoning. In Brown, S. S. (Ed.) *Clinical Chemistry and Chemical Toxicology of Metals*, pp. 189–200, Elsevier/North-Holland Biomedical Press, Amsterdam.

Cooper, E. H., and Plesner, T. (1980). Beta-2-microglobulin review: its relevance to clinical oncology. *Med. Pediatr. Oncol.*, **8**, 323–334.

Cutler, M. G., Moore, M. R., and Ewart, F. G. (1979). Effects of delta-aminolaevulinic acid administration on social behaviour in the laboratory mouse. *Psychopharmacology*, **61**, 131–135.

EPA (1977). Lead: ambient air quality standard. *Fed. Regist.*, **42**, 63076.

EPA (1983). *Independent Peer Review of Selected Studies Concerning Neurobehavioural Effects of Lead Exposures in Nominally Asymptomatic Children: Official*

Report of Findings and Recommendations of an Interdisciplinary Expert Review Committee. EPA-600/8-83-028A US Environmental Protection Agency, North Carolina.

Flanagan, P. R., McLellan, J. S., Haist, J., Cherian, G., Chamberlain, H. J., and Valberg, L. S. (1978). Increased dietary cadmium absorption in mice and human subjects with iron deficiency. *Gastroenterology*, **74**, 841–846.

Friberg, L., Piscator, M., Nordberg, G. F., and Kjellström, T. (1974). *Cadmium in the Environment,* 2nd edition, CRC Press, Cleveland, Ohio.

Hernberg, S., and Nikkanen, J. (1970). Enzyme inhibition by lead under normal urban conditions. *Lancet,* **1**, 63–64.

Hutton, M. (1983). Evaluation of the relationships between cadmium exposure and selected indicators of renal function. In *Heavy Metals in the Environment, Heidelberg, September 1983,* pp. 274-277, CEP Consultants Ltd, Edinburgh.

IARC (1980). *Monographs. Arsenic and Arsenic compounds; Volume 23,* Lyons, International Agency for Research on Cancer, pp. 39-141.

Kjellström, T. (1980). Epidemiological aspects of the dose-response relationship of cadmium induced renal damage. In *International Cadmium Conference, Cannes,* pp. 118–122, Metal Bulletin Ltd, London.

Lauwerys, R., and De Wals, Ph. (1981). Environmental pollution by cadmium and mortality from renal diseases. *Lancet,* **1**, 383.

Lin-Fu, J. (1980). Lead poisoning and undue lead exposure in children: history and current status. In Needleman, H. L. (Ed.), *Low Level Lead Exposure: The Clinical Implications of Current Research,* pp. 5–16, Raven Press, New York.

McGillion, F. B., Thompson, G. G., Moore, M. R., and Goldberg, A. (1974). The passage of δ-aminolaevulinic acid across the blood-brain barrier of the rat: effect of ethanol. *Biochem. Pharmacol.*, **23**, 472–474.

Marsh, D. O., Myers, G. J., Clarkson, T. W., Amin-Zaki, L., Tikriti, S., Majeed, M. A., and Dablagh, A. R. (1979). Dose-response relationship for human fetal exposure to methylmercury. International Congress of Neurotoxicology, Varese, Italy.

Methylmercury Study Group (1980). *Methylmercury Study.* McGill University, Montreal, Canada.

Milham, S. (1977). Studies of morbidity near a copper smelter. *Environ. Health Perspect.*, **19**, 131–132.

Moore, M. R. (1980). Prenatal exposure to lead and mental retardation. In Needleman, H. L. (Ed.), *Low Level Lead Exposure: The Clinical Implications of Current Research,* pp. 53–65, Raven Press, New York.

Moore, M. R., and Meredith, P. A. (1976). The association of δ-aminolevulinic acid with the neurological and behavioral effects of lead exposure. In Hemphill, D. D. (Ed.), *Trace Metals in Environmental Health,* pp. 363–371, University of Missouri, Columbia.

Nakagawa, Y., and Ibuchi, Y. (1970). On the follow-up investigation of Morinaga milk arsenic poisoning. *Igaku no Ayumi,* **74**, 1–3.

Needleman, H. L. (1980). Lead and neuropsychological deficit: finding a threshold. In Needleman, H. L. (Ed.), *Low Level Lead Exposure: The Clinical Implications of Current Research,* pp. 43–51, Raven Press, New York.

Needleman, H. L. (1984). Comments on chapter 12 and appendix 12C, Air Quality Criteria for Lead (external review draft no. 1). Available for inspection at: US Environmental Protection Agency, Central Docket Station, Washington, DC. Docket No. ECAO-CD-81-2 IIA. E. C.1.20.

Needleman, H. L., Gunnoe, C., Leviton, A., Reed, R., Peresie, H., Maher, C., and

Barrett, P. (1979). Deficits in psychologic and classroom performance of children with elevated dentine lead levels. *New Eng. J. Med.*, **300**, 689–695.
Newman, J. A., Archer, V. E., Saccomanno, G., Kuschner, M., Auerbach, O., Grondahl, R. D., and Wilson, J. C. (1976). Histologic types of bronchogenic carcinoma among members of copper mining and smelting communities. *Ann. N. Y. Acad. Sci.*, **271**, 260–268.
Nogawa, K., Kawano, S., and Nishi, M. (1981). Mortality study of inhabitants in a cadmium polluted area with special reference to low molecular weight proteinuria. In *Heavy Metals in the Environment, Amsterdam*, pp. 538–540, CEP Consultants Ltd, Edinburgh.
Nordström, S., Beckman, L., and Nordenson, I. (1978). Occupational and environmental risks in and around a smelter in northern Sweden. III: Frequencies of spontaneous abortion. *Hereditas*, **88**, 51–54.
Pershagen, G., (1983). The epidemiology of human arsenic exposure. In Fowler, B. A. (Ed.), *Biological and Environmental Effects of Arsenic*, pp. 199–232, Elsevier/North-Holland Biomedical Press, Amsterdam.
Pershagen, G., Elinder, C-G., and Bolander, A. M. (1977). Mortality in a region surrounding an arsenic emitting plant. *Environ. Health Perspect.*, **18**, 133–137.
Piomelli S. (1980). The effects of low-level lead exposure on heme metabolism. In Needleman, H. L. (Ed.), *Low Level Lead Exposure: The Clinical Implications of Current Research*, pp. 67–74, Raven Press, New York.
Piomelli, S., Seaman, C., Zullow, D., Curran, A., and Davidow, B. (1982). Threshold for lead damage to heme synthesis in urban children. *Proc. Natl. Acad. Sci.*, **79**, 3335–3339.
Piotrowski, J. K., and O'Brien, B. J. (1980). *Analysis of the Effects of Lead in Tissue upon Human Health Using Dose-Response Relationships.* MARC Report No. 17, Monitoring and Assessment Research Centre, Chelsea College, University of London: 88 pages.
Piotrowski, J. K., and Inskip, M. J. (1981). *Health Effects of Methylmercury.* MARC Report No. 24, Monitoring and Assessment Research Centre, Chelsea College, University of London: 82 pages.
Roels, H. A., Lauwerys, R. R., Buchet, J. P., and Bernard, A. (1981). Environmental exposure to cadmium and renal function of aged women in three areas of Belgium. *Environ. Res.*, **24**, 117–130.
Rutter, M. (1983). Low level lead exposure: sources, effects and implications. In Rutter, M., and Russell Jones, R. (Eds.), *Lead versus Health: Sources and Effects of Low Level Lead Exposure*, pp. 333–370, J. Wiley, Chichester.
Selander, S., and Cramer, K. (1970). Interrelationships between lead in blood, lead in urine and ALA during lead work. *Br. J. Ind. Med.*, **27**, 28–39.
Tseng, W. P. (1977). Effects and dose-response relationships of skin cancer and Blackfoot disease with arsenic. *Environ. Health Perspect.*, **19**, 109–119.
Tsubaki, T., Hirota, K., Shirakawa, K., Kondo, K., and Sato, T. (1978). Clinical, epidemiological and toxicological studies on methylmercury poisoning. In Plaa, G. L., and Duncan, W. A. M. (Eds.), *Proceedings of 1st International Congress on Toxicology*, pp. 339–357, Academic Press, New York and London.
Waldron, H. A. (1973). Lead poisoning in the ancient world. *Med. Hist. (Lond.)*, **17**, 391–399.
WHO (1976). *Environmental Health Criteria 1. Mercury.* World Health Organization, Geneva: 131 pages.
WHO (1981). *Environmental Health Criteria 18. Arsenic.* World Health Organization, Geneva: 174 pages.

Winneke, G. (1983). Neurobehavioural and neuropsychological effects of lead. In Rutter, M., and Russell Jones, R. (Eds), *Lead versus Health: Sources and Effects of Low Level Lead Exposure*, pp. 219–266, J. Wiley, Chichester.
Yamashita, N., Doi, M., Nishio, M., Hojo, H., and Tanaka, M. (1972). Current state of Kyoto children by arsenic tainted Morinaga dry milk. *Jpn. J. Hyg.*, **27**, 364–399 (in Japanese).
Yule, W., and Landsdown, R. (1983). Lead and children's development: recent findings. In *Heavy Metals in the Environment*, Heidelberg, September 1983, pp. 912–916, CEP Consultants Ltd, Edinburgh.
Zielhuis, R. L. (1975). Dose–response relationships for inorganic lead. 1 Biochemical and haematological responses. *Int. Arch. Occup. Health*, **35**, 1–18.

Lead, Mercury, Cadmium and Arsenic in the Environment
Edited by T. C. Hutchinson and K. M. Meema

CHAPTER 7

Atmospheric Emissions of Arsenic, Cadmium, Lead and Mercury from High Temperature Processes in Power Generation and Industry

JOZEF M. PACYNA

Norwegian Institute for Air Research,
P. B. 130, N-2001 Lillestrom,
Norway

ABSTRACT

High-temperature processes, such as coal and oil combustion in electric power stations and industrial plants, roasting and smelting of ores in non-ferrous metal smelters, melting operations in ferrous foundries, refuse incineration, and kiln operations in cement plants, emit several pollutants, especially those being easily volatile. Among the volatile pollutants are As, Cd, Pb and Hg.

In this chapter, atmospheric fluxes of As, Cd, Pb and Hg are calculated for the above-mentioned sources in a quantitative and qualitative sense, and are based on measurements and estimates. The measurements were carried out around coal-fired power plants and copper and lead smelter complexes in Poland during the period 1975–80. Estimates were based on: (1) the chemical composition of source materials, (2) the physical–chemical properties of metals, (3) the technology employed, and (4) the efficiency and type of emission control devices utilized. Non-ferrous metal smelters appear to be a predominant source of the As and Cd emissions. Traffic is a major contributor to Pb emission, however, this source is not considered here. Significant amounts of Pb are also emitted from lead smelters. The most important sources of Hg emission include mining and smelting of non-ferrous metals, burning of fossil fuels (mainly coal), and chloralkali process for production of chlorine and caustic soda.

Combustion temperature in boilers, furnaces, roasters, etc. is the key parameter affecting the amounts of As, Cd, Pb and Hg released into the atmosphere, as well as their chemical form and particle size distribution. The metallic As, Cd and Pb and their oxides and sulphides are commonly found in stack dust. Hg is released mainly in a gaseous form. Mercury associated with aerosol particles normally contributes only marginally to the total atmospheric load. High-temperature processes produce fine fraction particles with a significant portion of particles below 2 μm in diameter. These particles can penetrate through emission control devices, enter the atmosphere, and travel for long distances. Penetration of As, Cd and Pb through the boilers of two coal-fired power plants in Poland, equipped with electrostatic precipitators of 99.5% efficiency was between 2 to 3%. More than 95% of Hg was released in a vapour phase. Penetration rates 2–3 times lower for the above elements were measured through wet scrubbers.

Atmospheric fluxes of As, Cd, Pb and Hg were calculated on global, regional, and local scales. Estimates in this chapter are then compared with emission surveys by other authors.

Finally, recommendations for further efforts are presented emphasizing the role of emission surveys for planning control strategies to achieve ambient air quality goals. Gaps in the knowledge and uncertainties of emission assessments have been also taken into account.

INTRODUCTION

The processing of mineral resources at high temperatures, such as coal and oil combustion in electric power stations and industrial plants, roasting and smelting of ores in non-ferrous metal smelters, melting operations in ferrous foundries, refuse incineration and kiln operations in cement plants result in the release of volatile elements into the atmosphere. Among the volatile pollutants, arsenic (As), cadmium (Cd), lead (Pb) and mercury (Hg) are widely studied because of the serious health concern associated with them. The emissions of these elements by the above processes alter their biogeochemical cycles in the whole human environment. To assess these perturbations, extensive emission inventories are required. During the last decade the growing interest in this field has been documented in the literature. Emission strengths of As, Cd, Pb and Hg were estimated by several international and national communities and agencies. As an example, the US National Academy of Sciences, the US Environmental Protection Agency, and the National Research Council of Canada have produced emission inventories for all four elements in the USA and Canada (e.g. NAS, 1977; US EPA, 1979; NRCC, 1979). In Europe, special attention has been paid to the anthropogenic fluxes of Cd (e.g. van Enk, 1980; Hutton 1982; OECD, 1984). This does not mean that the evaluation of anthropogenic fluxes of other elements has been neglected. The As, Cd, Pb and Hg emission sources and their strengths were also discussed during several workshops (e.g. Lindqvist *et al.*, 1983; Nriagu, 1984) and by individual authors in scientific papers (e.g. Nriagu, 1979; Lantzy and Mackenzie, 1979; Walsh *et al.*, 1979.

In this chapter, atmospheric fluxes of arsenic, cadmium, lead, and mercury are presented with emphasis on: (1) the chemical composition of source materials, (2) the physical-chemical properties of metals, (3) the technology employed, and (4) the efficiency and type of emission control devices utilized. Special consideration is given to metal releases from power generation and non-ferrous metal production. The atmospheric transport of As, Cd, Pb and Hg is then discussed considering the information on their chemical nature and tendency to concentrate on fine particles.

EMISSIONS OF As, Cd, Pb AND Hg DURING COAL AND OIL COMBUSTION

Coal and oil are the main fossil fuels used to produce electricity. Apart from many unquestionable advantages, such as the relatively low cost of energy generation and easy availability of the fuel required, coal- and oil-power plants are known to create serious environmental hazards due to the emission of such pollutants as arsenic, cadmium, lead and mercury.

The trace element behaviour during combustion of fuels depends mainly on: (1) affinity of elements for pure coal and mineral matter, (2) physical-chemical properties of elements and their concentrations in coal and oil, and (3) combustion conditions. Kuhn *et al.* (1980), in their thorough review of the abundance of trace and minor elements in organic and mineral fractions of coal, found As, Cd, Pb and Hg as the sulphide-forming elements. They concluded that information on the affinity of trace elements for pure coal and mineral matter could be used, in conjunction with concentrations, for determining the chemical forms of elements, for estimating the theoretical percentage of an element that can be removed by coal cleaning, and for predicting material balances in the coal products and wastes. The needed information on trace element concentrations in coal has been collected by this author (Pacyna, 1986a). The As, Cd, Pb and Hg concentrations in coals, taken from this review, are presented in Table 7.1, together with the metal concentrations in crude oils. As can be seen, there are large differences between metal concentrations in coals from several producing fields of the world. As a result, it is difficult to generalize on impurities in coal. However, the literature (Gluskoter *et al.*, 1977; US EPA, 1980; Pacyna, 1980) suggests that lignite and sub-bituminous coals are less contaminated by these metals than medium-, low-, and high-volatility bituminous coals. As an example, coal from the Western Basin in the United States is much less contaminated than the coals from Eastern and Illinois basins. In Europe, German brown coal (lignite) seems to be cleaner than bituminous and sub-bituminous coals from Czechoslovakia and Poland (Pacyna, 1980; Heinrichs, 1982). On the other hand, the production of a certain amount of energy (i.e. electricity) requires a lignite charge almost two times as high as that of bituminous coal (Dvorak and Lewis, 1978), resulting in enhanced emissions of As, Cd, Pb and Hg.

Table 7.1 Concentrations of As, Cd, Pb and Hg in coals and crude oils (in μg/gm)

Metal	Coal	Crude oil
As	0.34–130.0	0.0024–1.63
Cd	0.01–300.0	0.03–2.10
Pb	0.7–220.0	0.001–0.31
Hg	0.01–1.6	0.014–30.0

During combustion, the volatile species in the coal evaporate in the boiler and recondense as submicrometre aerosol particles, or on the surfaces of ash particles as the flue gas cools in the convective sections. The concentrations of arsenic, cadmium and lead increase markedly with decreasing particle size from bottom ash through fly ash from control devices to stack fly ash (stack dust) emitted into the atmosphere. Apparently, more than 90% of mercury in coal is released as vapor (Billings *et al.*, 1973; Kaakinen *et al.*, 1975; Pacyna, 1980). Combustion temperature in the boiler is one of the key parameters affecting the amounts of metals released. The higher the temperature in a boiler, the larger the discharges of volatile elements. Thus, larger amounts of As, Cd and Pb are emitted into the atmosphere from conventionally-fired boiler systems (e.g. stoker-boiler or cyclone-boiler) burning fuel at temperatures higher than 1650 K, as compared with fluidized-bed systems with temperatures between 1100 and 1200 K. The latter systems are now being employed in electric utilities together with cyclone furnaces. The stoker-type boiler is the dominant unit used in industrial plants and in atypically small (<40 MW) power plants. The type of fly ash control system and its efficiency also influence the As, Cd and Pb emissions. Electrostatic precipitators (ESP), wet scrubbers, mechanical collectors and fabric 'bag-house' filters are the most commonly used systems. The two former are installed in coal-fired power plants. Generally, a venturi wet scrubber system is more efficient in removing As, Cd and Pb from a flue gas stream than electrostatic precipitators. Ondov *et al.* (1979) have studied the penetration of several elements contained in particles from coal-fired plants equipped with both control systems. They calculated the following values for penetration through ESP: 4.3–11.5% for As, 3.3–8.8% for Cd and 2.2–5.5% for Pb. The arsenic penetration through a venturi wet scrubber ranged from 2.5–7.5%. No information was presented for cadmium and lead.

Oil-fired power plants are also important sources of As, Cd, Pb and Hg emissions into the atmosphere. During refining of oil, these metals concentrate in the heavy distillate residuals, such as residual fuel oils, asphalts, and in the liquid and solid waste streams. The limited information available indicates that probably 30% of the metals in crude oil are retained in residual fuel oil and asphalt (Smith *et al.*, 1975). The amounts of As, Cd, Pb and Hg discharged during combustion of residual fuel oil in power plants and industrial boilers depend mainly on their concentrations in crude oils (Table 7.1), the efficiency of combustion, and the build-up of boiler deposits. Generally, the most contaminated crude oils are those from the Middle East and Venezuela (Pacyna, 1982), while crude oils from the Far East (Singapore, Indonesia), Africa (Algeria, Gabon, Nigeria) and North America seem to be lowest in impurities. Poor mixing in a boiler, low flame temperatures, and short residence time in the combustion zone result in larger particles with a higher content of combustible matter and higher particulate loadings. Considering boiler design, the As, Cd and Pb emission rates from two commonly used types of boilers, tangential and horizontal, working under similar conditions, are comparable. Oil-fired boilers do not require fly ash removal equipment and, under such circumstances, the emissions of elements are comparable with the emissions from a coal-fired unit of better than 99% collection efficiency. Generally, the stack dust loadings for industrial oil combustion units are higher than those for electricity generating plants.

The above information has been used to calculate the trace element emissions from coal and oil combustion in Europe (Pacyna, 1983). Emission factors have been estimated for each European country separately. The differences were very significant: from 15.4 μg As/MJ for hard coal-power plants in West Germany to 95.5 μg As/MJ for lignite-power plants in Poland, and from 8.8 μg As/MJ for Switzerland to 58.1 μg As/MJ for Denmark in the case of oil-fired power plants (Pacyna, 1982). Similar differences in emission factors were estimated for lead, cadmium and mercury emitted from coal and oil combustion for electric utility, industrial, commercial and residential

Table 7.2 Emissions of As, Cd, Pb and Hg from coal and oil combustion in Europe in 1979 (in tons/year)

Source	As	Cd	Pb	Hg
Coal combustion				
power plants	205	64	733	85.5
industry	240	77	870	135.0*
commercial and residential units	16	5	73	
Oil combustion				
power plants	79	37	450	v.l.
industry and commercial and residential units	138	73	709	v.l.
Total	678	256	2835	221

* Industry + commercial and residential units.
v.l.—very low.

furnaces. The emissions of As, Cd, Pb and Hg for Europe are presented in Table 7.2. The Hg emissions in Table 7.2 include both the metal releases in the vapor phase and as particles. Oil firing contributes insignificantly to the mercury emissions from fossil fuel combustion. The metal emissions due to burning of fossil fuels in the United States were the following: 497 tons of Cd in 1979 (US EPA, 1981), 120 tons of Hg in 1975 (Watson, 1979), and 650 tons of As in 1974 (NAS, 1977). Fuel combustion includes not only stationary combustion but also mobile sources, particularly gasoline combustion engines. The fate of lead in gasoline has been studied extensively, and is not reviewed here in detail. For statistical purposes only, the author's calculations of Pb emissions from gasoline combustion in Europe give a value of 74 300 tons in 1979, compared with 128 000 tons in the United States in 1975 (US EPA, 1977).

NON-FERROUS METAL PRODUCTION AS A SOURCE OF ATMOSPHERIC As, Cd, Pb AND Hg

A major source of arsenic and cadmium and an important source of lead and mercury is smelting and refining of non-ferrous metals. Both primary processing from mineral concentrates and secondary processing from scrap and residual materials are considered. The first process in non-ferrous metal production is mining and beneficiation of sulphide ores. Most of the ores are extracted by underground mining methods, but open-pit mining is also used. The exploitation of lead–zinc ores is particularly important due to their contamination by Cd and As. The latter element is also enriched in copper-nickel ores. The concentrations of Cd and As in non-ferrous metal ores vary widely from one field to another. As an example, Wedepohl (1969) gives the arsenic concentration range from 10 to 10 000 μg/gm.

Primary metal production converts the concentrated ore into metal by several processes. Roasting of ore concentrates and low grade ore enrichment, the initial steps in smelters, are common in copper, nickel, lead, zinc and cadmium production. The processes occur at about 1500 K, and volatile elements, such as arsenic, cadmium and mercury, are partly removed from the concentrates, and remain in the flue gases. As much as 95% of arsenic contained in copper ores may be driven off in the gases; however, 50% is more typical. Most of this is recycled. All the mercury in a smelter will vaporize early in the process and enter the gas stream, which is usually passed through an acid plant to remove mercury.

The next process, smelting, can be either thermal (in horizontal or vertical retort) or electrolytic extraction. The old, pyrometallurgic zinc plants employing thermal smelting are a major cadmium emission source. Hutton (1982), studying cadmium behaviour in the environment, suggests metal emission factors from 10–100 gm Cd/ton, in pyrometallurgic zinc production, to 0.20 gm Cd/ton in the electrolytic process. At present, five of 18 primary zinc plants within the European Community use thermal facilities while two pyrometallurgic plants operate in the United States. All the primary zinc in Canada is produced by the electrolytic process. In 1977 the thermal process in horizontal retorts was used in three countries: Mexico, Poland and the USSR. As in zinc smelting, the thermal process in copper and lead production results in cadmium emissions. Based on the literature reviewed by Hutton (1982), the Cd emissions from primary copper and lead production are lower than the Cd emission from pyrometallurgic zinc plants.

Smelting and converting in copper–nickel smelters are important sources of As emissions. Behaviour of the metal depends primarily upon the process itself. Several smelting technologies are currently used in the copper–nickel industry, including reverberatory smelting, flush smelting (two processes are currently in commercial use: the INCO process and the OUTOKUMPU process), and the Noranda and Mitsubishi Continuous Copper Smelting Process. About 90% of copper is presently being produced by these processes. The rest comes from hydrometallurgic techniques. Smelting technology has been widely reviewed in the literature (e.g. Environment Canada, 1982). Reimers *et al.* (1977) concluded that the portion of arsenic volatilized during the smelting operation ranges from 20% of the arsenic input for the reverberatory furnace to 85% for the Noranda reactor for making high grade matte. The effluent gases include also about 85% of the lead input in the copper–nickel ores and 80% of the cadmium input. Most of the metal is usually recycled in flue dusts, but the remainder is emitted into the atmosphere. Lead is produced in a similar way to zinc through processes such as roasting, sintering, smelting and refining. Most of the airborne trace element emissions in lead ore processing are due to the roasting operations. Reduction in the blast furnaces is also a source of air pollution by trace elements. The emission rates of As, Cd, Pb and Hg from primary non-ferrous metal production are highly sensitive to differences in the efficiency of control devices. High-efficiency control devices are often employed in smelters. Emissions from roaster facilities, as well as from sintering processes, are mainly controlled with fabric filters and electrostatic precipitators.

The share of non-ferrous metals produced by secondary processing in the total metal production varies from 20% for zinc and nickel to 37% for lead and 38% for copper (Barbour *et al.*, 1978). Besides, many 'primary' lead plants are fed with a mixed charge of roasted ores and scrap materials (Hutton, 1982). The chemical composition of input scrap and the type and efficiency of control devices are two factors which affect the Cd and Pb emissions most from this source.

To assess quantitatively atmospheric fluxes of As, Cd, Pb from non-ferrous metal production, the metal emission factors have been calculated for the above-mentioned processes, and used to estimate the metal releases in Europe in 1979. The results are listed in Table 7.3. Similar inventories have been prepared for the US emissions, indicating releases of about 1000 tons/year for Cd (several US EPA reports), 70 tons/year for Hg (Watson, 1979), and 4800 tons/year for As from copper smelting only (NAS, 1977).

Table 7.3 Emissions of As, Cd, Pb and Hg from non-ferrous metal production in Europe in 1979 (in tons/year)

Source	As	Cd	Pb	Hg
Primary non-ferrous metal production*				
copper–nickel	4 500	600	9 250	61
zinc–cadmium	900	1 050	7 880	13
lead	300	8	10 450	3
Secondary non-ferrous metal production				
copper		v.l.	55	
lead		v.l.	400	
Mining		v.l.	1 090	
Total	5 700	1 658	29 125	77

* Data include both pyrometallurgical and hydrometallurgical processes.
v.l.—very low.

OTHER ANTHROPOGENIC SOURCES OF As, Cd, Pb AND Hg EMISSIONS

Other anthropogenic sources releasing arsenic, cadmium, lead, and mercury include high-temperature processes in steel and iron manufacturing, municipal and sewage sludge incineration, and cement production, as well as industrial application of the metals. Recently, the trace element emissions from the above sources have been analysed in detail (Pacyna, 1986a). Briefly, the cadmium and lead emissions from steel works depend significantly on the type of technology used. Electric arc and basic oxygen furnaces are much safer than older open-hearth ones.

Large quantities of coke are produced for use in the iron and steel industries, but the release of metals from this process is considered when analysing coal combustion for industrial purposes.

Only limited information is available on trace element emissions from refuse incineration. As, Cd, Pb and Hg emissions are mainly affected by the chemical composition of the refuse input, the incinerator chamber design (combustion temperature), and the efficiency of control devices (if any).

The high-temperature kiln operations in cement plants emit several trace elements, among them Pb and Cd.

Industrial applications of As, Cd, Pb and Hg often involve their discharges as air pollutants. Arsenic is used in agriculture (herbicides, insecticides, feed additives), in glass manufacturing, and in wood preservatives. Approximately 10% of the As used to produce pesticides may be volatilized to the atmosphere by aerobic and anaerobic microbial activity. An important source of arsenic emissions is the cotton ginning process, removing the seeds from raw cotton. Cadmium is used in the plating of iron, steel, and other metals, as pigments for glass and paint, and in insecticide production. The metal is also used to produce nickel–cadmium batteries. The chloralkali industry is the largest consumer of mercury, followed by the electrical and paint industries, agriculture, dentistry and the chemical industry. Industrial applications of lead include lead alkyl manufacture, storage battery production, and production of can solder.

Table 7.4 shows atmospheric fluxes of As, Cd, Pb and Hg from the above sources in Europe in 1979 (Pacyna, 1983). The metal emissions from relevant sources in the United States are as follows: 1100 tons As/year (NAS, 1977), 135 tons Cd/year (US EPA, 1979), 40 tons Hg/year (Watson, 1979), and 10 130 tons Pb/year (US EPA, 1977).

Table 7.4 Emissions of As, Cd, Pb and Hg from sources other than fossil fuel combustion and non-ferrous metal production in Europe in 1979 (in tons/year)

Source	As	Cd	Pb	Hg
Iron, steel, and ferro-alloy manufacturing		60	14 660	
Refuse incineration	10	85	800	20
Cement production		15	750	
Industrial application of metals	140*	20	2 000	20
Total	150	180	18 210	40

* Without pesticide production.

It should be noted that for some sources, as well as for certain countries in Europe, the information necessary for calculations of trace element emissions was difficult to obtain or was simply lacking. The uncertainties included data on: (1) metal concentrations in fuels and other raw materials, (2) industrial technologies, and (3) the efficiencies of control devices used. A simple trajectory model has been used to verify the above estimates for As, Cd and Pb emissions from anthropogenic sources in Europe. The metal concentrations calculated by the model were compared with measured daily ambient concentrations at Birkenes in Southern Norway. The measured concentrations exceeded the calculated by 24, 34 and 38% for Pb, As, and Cd, respectively, thus in reasonable agreement.

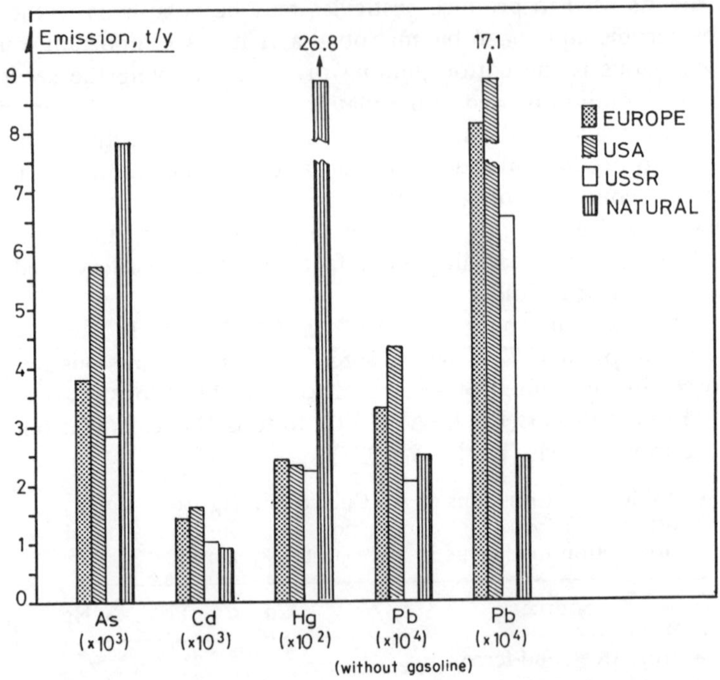

Figure 7.1 Total atmospheric emissions of As, Cd, Hg, and Pb in Europe, USA, USSR, and from natural sources

It is of interest to compare the anthropogenic emissions of the As, Cd, Pb and Hg to the metal releases from natural sources, shown in Figure 7.1. The trace metal emissions from sources in the Soviet Union have been estimated at the Norwegian Institute for Air Research (NILU, 1984). The data for natural sources in Figure 7.1 are taken from a literature review by Pacyna (1986b), except for the Hg emissions, given by NAS (1978). The natural emissions of arsenic and mercury* contribute very significantly to the total releases of the metals. The latter element is discharged mainly from natural sources, such as volatilization of gaseous mercury from soils and vegetation, as well as from the ocean and other water bodies. The predominant source of

* It should be noted that the mercury study groups suggest a global ratio of from 2:1 to 4:1 for natural versus anthropogenic contributions of Hg (*Editors*).

natural As and Cd appears to be volcanoes, whereas for Pb it is windblown dust. The natural Cd and Pb contributions to the total emissions of the metals seem to be less significant than the contribution for As and Hg, but cannot be neglected. Generally, very little is known about the magnitude of natural fluxes of Cd and Pb, or about the species involved.

The total anthropogenic emissions of the elements in Europe, USA and USSR, are broken up in source categories in Figure 7.2. The largest amounts of As, Cd and Pb are emitted in all these areas from non-ferrous metal smelters (excluding Pb from gasoline combustion), while fossil fuel combustion seems to be the most important anthropogenic source of mercury. Thus, the high-temperature processes emit a major part of As, Cd, Pb and Hg into the atmosphere. The magnitude of the metal fluxes is important information in the assessment of their potential threat to human health. However, the environmental behaviour and toxicity of an element are also dependent on its chemical speciation and the tendency to concentrate in fine particles. In high-temperature processes these two factors are closely related to the combustion temperature and volatility of the metal.

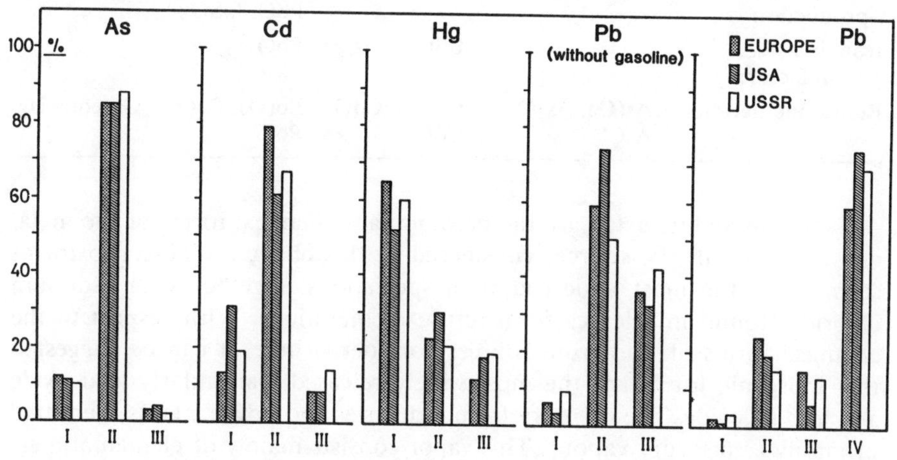

Figure 7.2 Atmospheric emissions of As, Cd, Hg, and Pb from fossil fuel combustion (I), non-ferrous metal production (II), other anthropogenic sources (III), and gasoline combustion (IV) in Europe, USA, and USSR

CHEMICAL SPECIATION OF As, Cd, Pb AND Hg FROM HIGH-TEMPERATURE PROCESSES

There is a very limited information in the literature on the chemical speciation of metals emitted into the atmosphere. Based on the results from the EPA's National Air Sampling Network, it can be suggested that for arsenic inorganic forms dominate in the air over emission areas. Major chemical species of arsenic and of the other elements discussed which evolve during high-temperature processes are presented in Table 7.5. Trivalent arsenic is the most common form of the metal. Air samples containing arsenic, of either smelter or coal-fired power plant origin, consist largely of trivalent arsenic in both vapour and particulate states. This is very important, because, as a general rule, inorganic arsenicals exhibit greater toxicity than organic arsenicals, and the trivalent state more so than the other states. Methylated forms of arsenic are probably of minor significance.

Table 7.5 Major chemical species evolved during fossil fuel combustion and industrial processes

Process	As	Cd	Pb	Hg
Coal combustion	As(O), As_2O_3 As_2S_3	Cd(O), CdO CdS	$PbCl_2$, PbO, PbS, Pb	Gaseous Hg.
Oil combustion	As(O), As_2O_3 Organic arsines	Cd(O), CdO	PbO	Gaseous Hg
Non-ferrous metal production	As_2O_3	CdO, CdS	PbO, $PbSO_4$, $PbO.PbSO_4$	
Iron and steel manufacturing		CdO	PbO	
Refuse incineration	As(O), As_2O_3, $AsCl_3$	Cd(O), CdO, $CdCl_2$	Pb(O), PbO, $PbCl_2$	Gaseous Hg

Cadmium and its oxide are the predominant chemical forms of the metal emitted from all the sources considered in this chapter. These two forms seem to be the most toxic cadmium speciations, together with cadmium chloride (found in releases from refuse incineration). With respect to the chemical forms of lead from sources considered here, it can be suggested that inorganic forms are the most widely released, particularly lead oxide and lead chloride. The mercury from industries and power plants is emitted primarily as mercury vapour. This vapor consists mainly of elemental mercury and dimethyl mercury. It is difficult to say which volatile compound dominates the discharge process. Mercury species other than elemental Hg and $(CH_3)_2Hg$ can also contribute. Seiler (in Lindqvist *et al.*, 1983) suggests that most mercury is emitted as dimethyl mercury with a relatively fast degradation to elemental mercury taking place in the air. Brosset (1981) concludes that Hg(O) is mobilized to the atmosphere where it is subjected to atmospheric oxidation processes to yield water soluble forms, subsequently scavenged by wet or dry deposition.

Considering the volatility of As, Cd, Pb and Hg compounds in Table 7.5, elemental arsenic (As(O)), As_2O_3, $AsCl_3$, elemental Cd (Cd(O)), and

$CdCl_2$ are volatile, $PbCl_2$ is intermediate, and CdO, elemental lead (Pb(O)) and PbO are non-volatile (Gerstle and Albrinck, 1982). The volatile species would be in the vapor state during the high-temperature processes, whereas the non-volatile compounds would continue to be emitted largely as fly ash, even at the higher temperature.

ATMOSPHERIC TRANSPORT

Volatile elements evaporated in a furnace condense on small particles in the flue gas stream. The relationship between volatile metal concentration and particle size during high-temperature processes has already been mentioned, when discussing emission from coal combustion. The condensation of trace elements continues in the atmosphere. The outermost layer on the surface of emitted ashes is often composed of H_2SO_4, which permits surface condensation of several elements including As, Cd, Pb and partly Hg. The condensation takes place preferentially on fine particles in the 0.1 to 1.0 μm diameter range. Additionally, the presence of sulphuric acid on some particle surfaces may give rise to post-combustion crystal formation of metal and/or ammonium sulphates, but how this affects the physico-chemical form of trace elements has not been studied. Small particles from high-temperature processes are very easily transported by air masses over long distances. Gaseous mercury behaves in a similar manner to As, Cd and Pb on fine particles. The concentrations of As, Cd and Pb on the larger particles will be low and decrease rapidly with distance from a power plant or smelter. The results from several studies seem to indicate that at a distance of 30 km from the high-temperature source the As, Cd and Pb concentrations reach the background level (e.g. Rozenshtein, 1970; Yankel *et al.*, 1977; Glowiak *et al.*, 1977; Pacyna, 1980). Lead concentrations around power plants, smelters or other industrial plants are often enhanced by emissions from gasoline combustion. Several studies in the surroundings of copper and lead smelters in Poland (Glowiak *et al.*, 1977; Kubacka *et al.*, 1977; Pacyna *et al.*, 1981) have been carried out to assess the 'local deposition' of trace elements. It was found that about 11% of the mass of trace element is deposited in the same area as it is released. Pacyna *et al.* (1984) have studied long range transport of several trace elements in Europe. Based on the literature sources (OECD, 1979), they have inferred local deposition to be 5% of the emissions.

A major portion of the arsenic, cadmium, and lead emissions from high temperature sources with tall stacks is, however, transported over long distances. The behaviour of these elements during transport depends on their physical-chemical properties, the particle size-distribution, and meteorological conditions, such as the rate of turbulent vertical air exchange and wind speed. The two former parameters seem to dominate the long range

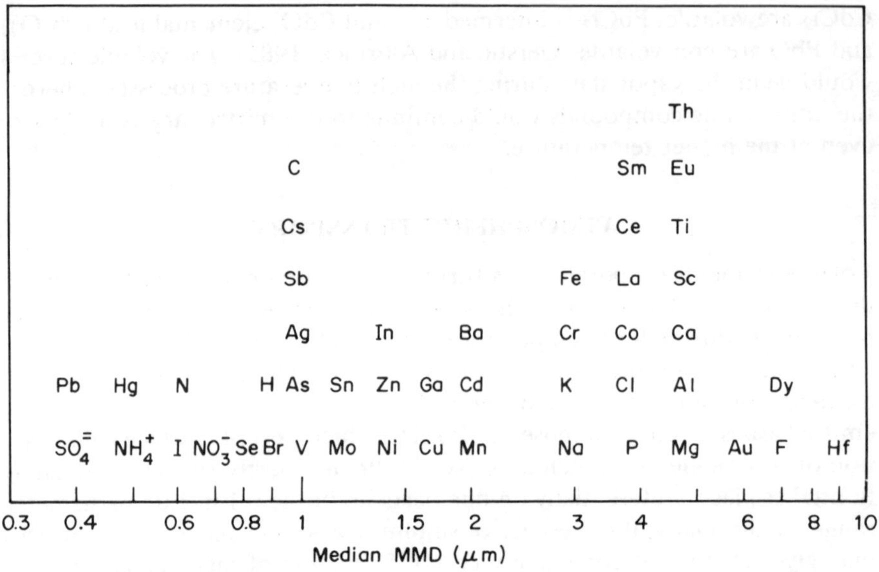

Figure 7.3 Mass median diameter of trace elements. Mean values (after Rahn, 1976)

transport of As, Cd and Pb. According to the literature review by Davidson and Osborn (1984), lead is generally associated with submicrometre particles, while cadmium size distribution contains submicrometre as well as supermicrometre aerosol. Rahn (1976) has compiled information on mass-median diameters of As, Cd, Pb, Hg and other trace elements, shown in Figure 7.3. The residence time of As, Cd, and Pb in the atmosphere was found to be about 7 days. This time is sufficient for transport over thousands of kilometres.

Trace elements can be removed from the atmosphere by dry deposition (sedimentation, interception, impaction and turbulent diffusion) and by wet (precipitation) deposition (rainout, washout). Dry deposition velocities of trace metals were measured by Cawse (1981) and estimated by Davidson (1980). A value of 0.1 seems to be representative for the mean Cd dry deposition velocity, and 0.2 for lead and arsenic. Generally, deposition models consider dry deposition velocity to be a function of aerosol aerodynamic diameter, state of the atmosphere (the variation of wind speed with height and particularly the turbulence of the atmosphere close to the ground), and surface conditions (type and status of the vegetation, open waters, snow, etc.).

Aerosol solubility is one of the key parameters for the wet precipitation process. However, the concentration of pollutants in precipitation is the result of several complex processes both within and below clouds, and uncertainties still remain as to a detailed description of the scavenging efficiency of precipitation. Generally arsenic and cadmium were found to be 'soluble' in the sense that these two elements are likely to be transported over significant distances dissolved in natural waters (Asami *et al.*, 1984). Lindberg and Harriss (1980) have studied solubility of trace elements in aerosols from a power plant plume. They found that coal combustion sources may be a major contributor of soluble trace elements (including Cd and Pb) to the regional tropospheric aerosol. However, our knowledge of complicated physical and chemical processes leading to wet deposition of trace elements is very incomplete and data of this type are largely absent from the literature.

The relative importance of dry and wet deposition depends on the transport distance from the sources. In a heavily polluted area, the amounts removed by dry deposition seem to be larger than by wet deposition, while in remote areas with much precipitation wet deposition may be at least equally important.

The atmospheric behaviour of mercury appears to be a combination of deposition and resuspension or vaporization (Kothny, 1973). After emission into the atmosphere, a part of mercury can be scavenged by even light precipitation, deposited on land, and re-vaporized within 1–2 days (NRCC, 1979). There is no evidence, however, for Hg re-vaporization over water. Most of the Hg in the atmosphere (in particular the insoluble fraction) is transported over long distances (Lindqvist *et al.*, 1983). The average residence time of this fraction of mercury in the atmosphere is estimated to be more than 1 year. The water-soluble fraction and the fraction associated with particulate matter are readily removed from the atmosphere and have a residence time of perhaps a few days or weeks (Lindqvist *et al.*, 1983). Lindqvist *et al.* (1983) suggest that dispersion parameters of these fractions of mercury are similar to those of pollutants occurring as soluble gases, or associated with submicrometre-size aerosol particles (e.g. Pb, As and Cd). This results in transport scales of several hundred to a few thousand kilometres for the soluble fraction of mercury. These authors also point out that a significant portion of Hg emissions may be deposited close to point sources.

CONCLUSION

A number of studies and reviews cited in this chapter demonstrate the progress in evaluation of trace element emissions. Emission surveys of As, Cd, Pb and Hg from anthropogenic sources are presented in the literature, with measurements and calculations done either for a certain country or for an individual source. Fuel combustion and metal smelting are the most investigated types of emission sources. However, a number of uncertainties have emerged during these investigations concerning the content of the metals in fuels and other raw materials, the efficiencies of control devices, and the uses of different technologies to produce industrial goods.

Future investigations should improve the accuracy of existing inventories, and extend the list of important sources for processes such as waste incineration, biomass burning, phosphate fertilizers, etc. In practice, to improve the accuracy of emission estimates sensitive analytical methodologies for metal and metalloid specification (chemical form and complexation) in emitted dust samples are required. This is particularly important when considering the fate of atmospheric emissions of the elements discussed with respect to the residence times in the air and scavenging processes during atmospheric transport.

To understand the impact of anthropogenic sources of As, Cd, Pb and Hg and their cycles in the environment, we need to know more accurately natural source strengths. The most extensively studied natural source appears to be volcanic eruptions. Information is very scarce on the As, Cd, Pb and Hg releases from wind-blown dust erosion, forest wild-fires, airborne sea-salt, and biogenic emissions from vegetation.

ACKNOWLEDGEMENTS

The author thanks Dr Val Vitols for helpful discussions.

REFERENCES

Asami, T., Andrea, M. O., Bertine, K. K., Buat-Ménard, P. E., Duce, R. A., Filip, Z., Forstner, U., Goldberg, E. D., Heinrichs, H., Jernelov, A. B., Pacyna, J. M., Thornton, J., Tobschall, H. J., and Zoller, W. H. (1984). Changing biogeochemical cycles. In Nriagu, J. O. (Ed.), *Changing Metal Cycles and Human Health*, pp. 359–373, Dahlem Konferenzen, Springer Verlag, Berlin, Heidelberg, New York, Tokyo.

Barbour, A. K., Castle, J. F., and Woods, S. E. (1978). Production of non-ferrous metal. In Parker, A. (Ed.), *Industrial Air Pollution Handbook*, McGraw-Hill Book Company (UK) Ltd., London.

Billings, C. E., Sacco, A. M., Matson, W. R., Griffin, R. M., Coniglio, W. R., and Harley, R. A. (1973). Mercury balance on a large pulverized coal-fired furnace. *J. Air Pollut. Control Assoc.*, **23**, 773–777.

Brosset, C. (1981). The mercury cycle. *Water, Air, and Soil Pollut.*, **16**, 253–255.

Cawse, P. A. (1981). *A Survey of Atmospheric Trace Elements in the U. K. Results for 1979*. AERE Report-R 9886. United Kingdom Atomic Energy Authority, Harwell, England.

Davidson, C. J. (1980). Dry deposition of cadmium from the atmosphere. In Nriagu, J. O. (Ed.), *Cadmium in the Environment*, pp. 115–140. John Wiley and Sons, Inc., New York.

Davidson, C. J., and Osborn, J. F. (1986). The sizes of airborne trace metal-containing particles. In Nriagu, J. O., and Davidson, C. (Eds), *Toxic Metals in the Atmosphere*, Advances in Environmental Science and Technology, pp. 355–390, John Wiley and Sons, Inc., New York.

Dvorak, A. J., and Lewis, B. G. (1978). *Impacts of Coal-fired Power Plants on Fish Wildlife and their Habitats.* US Dept. of Interior, Report FWS/OBS-78/29, Washington, DC.

van Enk, R. H., (1980). *The Pathway of Cadmium in the European Community.* Commission of the European Communities, Joint Research Centre Ispra Establishment, Report EUR 6626 EN, Ispra, Italy.

Environment Canada (1982). *Air Pollution Emissions and Control Technology: Primary Copper Industry.* Air Pollution Control Directorate, Mineral and Metal Processes Division, EPS 3-AP-82-4, Ottawa.

Gerstle, R. W., and Albrinck, D. N. (1982). Atmospheric emissions of metals from sewage sludge incineration. *J. Air Pollut. Control Assoc.*, **32**, 1113–1123.

Glowiak, B. J., Palczynski, R. J., and Pacyna, J. M. (1977). *Environmental Lead and Mercury in the Area of a Copper Smelter.* US ERDA, Environmental Quarterly Report HASL-321, Health and Safety Laboratory, New York.

Gluskoter, H. J., Ruch, R. R., Miller, W. G., Cahill, R. A., Dreher, G. B., and Kuhn, J. K. (1977). Trace elements in coal: occurrence and distribution. Illinois State Geological Survey. Circular 499, Urbana, Il.

Heinrichs, H. (1982). Trace element discharge from a brown coal fired power plant. *Environ. Technology Letters,* **3**, 127–136.

Hutton, M. (1982). *Cadmium in the European Community.* Monitoring and Assessment Research Centre, MARC Report No. 26, London.

Kaakinen, J. W., Jorden, R. M., Lawasani, M. H., and West, R. E. (1975). Trace element behavior in coal-fired power plant. *Environ. Sci. Technol.*, **9**, 862–863.

Kothny, E. L. (1973). The three-phase equilibrium of mercury in nature. In Kothny, E. L. (Ed.), *Trace Elements in the Environment.* Adv. Chem. Ser., pp. 48–80, Amer. Chem. Soc., Washington, D. C.

Kubacka, J., Pacyna, J. M., and Palczynski, R. J. (1977). *Chemical Composition of Metalloferrous Dust Discharged from Stacks of the Copper Smelters.* Scientific Papers of Environmental Protection Engineering Institute, Technical University of Wrocław, Poland.

Kuhn, J. K., Fiene, F. L., Cahill, R. A., Gluskoter, H. J., and Shimp, N. F. (1980). *Abundance of Trace and Minor Elements in Organic and Mineral Fractions of Coal.* Illinois Institute of Natural Resources. Environ. Geology Notes No. 88, Urbana, Il.

Lantzy, R. J., and Mackenzie, F. T. (1979). Atmospheric trace metals: global cycles and assessment of man's impact. *Geochim. Cosmochim. Acta,* **43**, 511–523.

Lindberg, S. E., and Harriss, R. C. (1980). Trace metal solubility in aerosols produced by coal combustion. In Singh, J. J., and Deepak, A. (Eds.), *Environmental and Climatic Impact of Coal Utilization,* pp. 589–608. Academic Press, New York.

Lindqvist, O., Jernelov, A., Jonhansson, K., and Rodhe, H. (1983). *Mercury Pollution of Swedish Lakes. Global and Local Sources.* Swedish Environmental Protection Board, Solna, Sweden.

NAS (1977). *Arsenic.* National Academy of Science, National Research Council, Committee on Medical and Biological Effects on Environmental Pollutants, Washington, DC.

NAS (1978). *An Assessment of Mercury in the Environment.* National Academy of Science, National Research Council, Committee on Medical and Biological Effects on Environmental Pollutants, Washington, DC.

NILU (1984). *Emission Sources in the Soviet Union.* Technical Report 4/1984. The Norwegian Institute for Air Research, Lillestrom, Norway.

NRCC (1979). *Effects of Mercury in the Canadian Environment.* National Research Council of Canada. NRCC Report No. 16739, Ottawa.

Nriagu, J. O. (1979). Global inventory of national and anthropogenic emissions of trace metals to the atmosphere. *Nature (London)*, **279**, 409–411.

Nriagu, J. O. (Ed.) (1984). *Changing Metal Cycles and Human Health.* Springer-Verlag, Berlin, Heidelberg, New York, Tokyo, 445 pages.

OECD (1979). *The OECD Programme on Long Range Transport of Air Pollutants. Measurements and Findings.* Second Edition, Organization for Economic Co-operation and Development, Paris.

OECD (1984). *Control of Specific Toxic Substances in the Atmosphere.* Cadmium. Organization for Economic Co-operation and Development, Air Management Policy Group, Paris.

Ondov, J. M., Ragaini, R. C., and Bierman, A. H. (1979). Elemental emissions from a coal-fired power plant. Comparison of a venturi wet scrubber system with a cold-side electrostatic precipitator. *Environ. Sci. Technol.*, **13**, 588–601.

Pacyna, J. M. (1980). Coal fired power plants as a source of environmental contamination by trace metals and radionuclides. Habilitation thesis. Technical University of Wrocław, Poland.

Pacyna, J. M. (1982). *Trace element emissions from coal-and oil-power plants in Europe. Methodology of calculations.* Technical Report 5/1982. The Norwegian Institute for Air Research, Lillestrom, Norway.

Pacyna, J. M. (1983). *Trace element emission from anthropogenic sources in Europe.* Technical Report 10/1982. The Norwegian Institute for Air Research, Lillestrom, Norway.

Pacyna, J. M. (1986a). Emission factors of atmospheric elements. In Nriagu, J. O., and Davidson, C., (Eds), *Toxic Metals in the Atmosphere.* Advances in Environmental Science and Technology, pp. 1–32. John Wiley and Sons, Inc., New York (in press).

Pacyna, J. M. (1986b). Atmospheric trace elements from natural and anthropogenic sources. In Nriagu, J. O., and Davidson, C., (Eds.). *Toxic Metals in the Atmosphere.* Advances in Environmental Science and Technology, pp. 33–52. John Wiley and Sons, Inc., New York (in press).

Pacyna, J. M., Zwozdziąk, A., Zwozdziak, J., Matyniak, Z., Kuklinski, A., and Kmiec, G. (1981). *Zagadnienia ochrony powietrza atmosferycznego dla stanu istniejacego i perspektywicznego na obszarze pilotowym wojewodztwa legnickiego.* (Air pollution problems caused by the LGOM copper smelter complex—present situation and perspectives.). Report SPR 14–81, Technical University of Wrocław, Wrocław, Poland (in Polish).

Pacyna, J. M., Semb, A., and Hanssen, J. E. (1984). Emission and long-range transport of trace elements in Europe. *Tellus (Series B)*, **36B**, 3, 163–178.

Rahn, K. A. (1976). *The Chemical Composition of the Atmospheric Aerosol.* Technical Report, Graduate School of Oceanography, University of Rhode Island, Kingston, RI.

Reimers, J. H., Snelgrove, S. R., and Taylor, J. C. (1977). Study on emission control technology for arsenic in the non-ferrous metallurgical industry. Environment Canada, Air Pollution Control Directorate, Mining, Mineral and Metallurgical Division, unpublished report.

Rozenshtein, J. S. (1970). Sanitary toxicological assessment of low concentrations of As_2O_3 in the atmosphere. *Gig. Sanit.*, **35**, 1, 15–20.

Smith, J. C., Ferguson, T. L., and Carson, B. L. (1975). Metals in new and used petroleum products and by-products. Quantities and consequences. In Yen, T. F.

(Ed.), *The Role of Trace Metals in Petroleum*, pp. 123–148, Ann Arbor Press, Ann Arbor, Michigan.

US EPA (1977). *Control Techniques for Lead Air Emissions.* US Environmental Protection Agency, EPA-450/2-77-012, Research Triangle Park, NC.

US EPA (1979). *Sources of Atmospheric Cadmium.* US. Environmental Protection Agency, EPA-450/5-79-006, Research Triangle Park, NC.

US EPA (1980). *Experimental/Engineering Support for EPA's FBC Program: Final Report—Vol. II. Particulate, Nitrogen Oxide and Trace Element Control.* US. Environmental Protection Agency, EPA-600/7-80-015b, Research Triangle Park, NC.

US EPA (1981). *Survey of Cadmium Emission Sources.* US Environmental Protection Agency, EPA-450/3-81-013, Research Triangle Park, NC.

Walsh, P. R., Duce, R. A., and Fashing, J. L. (1979). Considerations of the enrichment, sources, and fluxes of arsenic in the troposphere. *J. Geophys. Res.*, **84**, 1719–1726.

Watson, W. D., Jr. (1979). Economic considerations in controlling mercury pollution. In Nriagu, J. O. (Ed.), *The Biogeochemistry of Mercury in the Environment*, pp. 41–78. Elsevier/North-Holland Biomedical Press.

Wedepohl, K. H. (1969). *Arsenic. Handbook of Geochemistry.* Volume II/2, Springer-Verlag, Berlin.

Yankel, A. J., von Lindern, J. H., and Walter, S. D. (1977). The Silver Valley Lead Study. The relationship between childhood blood lead and environmental exposure. *J. Air Pollut. Control Assoc.*, **27**, 763–7.

Lead, Mercury, Cadmium and Arsenic in the Environment
Edited by T. C. Hutchinson and K. M. Meema

CHAPTER 8

Emission and Deposition of Atmospheric Mercury Vapor

S. E. LINDBERG

Environmental Sciences Division,
Oak Ridge National Laboratory,
Oak Ridge, Tennessee,
37831 USA

ABSTRACT

The global biogeochemical cycle of mercury is complicated by the fact that the atmospheric burden is dominated by various forms of mercury vapor, not all of which have been identified. We have studied three of the most important industrial sources of atmospheric mercury: the chloralkali, cinnabar mining, and electric power generation industries. These studies are reviewed here in the light of more recent work on the global cycle of mercury.

Our work at a commercial coal-fired power plant in the United States involved collection of samples in the stack, plume, and ambient air. Mercury vapor was the dominant form of mercury in all samples, accounting for 92–99% of the total, and was not converted to particle-associated forms during plume interactions. Because of these characteristics, mercury is inefficiently removed by conventional means during fossil fuel combustion, with nearly 99% of the feed coal mercury being discharged to the atmosphere.

Contamination of aquatic environments by direct discharges from chloralkali plants is well known. Our work at a non-operating plant in the United States identified a previously unrecognized source of mercury to the environment, atmospheric emission of the vapor from stored waste deposits. Our studies indicated that atmospheric and aquatic emissions were comparable, and that losses from defunct plants could approach those from active plants. The legacy of continued mercury pollution from such sites must be considered.

Studies at the Almadén mercury mine in Spain indicated that mercury vapor is also emitted from mercury-rich soils, and that the emission rates are temperature dependent and influenced by vegetation cover. Plants grown on these soils accumulate mercury by two means: the roots from soil mercury, and the leaves by direct absorption of emitted mercury vapor.

Predominance of the vapor form of mercury is conducive to long range transport, global dispersion, and direct uptake by plants and animals. On a global scale, precipitation scavenging appears to be the major removal process, but dry deposition measurements are virtually non-existent. Recent reports of relationships between mercury in fish and lake acidification suggest that air pollution may be having an indirect effect on the biogeochemical cycle of mercury.

INTRODUCTION

Reports of potential relationships between acidic deposition, lake acidification, and concentrations of Hg in fish (Brouzes *et al.*, 1977; Brosset and Svedung, 1977) have prompted a recent resurgence of interest in the environmental behavior of Hg. This has been demonstrated by recent publications on global Hg cycling (Nriagu, 1979; Slemr *et al.*, 1981; Fitzgerald *et al.*, 1983; Lindqvist *et al.*, 1984). In the early 1970s, the amount of Hg discharged to the environment from anthropogenic sources was estimated to be comparable with quantities derived from continental weathering processes (Klein and Goldberg, 1970). Following new regulations, most large industrial sources in the developed countries were identified and the aquatic discharges of Hg reduced to amounts that were negligible relative to natural sources. However, current anthropogenic emissions of Hg to the atmosphere exceed direct releases to surface waters by more than an order of magnitude (Harriss and Hohenemser, 1978), and are considered to be of the same order of magnitude as atmospheric emissions from natural processes (Lindqvist *et al.*, 1984). Mercury released from both high-and low-temperature processes enters the atmosphere primarily as a vapor. Ultimately, atmospheric deposition processes return airborne Hg to the Earth's surface, where it undergoes the same reactions as Hg discharged directly into aquatic and terrestrial environments.

This chapter is not intended to be a comprehensive review of the biogeochemical cycle of Hg, which has been published elsewhere (Nriagu, 1979). Rather, it addresses the implications of the fact that vapor-phase Hg dominates the atmospheric pool of Hg by summarizing several of our own and other recent field studies in the areas of atmospheric emission, deposition, and plant uptake of Hg vapor. The reader is referred to more extensive reviews of the behavior of Hg in the environment by Lindqvist *et al.* (1984) and reviews of recent developments in analytical methods for speciation of airborne Hg by Schroeder (1982) and Schroeder and Jackson (1983).

EMISSION AND DEPOSITION

Estimates of global anthropogenic emissions of Hg to the atmosphere are on the order of 2000 to 3000 metric tons/year (Harriss and Hohenemser, 1978; Fitzgerald *et al.*, 1984; Lindqvist *et al.*, 1984). The major contributors to this flux are mining and smelting of Hg ores, industrial processes involving mercury compounds, and fossil fuel combustion (particularly coal). Over the past several years we have studied the mechanisms of release of Hg vapor from three of the most important industrial sources: electric power generation, chlorine/caustic soda production, and cinnabar mining. The details of these studies and analytical/sampling methods used are in the literature (Lindberg and Turner, 1977; Lindberg *et al.*, 1979; Lindberg, 1980).

Coal Combustion

Our research on Hg emission from power generation involved an investigation of Hg partitioning between particle and vapor forms in the plume of a modern utility coal-fired power plant (Lindberg, 1980). Based on samples collected in the stack and in the combustion plume, we determined the relationship between plume age and atmospheric Hg (Figure 8.1). As expected, Hg collected in the plume 0.25 km from the stack was dominated by the vapor phase, most likely Hg^o (elemental mercury vapor). Approximately 92% of the total Hg at this point in the plume was present as Hg^o at a measured concentration of 1700 ng/m^3. During plume travel from 0.25 to 7 to 22 km, the measured concentrations of Hg^o decreased to 1000 ng/m^3

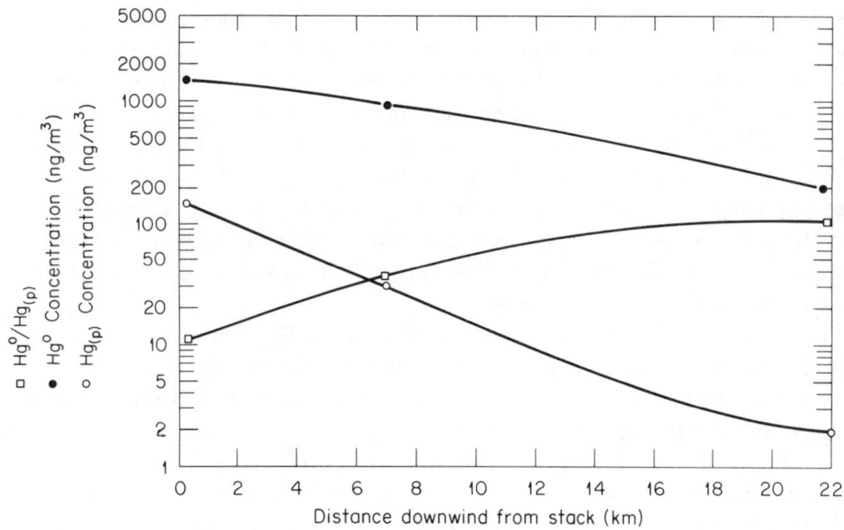

Figure 8.1 Concentration of mercury vapor (Hg^o), particulate mercury [$Hg_{(p)}$], and Hg vapor to particulate Hg concentration ratio [$Hg^o/Hg_{(p)}$] in the plume of a coal-fired power plant. □, $Hg^o/Hg_{(p)}$; •, Hg^o concentration (ng/m^3); o, $Hg_{(p)}$ concentration (ng/m^3) (from Lindberg, 1980)

at 7 km and 200 ng/m^3 at 22km. (Background concentrations were on the order of 10–20 ng/m^3.) Because of some uncertainty in the plume dimensions, the absolute concentrations of airborne Hg in the plume itself should be regarded cautiously, particularly at the point closest to the smokestack. It is best to consider the ratios of the measured concentrations.

If significant adsorption or condensation of Hg^0 vapor onto particles were occurring as the plume cooled and mixed with ambient air, as suggested elsewhere (Williston, 1968; Staff, Environmental Magazine, 1971; Billings and Matson, 1972; Lockeretz, 1974), the vapor to particle ratio should decrease, assuming little settling loss for particles in the size range encountered in the plume (mass median diameters of ~0.1 to 0.5 μm, Lindberg and Harriss, 1980). However, this ratio increased with distance from 11 at 0.25 km, to 33 at 7 km, to 100 at 22 km. Similarly, the fraction of the total Hg present as a vapor increased to 97% at 7 km and 99% at 22 km. This increase in the vapor to particle ratio indicates not only the absence of any measurable gas-to-particle conversion but also the loss of some fraction of the initial particulate Hg population. Considering the measured concentrations of Hg^0 and that the vapor pressure of Hg at ambient temperatures is ~20 mg/m^3, one would not expect gas to liquid phase condensation to influence the gas-phase concentrations. The loss of particulate Hg could be the result of desorption or displacement of Hg vapor from the aerosol, or physical removal of some fraction of the particles during plume travel. Because settling of these particles is negligible over these distances, displacement is more likely. This displacement may be the result of reactions between various mercury species (e.g. HgO) on the suspended particulate matter and gaseous oxides of S and N in the plume. Reactions between SO_2, for example, and surfaces of metal-containing particles are well known (Fennelly, 1975), although the effects of these interactions on adsorbed species have not been documented. Reactions between HgO and SO_2 to yield Hg^0 vapor have been recently suggested as an important Hg reduction process in the atmosphere (Stevens *et al.*, 1982). Such interactions could explain the observations in our plume study.

The fraction of the total incoming feed coal Hg which is discharged to the atmosphere from the power plant and the source strength can be estimated by using the 0.25-km downwind plume data to represent the maximum particle to vapor concentration ratio and by knowing the coal feed rate (4.8×10^5 kg/hour), the precipitator efficiency (99%), the composite coal Hg content, and the precipitator ash Hg concentration (Lindberg, 1980). Less than 1% of the initial feed coal Hg is retained in the plant (in collected ash), and the remainder is released in the stack emissions, ~7% in particulate form and 92% in vapor form. The estimated source strength for the unit sampled was ~3.5 kg Hg/day as a vapor and ~0.3 kg Hg/day in particulate form. Because nearly all of the feed coal Hg is emitted from the stacks, total Hg emission rates should be proportional to coal usage rates of power plants as reflected by megawatt capacity. As shown in Table 8.1 this is generally true, thus providing a simple method of estimating total Hg emissions from coal-fired power plants. Estimates of total atmospheric emissions from coal burning for the early 1970s were on the order of 100 metric tons per year (Harriss and Hohenemser, 1978; National Research Council, 1978). Our 'model' (atmospheric Hg loss $\simeq 5 \times 10^{-3}$ kg/day per MW of electricity generated, determined from the mean ratio of the MW and emission values in Table 8.1), combined with statistics for the United States (National Coal Association, 1980), suggests that there has been an increasing emission trend as coal use has increased (values in metric tons Hg/year): 1976 = 190, 1978 = 200, 1980 = 230.

Table 8.1 Emission rates of Hg from modern coal-fired power plants. Generally, more than 90% of the Hg emitted was in a vapor phase in each case

Electric generating capacity (MW)	Atmospheric Hg emission rate (kg/day)	Reference
200	1.5	Anderson and Smith, 1977
1000	4.3	Lockeretz, 1974
1300	3.8	Lindberg, 1980
2100	7.5	Billings and Matson, 1972
United States	270	Harriss and Hohenemser, 1978
World	1700	Harriss and Hohenemser, 1978

Chloralkali Wastes

Our research on the emission of Hg from the chloralkali industry involved measurement of the loss of Hg vapor from solid wastes stored near an inactive Hg-cell chloralkali plant in a rural area of the eastern United States (Lindberg and Turner, 1977). Although estimates have been published on the emission of Hg vapor from production processes in active plants (Caban and Chapman, 1972; Van Horn, 1975; Wallin, 1976; Högström *et al.*, 1979), none have considered the problem of atmospheric emissions from waste products stored in ponds or otherwise stored near either active or inactive plants. Based on laboratory studies, field data, and plume dispersion models, we estimated the emission rate of Hg vapor from solid wastes and the influence of these emissions on local air quality near two solid waste ponds of 4×10^5 m^2 surface area (Hildebrand *et al.*, 1980).

The volatility of elemental Hg and many of its compounds suggests that its presence in large-scale waste deposits such as these might have an influence on local air quality. In laboratory experiments, the emission of Hg vapor from the waste material was clearly related to surface temperature, as expected (Figure 8.2). The relationship is similar to that between temperature and the saturation concentration of Hg° in air (Wallace *et al.*, 1971), suggesting that the mercury occurs in the waste material to some extent in the elemental state. The mean total Hg content of these wastes was 150 μg/gm, but that present as Hg° was not specifically determined. Estimates of natural degassing rates of Hg on regional or global scales and even measured rates for mineralized soils are one to three orders of magnitude lower than those we have measured for the waste material (Table 8.2).

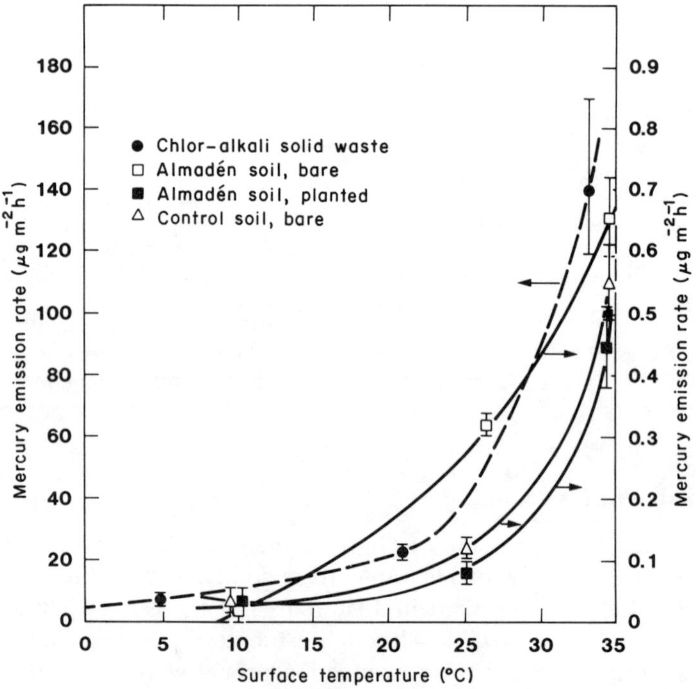

Figure 8.2 Relationship between emission of vapor phase mercury and surface temperature for chloralkali waste material (155 μg Hg/gm) and for mercuriferous soils (97 μg Hg/gm) and control soils (2.3 μg Hg/gm) with and without plant cover (from Lindberg and Turner, 1977; Lindberg *et al.*, 1979)

We used a Gaussian diffusion model (Gifford, 1968) to estimate 24-hour mean concentrations of Hg° vapor in the ambient air downwind of the waste area, based on emission rates measured under controlled conditions (Figure 8.2) and local meteorological measurements (Hildebrand *et al.*, 1980). For comparison, air sampling stations near the ponds measured Hg° in air and

Table 8.2 Natural emission rates of Hg vapor from solid materials

Material	Method	Hg emission rate (μg/m^2.hour)	Reference
Global land area	Input/output models	0.02–0.03	Weiss *et al.*, 1971; Kothny, 1973
Bedrock	Rock volatility	0.04–0.08	Desaedeleer and Goldberg, 1978
Soils 15 km from cinnabar mine	Chamber measurements	0.12–0.14	Lindberg *et al.*, 1979
Soils 1 km from cinnabar mine	Chamber measurements	0.32–0.34	Lindberg *et al.*, 1979
Mercuriferous soils	Chamber measurements	0.2–1.7	McCarthy *et al.*, 1969
Chloralkali wastes	Chamber measurements	35–40	Lindberg and Turner, 1977

total Hg on suspended particles, wind speed and direction, air temperature, and bulk deposition. At temperatures from 6 to 30°C, the measured and calculated air concentrations agreed to within ~30% (Lindberg and Turner, 1977), suggesting that the entire surface area of the waste pond was emitting Hg° vapor at a rate comparable with that measured in controlled experiments.

The levels of Hg° in air measured near the pond at 6°C (60 to 90 ng/m^3) were slightly elevated above those reported for large US cities and areas near active chloralkali plants (Table 8.3). However, these concentrations decreased to rural background levels within ~2 km of the waste area. At temperatures near 30°C the concentration of Hg measured in the pond vicinity approached the US Environmental Protection Agency guideline level for total Hg in ambient air (1000 ng/m^3) (*Federal Register*, 1973). This is comparable with levels in air over natural Hg deposits and near geothermal emissions (Table 8.3).

The concentration of particulate Hg during this sampling period was very low and comparable with that reported for urban areas. The indication that essentially 100% of the Hg in air is in vapor form agrees with reported results for ambient air in remote and urban areas (Johnson and Braman, 1974; Fitzgerald *et al.*, 1983; Lindqvist *et al.*, 1984).

The bulk deposition of Hg was also measured during one 24-hour period of air monitoring, during which it rained for 1 hour (0.25 cm of precipitation). The concentrations of Hg in replicate bulk precipitation samples were 1.1 and 1.4 μg/litre, comparable with values reported for urban areas (see the review by Galloway *et al.*, 1980) but considerably higher than concentrations of Hg in precipitation collected in generally rural areas (~0.01–0.1 μg/litre)

Table 8.3 Concentrations of Hg vapor and particulate Hg in various atmospheres

	Hg concentration (ng/m^3)		
Study area	Vapor	Particle	Reference
Remote marine	1–3	0.0004–0.002	Slemr *et al.*, 1981; Fitzgerald *et al.*, 1983
Rural	3–10		Kothny, 1973; Brosset, 1982; Lindberg and Turner, 1977
Urban	2–30	0.2–0.3	Ferrara *et al.*, 1982
Large cities	5–50		Johnson and Braman, 1974; Van Horn, 1975
Industrial	10–50		Högström *et al.*, 1979
Chloralkali waste ponds	60–1000	0.2–0.4	Lindberg and Turner, 1977
Cinnabar deposits	30–1600		Kothny, 1973
Power plant plume	200–1700	2–150	Lindberg, 1980
Volcanic and geothermal vents	10–40 000		Siegel and Siegel, 1975

(see the review by Lindqvist *et al.*, 1984). The bulk deposition rates of Hg during this period were 7.4 and 9.0 $\mu g/m^2$ per day, which are comparable with values measured near active chloralkali plants but much higher than those in rural areas not influenced by local emissions (Table 8.4).

Using regression equations and air temperature records, we estimated the total annual surface emission of Hg vapor from the waste area. The calculated flux is 36 $\pm$ 4 kg Hg/year (mean and SE). Interestingly, one estimate of annual aquatic loss of Hg by leaching and runoff from the waste ponds into the adjacent river was similar (39 $\pm$ 2 kg Hg/year; Turner and Lindberg, 1978). The estimated total annual flux of Hg from this inactive plant, 75 $\pm$ 5 kg/year, is less than 10% of the US Environmental Protection Agency standard for emission of Hg to air and water from active chlorine plants, but it is ~25% of the emission measured from active European plants (Högström *et al.*, 1979).

Emission of Hg vapor from waste deposits is likely to increase in relative importance because of modernization of the chloralkali industry. Reduction of direct losses results in an increase in landfill waste Hg; 60 to 90% of current Hg used by the industry is deposited in waste ponds ($>10^5$ kg Hg/year in the United States, Harriss and Hohenemser, 1978). In addition, conversion to the non-Hg diaphragm process suggests that existing waste ponds will be the major source of Hg vapor from this industry in the future. Because of the rate of generation of these wastes ($>10^9$ kg/year in the

Table 8.4 Atmospheric deposition rates of Hg at diverse locations

Study area	Hg deposition rate ($\mu g/m^2.d$)	Reference
Remote marine	0.02–0.06	Fitzgerald *et al.*, 1983
Rural	0.2–3	Schlesinger *et al.*, 1974, Galloway *et al.*, 1980
500–50 000 m from active chloralkali plant	0.5–1	Jernelöv and Wallin, 1973; Högström *et al.*, 1979
50–500 m from active chloralkali plant	1–40	Jernelöv and Wallin, 1973; Högström *et al.*, 1979
Chloralkali solid waste ponds	7.4–9.0	Lindberg and Turner, 1977

United States) and the residence time of Hg in such deposits (>100 years, Hildebrand *et al.*, 1980), these areas will continue to be sources of Hg vapor well into the future.

Since we completed our research at this plant, and in part because of our findings, considerable effort has been expended to seal the waste deposits to prevent continued losses. However, these efforts have been aimed primarily at reducing leaching rates of dissolved Hg into nearby ground and surface waters (Rucker, 1983). Because of the size of the waste ponds, complete sealing of the surface was considered to be impossible. Hence, although direct leaching of dissolved mercury should be reduced, the atmospheric source will be unaffected, allowing mercury vapor to continue to enter the atmosphere.

Mercuriferous Soils

Emissions of Hg to the atmosphere from ore-rich areas by mining and refining operations have been estimated to contribute ~12% of the global anthropogenic emissions of Hg to air (Van Horn, 1975). Local emissions can further increase the Hg burden of vegetation growing on mercuriferous soils; however, there are few measurements of Hg emission rates from such soils or of plant uptake of the emitted Hg. Our study of this industry examined agricultural soils and alfalfa (*Medicago sativa*) plants near the Almadén, Spain, mercury mine and smelter, the largest and oldest mercury mining/refining operation in the world. Emission rates and plant uptake of Hg vapor were measured in controlled chamber experiments as described elsewhere (Lindberg *et al.*, 1979).

Proximity to the mine had an obvious influence on the surface soil Hg levels, which ranged from 68 to 97 to 263 μg Hg/gm at points 0.5, 1.0, and 2.0 km downwind from the mine, respectively. Wind direction effects were also apparent further from the mine; the Hg concentrations at 15 km downwind decreased to 2–5 μg/gm, while those 15 km upwind were 0.3–0.5 μg/gm. All of these concentrations are elevated above the average crustal abundance of Hg of 0.07 μg/gm (Vinogradov, 1959) and above uncontaminated topsoils collected around the world (~0.02–0.07 μg/gm), but are comparable with soils collected near other mercury sources [2.5–10 μg/gm (US Geological Survey, 1970)].

Two of these soils (Almadén, Hg = 97 ± 7 μg/gm, and control, 2.3 ± 0.4 μg/gm) were used for detailed chemical analysis, plant uptake studies, and emission rate experiments. The soils were comparable in characteristics known to influence soil Hg behavior, such as pH, organic content, and cation exchange capacity. In addition, density gradient fractionation indicated that a major portion of the Hg in each soil (40% in Almadén, 80% in control) was associated with the most dense inorganic soil fractions, and likely represented cinnabar. The least dense soil fraction, including organo-clay complexes and surface deposited Hg^{0}, comprised 50% of the Almadén soil Hg (or 48 μg/gm), but only 10% of the control (0.2 μg/gm).

The results of some of our Hg emission experiments are shown in Figure 8.2. The emission rates were clearly related to surface temperature, as expected, and were also related to soil Hg content, as noted for soils experimentally amended with Hg (Landa, 1978). The measured Hg emission rates for bare Almadén and control soils at 25°C exceeded natural rates measured for non-mineralized regions by factors of 4 to 10, but were within the range reported for mineralized soils (Table 8.2). At 25°C, the emission rate of Hg from bare Almadén soils was significantly greater ($P < 0.05$) than the rate from these soils planted with alfalfa. Alfalfa cover may reduce mixing of Hg-rich air at the soil surface with air above the alfalfa canopy. However, because of the flow rate used, duration of the experiments, and small amount of plant cover present (Lindberg *et al.*, 1979), we believe this effect to be minor. A more likely possibility is direct absorption of emitted Hg^{0} by foliage.

Our measurements of Hg accumulation in alfalfa grown on these soils suggest two mechanisms of Hg uptake by plants under these conditions: the first controlling above ground, foliar levels of Hg, which is independent of soil Hg levels; and the second influencing root uptake, which is related to soil concentrations. The Hg concentrations in alfalfa foliage grown on both soil types were similar (pooled mean = 1.4 + 0.6 μg/gm, $n = 18$) and exceeded world average levels in grasses by an order of magnitude (Wallace *et al.*, 1971). On the other hand, concentrations of Hg in well-washed Almadén roots (8.0 ± 0.1 μg/gm) far exceeded those in control roots (0.38 ± 0.08 μg/gm) and represented 80% of the total Hg accumulated by plants in Almadén soils, while for control soils 80% of the accumulated Hg occurred in above ground foliage.

The ability of plants to absorb Hg vapor from the atmosphere is well known (Hitchcock and Zimmerman, 1957; Huckabee, 1973; Wallin, 1976; Lodenius and Laaksorvirta, 1979). Our alfalfa plants were exposed to elevated levels of Hg vapor emitted from the soils throughout the uptake experiments. In the plant growth chamber containing several pots of control and Almadén soils, the air concentration ranged from 50 to 220 ng/m^3, averaging 80 ng/m^3 during the 16-week plant uptake studies (compare background levels in Table 8.3). To quantify foliar uptake of Hg vapor a duplicate set of alfalfa plants was grown from seed on an inert silicate substrate in the same growth chamber and exposed to Hg vapor emitted from Almadén and control soils. Although the silicate contained essentially no Hg (<0.0005 μg/gm), the alfalfa grown on it exhibited a foliar Hg concentration of 2.0 ± 0.1 μg/gm, but a root level of only 0.35 ± 0.01 μg/gm. Hence, foliar Hg was comparable among plants grown on all three substrates, while root concentrations differed in the order Almadén > control $\simeq$ silicate. Translocation of foliar Hg to the roots, or direct deposition of Hg vapor to the silicate followed by root absorption, may explain why the silicate-grown roots had a higher Hg concentration than the silicate itself.

Previous plant uptake studies support the hypothesis of direct foliar absorption (and emission) of Hg vapor (Ross and Stewart, 1962; Gilmour and Miller, 1973; Kozuchowski and Johnson, 1978). Mechanisms of foliar uptake in wheat were studied by Browne and Fang (1978) using labeled Hg vapor. They found that: uptake was controlled by the stomata, but not influenced by air temperature; surface adsorption was negligible; uptake rates increased with increasing air concentration; total uptake increased with duration of exposure; and the absorbed Hg was confined solely to leaves.

IMPLICATIONS AND GLOBAL CYCLING

The activated charcoal adsorption traps used to sample airborne Hg in all our studies collect both elemental and chemically bound Hg vapor (Lindberg, 1981). However, our data and that of others suggest that the major or predominant species of airborne Hg vapor is elemental Hg, the source of which is thought to be direct industrial emissions plus chemically or biologically reduced mercury compounds in soil and water (Frear and Dills, 1967; Gilmour and Miller, 1973; Johnson and Braman, 1974; Nriagu, 1979; Fitzgerald *et al.*, 1984). Organic and inorganic Hg compounds react with soil organic matter to release Hg^o (Hitchcock and Zimmerman, 1957). Direct, abiotic methylation of divalent Hg in soils has been reported by Rogers (1977) and by Rogers and MacFarlane (1979), who found that Hg^o (not methyl Hg) was the predominant vapor species ($\sim$70%) emitted from these soils. Based on tests with autoclaved soils Landa (1978) suggested that soil micro-organisms mediate the volatilization of Hg^{o}. Following the reduction of various Hg compounds and organo-Hg complexes to Hg^{o} in soils, sediments, and surface waters, the vapor is lost to the atmosphere, where it apparently remains in this same form (Slemr *et al.*, 1981; Fitzgerald *et al.*, 1983).

Current estimates suggest that natural and anthropogenic emissions of Hg to air are of the same order of magnitude (Fitzgerald *et al.*, 1984; Lindqvist *et al.*, 1984). However, when emission rates are normalized to a unit surface area basis, it is apparent that atmospheric *fluxes* from the combustion of coal dwarf those from other sources (Figure 8.3). Despite the wide range in measured fluxes, the compositions of these emissions have one characteristic in common with the global atmosphere: dominance of total airborne Hg by vapor species. There are many important implications of this vapor-phase dominance. As discussed earlier, crop plants have been shown to absorb and retain Hg through leaf uptake, while the incorporation of particulate forms is considered less likely (Hosker and Lindberg, 1982). In addition, inhaled metallic Hg vapor is able to diffuse much more extensively into blood cells and various tissues than inorganic particle-associated Hg (Magos, 1968).

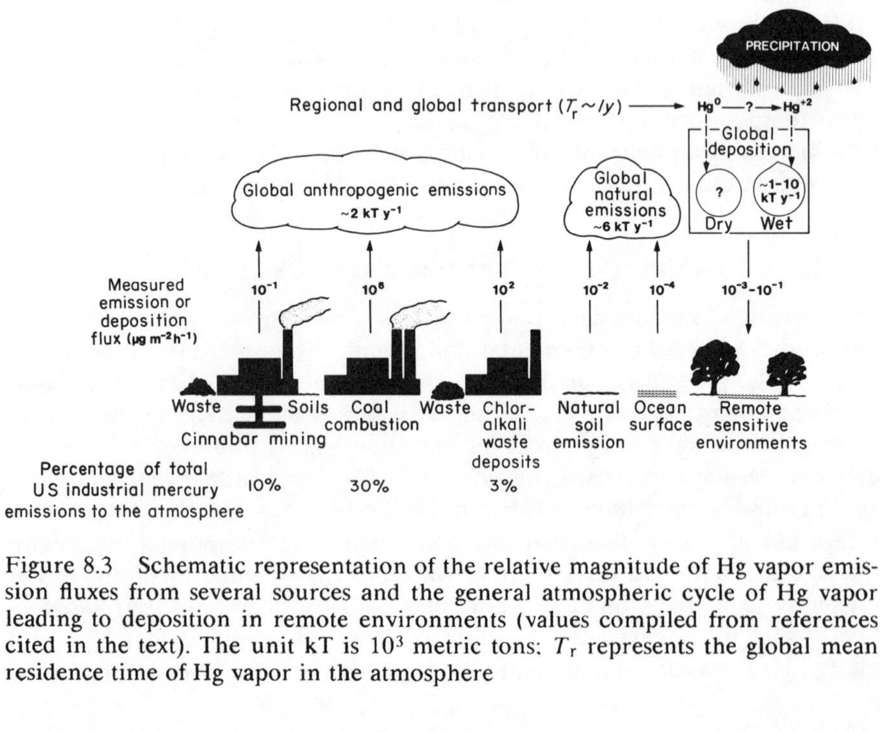

Figure 8.3 Schematic representation of the relative magnitude of Hg vapor emission fluxes from several sources and the general atmospheric cycle of Hg vapor leading to deposition in remote environments (values compiled from references cited in the text). The unit kT is 10^3 metric tons; T_r represents the global mean residence time of Hg vapor in the atmosphere

Other important implications involve atmospheric transport, residence time, and deposition. The occurrence of airborne Hg as a vapor is conducive to long-range transport from the source and a long atmospheric residence time. Several studies of the dispersion and deposition of Hg near point sources have confirmed this hypothesis (Jernelöv and Wallin, 1973; Lockeretz, 1974; Anderson and Smith, 1977; Crockett and Kinnison, 1979; Högström *et al.*, 1979). Slemr *et al.* (1981) recently reappraised earlier estimates of the mean global tropospheric residence time of atmospheric Hg, based on numerous measurements of Hg in the atmosphere over the Atlantic Ocean. Using box model calculations and the relationship between average residence time and the variability of gas concentrations in the troposphere proposed by Junge (1974), these authors calculate residence times in the range of 0.7 to 2 years, considerably larger than values of 0.03 to 0.1 year proposed earlier (Weiss *et al.*, 1971; Kothny, 1973; Wollast *et al.*, 1976; Andren and Nriagu, 1979). More recent estimates based on oceanic measurements (Fitzgerald *et al.*, 1983) and theoretical calculations (Lindqvist *et al.*, 1984) support these longer times. However, it should also be noted that recent analyses of some lesser known atmospheric reactions of Hg^0 (with O_3 for example) suggest that under certain conditions its half life could be on the order of minutes (Stevens *et al.*, 1982). Clearly more work in the area of atmospheric reactions of mercury and its compounds is needed.

Minimal data exist on the rates and mechanisms of Hg removal from the atmosphere, particularly by dry deposition. Theoretical estimates of the contribution of dry deposition to the overall flux of Hg to the Earth's surface are as follows: (1) 4 to 40% within 2 km of a power plant and 40 to 90% at 20 km from this same source (Lockeretz, 1974); (2) ~99% within 0.2 km of a chloralkali plant and 94% at 5 km from the plant (Högström *et al.*, 1979); (3) <20% to a forest canopy in the eastern United States (Andren and Lindberg, 1977); (4) 50% to the sea surface in the north Pacific (Fitzgerald *et al.*, 1983), and to the globe as a whole (Lindqvist *et al.*, 1984); and (5) from <1% to 'possibly significant' on the global scale (Lantzy and Mackenzie, 1979, and National Research Council, 1978, respectively). Clearly, empirical data on dry removal rates of atmospheric Hg are needed. A comparison of total global emission and deposition rates confirms the degree of uncertainty in dry fluxes (Figure 8.3).

Once Hg is dispersed from the source, precipitation scavenging may favor greater removal rates for the vapor than for the particles. We have found particulate Hg to be concentrated in the 0.6- to 1.1-μm size range in ambient air (Lindberg, unpublished data), a size range for which precipitation scavenging efficiencies are at a minimum (Beard, 1977). Because precipitation scavenging of particles is largely a physical process, a captured particle may or may not release Hg to solution. Thus, the initial composition of the raindrop has little influence on the scavenging efficiency for particulate

Hg, although it will influence the ultimate dissolved Hg concentration in the droplet. However, scavenging of the vapor is highly dependent on its solubility in the rain-drop, and any characteristic of the initial drop which increases the solubility of vapor-phase Hg can enhance the removal rate.

Reported solubilities of Hg^{o} in pure water are 30 to 60 μg/litre, for a saturated atmosphere (Onat 1974; Sanemasa 1975). Concentrations of Hg in precipitation (0.001 to 1.0 μg/litre) (Nriagu, 1979; National Research Council, 1978) do not approach these levels, suggesting that the solubility of Hg vapor in the raindrop is not a limiting factor in its removal from the atmosphere (McCarthy *et al.*, 1969). However, the equilibrium solubilities for ambient air concentrations of Hg^{o} are much lower ($\sim 3 \times 10^{-5}$ μg/litre for a background air concentration of 10 ng/m^3, Fogg and Fitzgerald, 1979). Hence, even at the highest air concentrations measured in our studies (1700 ng/m^3; Table 8.3), the equilibrium concentration of Hg^{o} in pure water (0.005 μg/litre) is considerably below reported levels in precipitation. Possible explanations for these discrepancies include the following: (1) Hg^{o} solubility is greatly enhanced by the presence of oxidants and H^+ in rain, (2) Hg^{o} is rapidly transformed into considerably more soluble vapor species in an oxidizing atmosphere saturated with water vapor (i.e. during precipitation), and (3) most of the Hg in rain originates from particle removal from the air. The last hypothesis seems least likely because of the low particle/gas Hg concentrations in air and the low particle removal efficiency by rain. Although testing of the other hypotheses awaits further analytical and field data, the existence of highly soluble vapor-phase molecular species [e.g. $Hg(OH)_2$] has been suggested (Brosset, 1982; Lindqvist *et al.*, 1984).

The global wet deposition of Hg from the atmosphere has been estimated to be from 10^6 to 10^8 kg/year (Weiss *et al.*, 1971; Garrels *et al.*, 1975; Fitzgerald, 1976; Slemr *et al.*, 1981); more recent estimates are in the range of 10^6 to 10^7 kg/year (Lindqvist *et al.*, 1984), indicating considerable uncertainty in wet removal rates. Lack of measurements of dry deposition rates of Hg hinders the assessment of the importance of this process as a removal mechanism. However, precipitation scavenging of the vapor appears to be the major removal process on a global scale and theoretically should increase in efficiency as precipitation acidity and airborne oxidant concentrations increase. Thus, enhanced atmospheric deposition of Hg should be considered as a possible contributing factor to recent reports of relationships between acid precipitation and elevated Hg levels in fish from remote locations, previously ascribed to regional soil/watershed acidification (Brouzes *et al.*, 1977; Brosset and Svedung, 1977; see also the review by Lindqvist *et al.*, 1984).

ACKNOWLEDGEMENTS

Financial support was provided in part by the National Science Foundation, the Electric Power Research Institute, and by the US Department of

Energy under contract no. DE-AC05-84OR21400 with Martin Marietta Energy Systems, Inc. Publication No. 2402, Environmental Sciences Division, ORNL.

REFERENCES

Anderson, W. L., and Smith, K. E. (1977). Dynamics of mercury at coal-fired power plant and adjacent cooling lake. *Environ. Sci. Technol.*, **11**, 75–80.

Andren, A. W., and Lindberg, S. E. (1977). Atmospheric input and origin of selected elements in Walker Branch Watershed, Oak Ridge, Tennessee. *Water Air Soil Pollut.*, **8**, 199–215.

Andren, A. W., and Nriagu, J. O. (1979). The global cycle of mercury. In Nriagu, J. O., (Ed.) *The Biogeochemistry of Mercury in the Environment*, pp. 1–21. Elsevier/North-Holland Biomedical Press, Amsterdam.

Beard, K. V. (1977). Rain scavenging of particles by electrostaticinertial impaction and Brownian diffusion. In Semonim, R. G., and Beadle, R. W. (Eds), *Precipitation Scavenging—1974*. ERDA Symposium Series 41. CONF-741003. Technical Information Center, Oak Ridge, Tennessee.

Billings, C. E., and Matson, W. R. (1972). Mercury emission from coal combustion. *Science*, **176**, 1232–1233.

Brosset, C., (1982). Total airborne Hg and its possible origin. *Water Air Soil Pollut.*, **17**, 37–50.

Brosset, C. and Svedung, I. (1977). *Preliminary Study of the Possibility of a Relationship Between High Acidity in Lakes and High Hg Content of Fish Populations.* Report B378, Swedish Water and Air Pollution Res. Lab., Gothenburg, Sweden.

Brouzes, R. J. P., McLean, R. A. N., and Tomlinson, G. H. (1977). Mercury—the link between pH of natural waters and the mercury content of fish. Paper presented at the meeting of the US National Academy of Sciences–National Research Council Panel on Mercury, Washington, DC.

Browne, C. L., and Fang, S. C. (1978). Uptake of mercury vapor by wheat: an assimilation model. *Plant Physiol.*, **61**, 430–433.

Caban, R., and Chapman, T. (1972). Losses of Hg from chlorine plants: a review of a pollution problem. *Amer. Inst. Chem. Eng. J.*, **18**, 892–903.

Crockett, A. B., and Kinnison, R. R. (1979). Mercury residues in soil around a large coal-fired power plant. *Environ. Sci. Technol.*, **13**, 465–476.

Desaedeleer, G., and Goldberg, E. D. (1978). Rock volatility—some initial experiments. *Geochem. J.*, **12**, 75–79.

Federal Register (1973). *National Emissions Standards for Hazardous Air Pollutants*, Fed. Reg. 38, 8820-8845, US Government Printing Office, Washington, DC.

Fennelly, P. F. (1975). Primary and secondary particulates as pollutants: a literature review. *J. Air Pollut. Control Assoc.*, **25**, 697–704.

Ferrara, R., Petrosino, A., Maserti, E., Seritti, A. and Bavghigiani, C. (1982). The biogeochemical cycle of mercury in the Mediterranean Part II. *Environ. Technol. Lett.*, **3**, 449–456.

Fitzgerald, W. F. (1976). Mercury studies of seawater and rain: Geochemical flux and implications. In Windom, H. L., and Duce, K. A. (Eds.), *Marine Pollutant Transfer*, pp. 121–134. Lexington Books, D. C. Health Co., Lexington, Massachusetts.

Fitzgerald, W. F., Gill, G. A., and Hewitt, A. D. (1983). Air-sea exchange of mercury. In Wong, C. S., Boyle, E., Bruland, K. W., Burton, J. D., and Goldberg, E. D. (Eds.), *Trace Metals in Sea Water*, pp. 297–316. Plenum Publishing Corp., New York.

Fitzgerald, W. F., Gill, G. A. and Kim, J. P. (1984). An equatorial Pacific Ocean source of atmospheric mercury. *Science*, **224**, 597–599.
Fogg, T. R. and Fitzgerald, W. F. (1979). Mercury in southern New England coastal rains. *J. Geophys. Res.*, **84**, 6987–6989.
Frear, D. E. H. and Dills, L. E. (1967). Mechanism of the insecticidal action of mercury and mercury salts. *J. Econ. Entomol.*, **60**, 970–974.
Galloway, J. N., Eisenreich, S. J. and Scott, B. C. (1980). *Toxic Substances in Atmospheric Deposition: A Review and Assessment.* Environmental Protection Agency Report EPA 560/5-80-001, USEPA, Washington, DC.
Garrels, R. M., McKenzie, F. T. and Hunt, C. (1975). *Chemical Cycles and the Global Environment: Assessing Human Influences.* Report published by W. Kaufman, Inc., Los Altos, California.
Gifford, F. A. (1968). *Meteorology and Atomic Energy.* United States Atomic Energy Agency Report 66116, Washington, D. C.
Gilmour, J. T., and Miller, M. S. (1973). Fate of a mercuric-mercurous chloride fungicide added to turfgrass. *J. Environ. Qual.*, **2**, 145–148.
Harriss, R. C., and Hohenemser, C. (1978). Mercury: measuring and managing the risk. *Environment*, **20**, 25–36.
Hildebrand, S. G., Lindberg, S. E., Turner, R. R., Huckabee, J. W., Lund, J. R. and Andren, A. W. (1980). *Biogeochemistry of Mercury in a River-Reservoir System: Impact of an Inactive Chloralkali Plant in the Holston River—Cherokee Lake.* ORNL/TM-6141. Oak Ridge National Laboratory, Oak Ridge, Tennessee.
Hitchcock, A. E. and Zimmerman, P. W. (1957). Toxic effects of vapors for mercury and compounds of mercury on plants. *Ann. N. Y. Acad. Sci.*, **65**, 474–497.
Högström, U., Enger, L. and Svedung, I. (1979). A study of atmospheric Hg dispersion. *Atmos. Environ.*, **13**, 465–476.
Hosker, R. P. and Lindberg, S. E. (1982). Review: atmospheric deposition and plant assimilation of gases and particles. *Atmos. Environ.*, **16**, 889–910.
Huckabee, J. W. (1973). Mosses: sensitive indicators of airborne Hg pollution. *Atmos. Environ.*, **7**, 749–754.
Jernelöv, A. and Wallin, T. (1973). Air-borne mercury fallout on snow and around five Swedish chloralkali plants. *Atmos. Environ.*, **7**, 209–214.
Johnson, D. L., and Braman, R. S. (1974). Distribution of atmospheric mercury species near ground. *Environ. Sci. Technol.*, **8**, 1003–1009.
Junge, C. E. (1974). Residence time and variability of tropospheric trace gases. *Tellus*, **26**, 477–488.
Klein, D. H. and Goldberg, E. D. (1970). Mercury in the marine environment. *Environ. Sci. Technol.*, **4**, 765–768.
Kothny, E. L. (1973). The three-phase equilibrium of Hg in nature. In Kothny, E. L. (Ed.), *Trace Elements in the Environment.* Advances in Chemistry Series 123, pp. 48–79, American Chemical Society, Reinhold, New York.
Kozuchowski, J. and Johnson, D. L. (1978). Gaseous emissions of Hg from an aquatic vascular plant. *Nature*, **274**, 467–469.
Landa, E. R. (1978). Microbial aspects of the volatile loss of applied mercury (II) from soils. *J. Environ. Qual.*, **7**, 84–86.
Lantzy, R. J. and Mackenzie, F. T. (1979). Atmospheric trace metals: Global cycles and assessment of man's impact. *Geochim. Cosmochim. Acta*, **43**, 511–525.
Lindberg, S. E. (1980). Mercury partitioning in a power plant plume and its influence on atmospheric removal mechanisms. *Atmos. Environ.*, **14**, 227–231.
Lindberg, S. E. (1981). Authors reply with details of sampling system for Hg vapor. *Atmos. Envir.*, **15**, 632–633.

Lindberg, S. E. and Turner, R. R. (1977). Mercury emission from chlorine-production solid waste deposits. *Nature (London)*, **268**, 133–136.
Lindberg, S. E. and Harriss, R. C. (1980). Trace metal solubility in aerosols produced by coal combustion. In Singh, J. J., and Deepak, A. (Eds.) *Environmental and Climatic Impact of Coal Utilization*, pp. 589–608, Academic Press, New York.
Lindberg, S. E., Jackson, D. R., Huckabee, J. W., Jansen, S. A., Levin, M. J. and Lund, J. R. (1979). Atmospheric emission and plant uptake of mercury from agricultural soils near the Almadén mercury mine. *J. Environ. Qual.*, **8**, 572–578.
Lindqvist, O., Jernelöv, A., Johansson, K. and Rodhe, H. (1984). *Mercury in the Swedish Environment, Global and Local Sources.* Report of the workshop held at Lerum, Sweden, November, 1983, SNV PM 1816, National Swedish Environment Protection Board, Solna, Sweden.
Lockeretz, W. (1974). Deposition of airborne mercury near point sources. *Water Air Soil Pollut.*, **3**, 179–193.
Lodenius, M. and Laaksorvirta, K. (1979). Mercury content of *Hypogymnia physodes* and pine needles affected by a chloralkali works at Kuusankoski, Finland. *Ann. Bot. Fennici*, **16**, 7–10.
Magos (1968), cited in Wallace, R. A., Fulkerson, W., Shultz, W. D. and Lyon, W. S. (1971). *Mercury in the Environment: The Human Element.* ORNL/NSF/EP-1. Oak Ridge National Laboratory, Oak Ridge, Tennessee.
McCarthy, J. R., Vaughan, W. W., Learned, R. E. and Mueschke, J. L. (1969). *Mercury in Soil, Gas, and Air–A Potential Tool in Mineral Exploration.* United States Geological Survey Circular 609, Washington, DC.
National Coal Association (1980). *Statistics on Electric Generating Plants.* Washington, DC.
National Research Council (1978). *An Assessment of Mercury in the Environment.* Printing and Publishing Office, National Academy of Sciences, Washington. DC 185 pages.
Nriagu, J. O. (1979). *The Biogeochemistry of Mercury in the Environment.* Elsevier/North Holland Biomedical Press, Amsterdam, 696 pages.
Onat, E. (1974). Solubility studies of metallic mercury in pure water at various temperatures. *J. Inorg. Nucl. Chem.*, **36**, 2029–2032.
Rogers, R. D. (1977). *A Biological Methylation of Mercury in Soil.* Environmental Protection Agency Report EPA-600/3-77-007, USEPA, Washington, DC 11 pages.
Rogers, R. D. and MacFarlane, J. C. (1979). Factors influencing volatilization of Hg from soil. *J. Environ. Qual.*, **8**, 255–260.
Ross, R. G. and Stewart, D. K. R. (1962). Movement and accumulation of mercury in apple trees and soil. *Can. J. Plant Sci.*, **42**, 280–285.
Rucker, D. (1983). Quicksilver cleanup. *Impact Magazine*, NRO–83/1, Tennessee Valley Authority, Chattanooga, Tennessee.
Sanemasa, I. (1975). The solubility of elemental mercury vapor in water. *Bull. Chem. Soc. Jpn.*, **48**, 1795–1798.
Schlesinger, W. H., Reiners, W. A., and Knopman, D. S. (1974). Heavy metal concentrations and deposition in bulk precipitation in montane ecosystems of New Hampshire *USA Environ. Pollut.*, **6**, 39–47.
Schroeder, W. H. (1982). Sampling and analysis of Hg and its compounds in the atmosphere. *Environ. Sci. Technol.*, **16**, 384A–400A.
Schroeder, W. H. and Jackson, R. A. (1983). A Method for the Separation of Several Volatile Mercury Species in Atmosphere. Presented at the 6th World Congress on Air Quality (IUAPPA), Paris, May, 1983.

Siegel, S. M. and Siegel, B. Z. (1975). Geothermal hazards: mercury emission. *Environ. Sci. Technol.*, **9**, 473–474.

Slemr, F., Seiler, W. and Schuster, G. (1981). Latitudinal distribution of Hg over the Atlantic Ocean. *J. Geophys. Res.*, **86**, 1159–1166.

Staff, Environmental Magazine. (1971). Mercury in the air. *Environment*, **13**, 24–35.

Stevens, R. D., Reid, N. W., Schroeder, W. H. and McLean, R. A. N. (1982). Chemical forms and lifetimes of Hg in the atmosphere. Presented at the 65th Canadian Chemistry Conference in Toronto, May 1982.

Turner, R. R. and Lindberg, S. E. (1978). Behavior and transport of mercury in a river-reservoir system downstream of an inactive chloralkali plant. *Environ. Sci. Technol.*, **12**, 918–923.

US Geological Survey (1970). *Mercury in the Environment.* Department of the Interior, Geol. Surv. Prof. Pap. 713/67 pages.

Van Horn, W. (1975). *Materials Balance and Technology Assessment of Mercury and its Compounds on National and Regional Bases.* Environmental Protection Agency EPA 560/3-75-007, USEPA, Washington, DC, 293 pages.

Vinogradov, A. (1959). *Geochemistry of Rare and Dispersed Chemical Elements in Soils.* Chapman and Hall Publishing, London, 209 pages.

Wallace, R. A., Fulkerson, W., Schults, W. D. and Lyon, W. S., (1971). *Mercury in the Environment.* ORNL/NSF-EPI. Oak Ridge National Laboratory, Oak Ridge, Tennessee, 61 pages.

Wallin, T. (1976). Deposition of airborne mercury from six Swedish chloralkali plants surveyed by moss analysis. *Environ. Pollut.*, **10**, 101–114.

Weiss, H. V., Koide, M. and Goldberg, E. D. (1971). Mercury in the Greenland ice sheet: evidence of recent input by man. *Science*, **174**, 692–694.

Williston, S. H. (1968). Mercury in the atmosphere. *J. Geophys. Res.*, **73**, 7051–7055.

Wollast, R., Billen, G. and McKenzie, F. T. (1976). Behavior of Hg in natural systems and its global cycle. In McIntyre, D., and Mills, C. F. (Eds.), *Ecological Toxicity Research*, Plenum, New York.

Lead, Mercury, Cadmium and Arsenic in the Environment
Edited by T. C. Hutchinson and K. M. Meema

CHAPTER 9

Impact of Long-range Atmospheric Transport of Heavy Metals to the Terrestrial Environment in Norway

EILIV STEINNES

*Department of Chemistry, University of Trondheim,
7055 Dragvoll, Norway*

ABSTRACT

The deposition of Pb, Cd, Zn, As, Sb and Se in southernmost Norway is shown on the basis of moss analysis to be about tenfold higher than in inland areas of Central and Northern Norway. Air-filter analyses show most of the supply to occur during short episodes from air masses transported from more heavily industrialized parts of Europe.

Investigations of surface soils demonstrate a very similar regional pattern of distribution for the above elements in surface soils. Soil profile studies exhibit large enrichments of Pb, Cd and As in the surface layer of soils in southernmost Norway. The greatest heavy metal concentrations in soils are found in largely uninhabited areas with very high annual precipitation levels. Long-distance atmospheric transport appears to have a profound influence on the supply of heavy metals to natural soils in Southern Norway. Further evidence of this is obtained from profile studies of ombrotrophic bogs from all over Norway.

A comparison of coniferous forest ecosystems in South Norway and Central Norway with respect to heavy-metal contents of common species of higher plants shows a similar trend as for soils. Plants from localities in southernmost Norway show substantially higher lead and cadmium contents than those from similar localities in Central Norway, the difference in many cases amounting to a factor of 5 or more. Copper and zinc do not exhibit a similar trend.

INTRODUCTION

The long-range atmospheric transport of pollution aerosols is well documented at present. Southern Scandinavia belongs to the areas that are appreciably affected by air pollutants derived from distant source regions. In

Norway most studies in this area have been confined to acidic components (e.g. Overrein *et al.*, 1981). Far less attention has been paid to other pollutants such as heavy metals in this respect.

Since the late 1970s, a substantial amount of work has been carried out in Norway in order to study the atmospheric supply of trace elements from other areas, the regional deposition patterns of these elements, and their significance with respect to contamination of terrestrial environments. The latter is mainly from investigations of soils and vegetation. In the present chapter this work is briefly reviewed.

LONG-RANGE TRANSPORT OF HEAVY METALS TO NORWAY

The first regular sampling of precipitation for trace element analyses started in 1978 at four different stations in Norway (Hanssen *et al.*, 1980), the locations of which are shown in Figure 9.1. Three of those stations (Birkenes, Kårvatn, Jergul) have been continuously in operation since then, and the elements Pb, Zn and Cd have been monitored on a regular basis. Recently As, Sb, and Se have been included in this programme. Wet deposition data for Pb and Cd during the period 1978–82 are listed in Table 9.1 (Hanssen *et al.*, 1980; State Pollution Control Agency, 1983).

Table 9.1 Deposition of lead and cadmium in Norway from precipitation data ($mg/m^2/y$)

Element	Location	1978/79	1980	1981	1982
Pb	Birkenes	11.4	8.4	9.1	12.9
	Narbuvoll	1.9	1.6	1.3	1.1
	Kårvatn	2.0	1.8	1.4	1.7
	Jergul	1.2	0.6	0.7	1.1
Cd	Birkenes	0.28	0.36	0.29	1.04*
	Narbuvoll	0.08	0.09	0.07	0.04
	Kårvatn	0.05	0.06	0.05	0.08
	Jergul	0.07	0.02	0.02	0.05

* Contamination problems suspected.

At three of the stations (Birkenes, Hummelfjell, Jergul), 24 hour or 48 hour air particulate samples were collected during 1978–79. These samples were analysed for 23 elements by a combination of instrumental neutron activation analysis and atomic absorption spectrometry (Steinnes, 1980). Samples for SO_2 analysis were collected simultaneously. Certain of these elements showed a very typical episodic occurrence. This group was comprised of V, Cr, Mn, Fe, Zn, As, Se, Br, Cd, Sb, Pb and also SO_2 and 'excess SO_4'. The episodic character is particularly pronounced at the southernly Birkenes site, where highly significant correlations occurred between the above

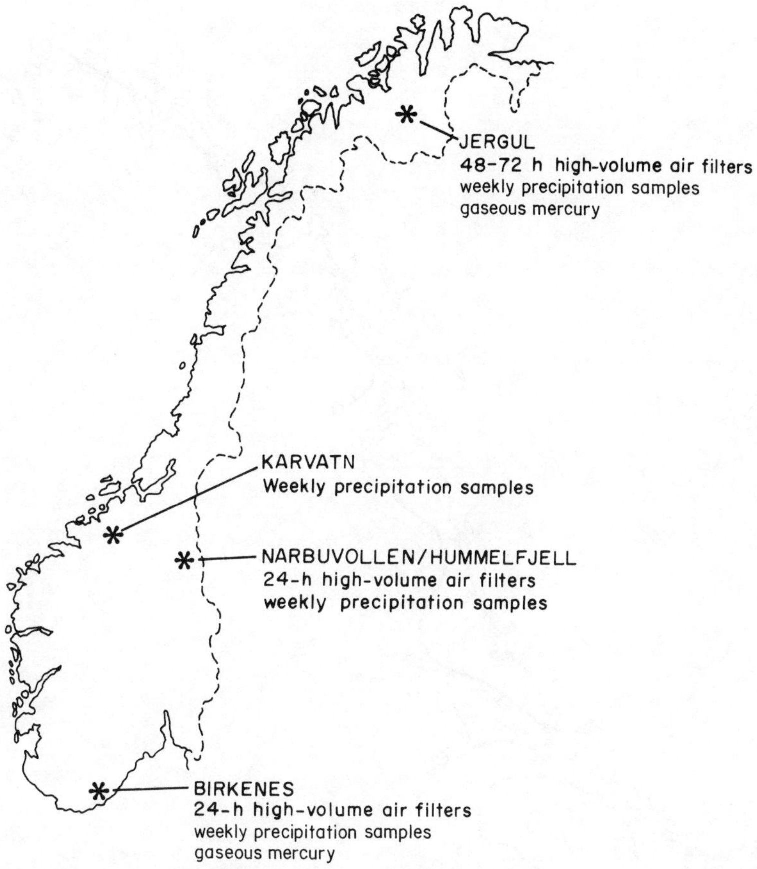

Figure 9.1 Sampling stations for air particulates and precipitation in Norway

components. Trajectory analysis showed all major episodes of this kind during the period concerned to be associated with air masses derived from heavily populated and industrial regions in other parts of Europe (Hanssen *et al.*, 1981). A sector analysis of the air concentration data at Birkenes based on air trajectories (Pacyna *et al.*, 1984) indicates that at least 80% of the supply of these elements to this region is due to emissions in areas outside of Norway.

TRACE ELEMENT DEPOSITION PATTERNS

Relative deposition patterns for a number of trace elements in Norway were established by means of moss analysis in a similar way to that originally

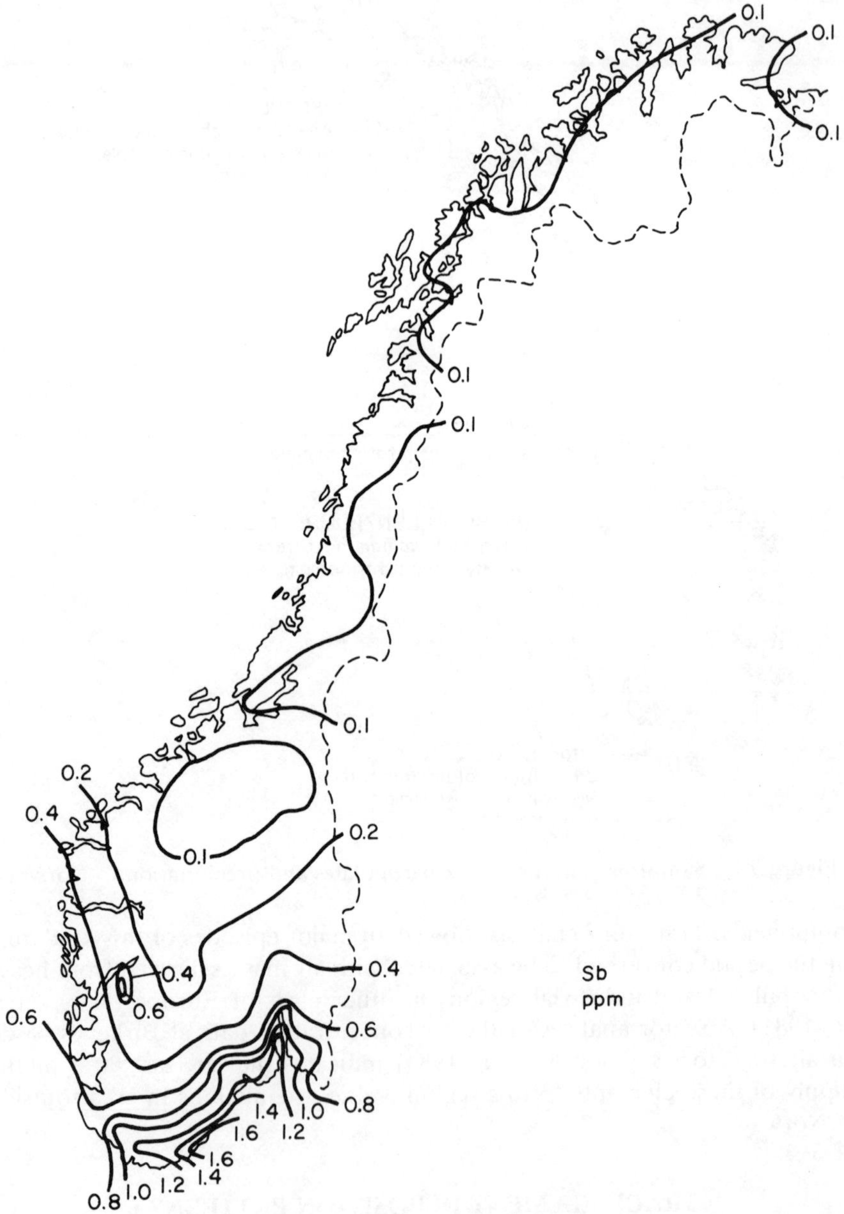

Figure 9.2 Relative deposition figures for antimony expressed as ppm Sb in moss samples

suggested by Rühling and Tyler (1971). Samples of the forest feather moss *Hylocomium splendens* which grows on the ground, were collected from about 500 sites all over Norway in 1977 (Rambaek and Steinnes, 1980). The samples were analysed in a manner similar to that indicated above for air filters. Elements known from the air filter studies to be mainly associated with long-range transport, tend to show very distinct regional differences in their deposition pattern. The map shown for Sb in Figure 9.2, exhibiting tenfold higher deposition in the far south than in many areas in the north, is quite typical for the group of elements also comprising Pb, As, V, Zn, Cd, Ag, Mo and Se. The areas exhibiting the highest metal deposition coincide with those previously shown to receive the maximal input of acid precipitation. Other heavy metals such as Cr, Cu and Ni show deposition patterns mainly dominated by local sources.

A second national moss survey is planned in 1985, concurrently with similar surveys in Sweden and Denmark.

HEAVY METALS IN SOILS

During the sampling for the 1977 national moss survey, natural surface soils were collected at the same 500 sites as for the moss samples. The samples were taken from the middle of the humus layer if the thickness was less than 10 cm, and at about 5 cm depth in cases of thicker humus layers. The majority of the soils were from podzol profiles. Further details on sampling and pre-analysis treatment are given elsewhere (Allen and Steinnes, 1980).

The samples were analysed for a number of trace elements mainly by atomic absorption and ICP emission spectrometry after digestion for 20 hours with hot concentrated nitric acid. The regional distribution of Pb in surface soils is shown in Figure 9.3, exhibiting a very distinct north–south gradient with values about ten times higher in the south and southwest than in the far north. Similar trends were observed for Cd, As, Sb and to a lesser extent for Zn and Cu (Allen and Steinnes, 1980). For other heavy metals such as Cr, Co, and Ni this kind of regional distribution was not evident (Bølviken, B. and Steinnes, E., unpublished). For the former group of elements, which is known to be associated with long-range transport, the regional distribution in the surface soils is rather similar to the regional deposition patterns shown by the moss survey. For these elements long-range atmospheric transport has apparently had a major impact on the concentrations in natural surface soils in the southern and south-western parts of Norway.

This conclusion is supported by data obtained from studies of metals in soil profiles from South Norway and Central Norway (Petersen and Steinnes, 1983; Solberg and Steinnes, 1983). A typical example is presented in Figure 9.4, showing the vertical distribution of Pb and Cu in two

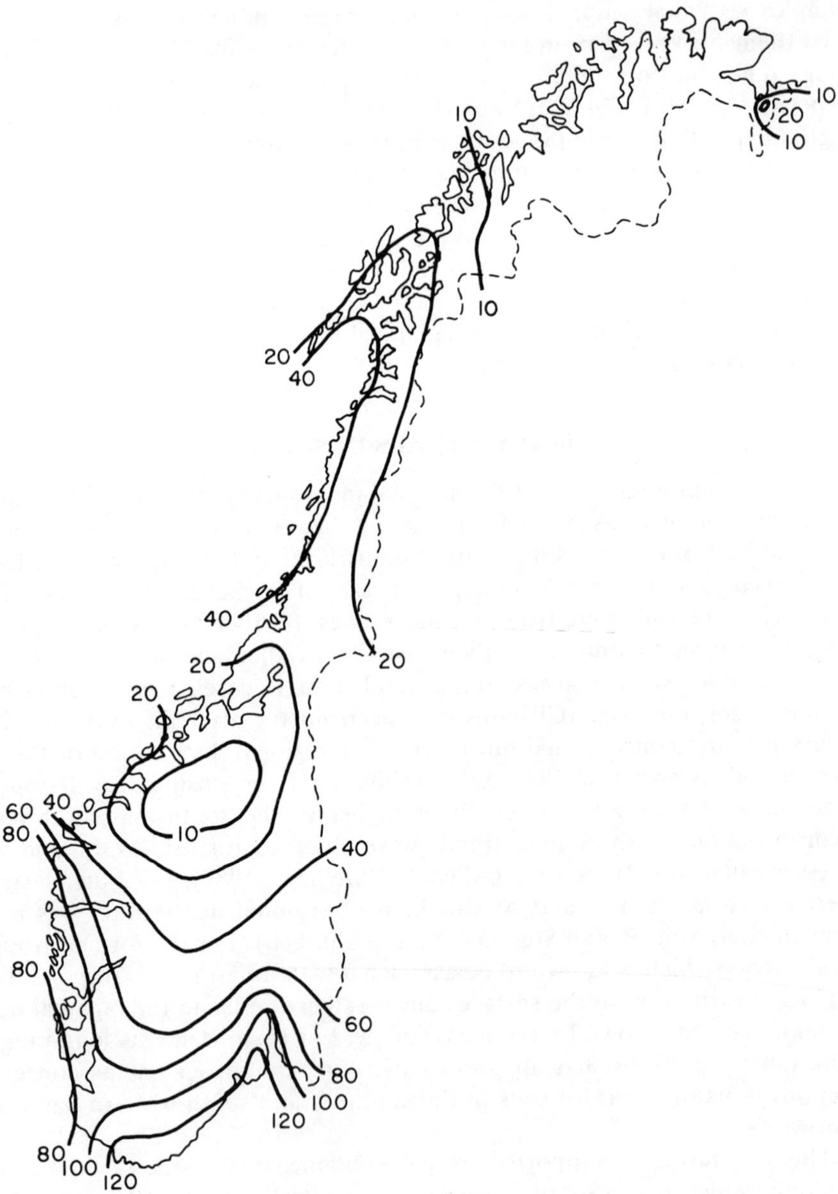

Figure 9.3 Geographical distribution of lead in natural surface soils in Norway (ppm)

brown earth profiles. Lead is strongly enriched in the surface in South Norway, whereas in Central Norway where the airborne supply is low, there is no such trend. For copper, the deposition of which is not very different in the two regions, there is no difference in surface enrichment corresponding to that apparent for Pb.

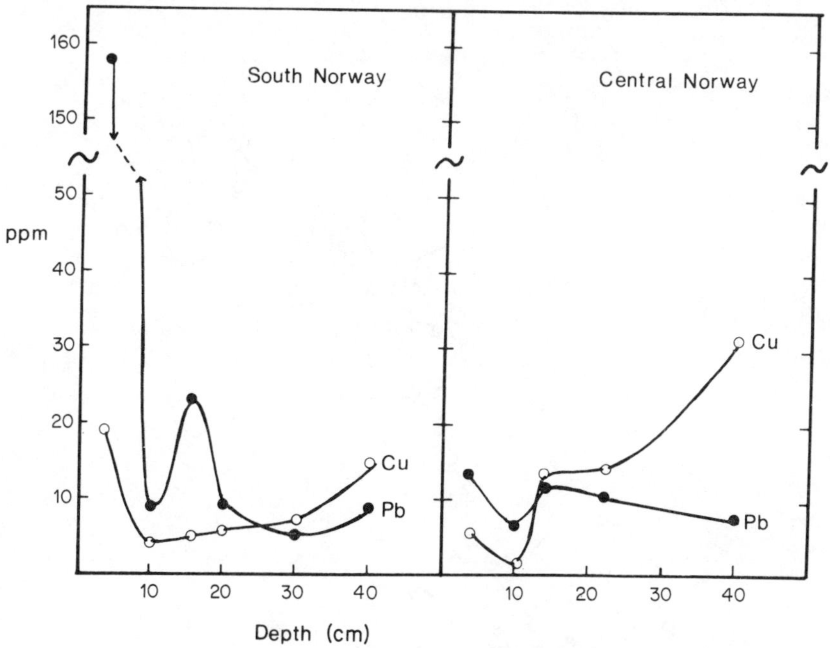

Figure 9.4 Examples of vertical distribution of copper and lead in brown earth profiles from South Norway and Central Norway, respectively (ppm)

The elements selenium and mercury require special comments as far as their regional distribution in surface soils is concerned. The Se content is considerably higher in soils from coastal areas than in corresponding soils inland even in Central and Northern Norway where the contribution from air pollution is presumably low (Allen and Steinnes, 1980). Selenium data from the moss survey show a similar coastal enrichment all along the coast. This supports the hypothesis put forward in earlier work (Låg and Steinnes, 1974, 1978a) that there is a natural source of atmospheric Se, possibly associated with the decomposition of Se-enriched marine organisms at the sea surface, which provides a relative excess of Se to soils in the coastal areas. In the case of Hg, work based on material collected during the period 1960–64 (Låg and Steinnes, 1978b) shows that there is no appreciable difference in soil concentration between the south-eastern and the central parts of the country, thus indicating that natural sources may be more important than anthropogenic ones in supplying Hg to surface soils in Norway.

In the southernmost areas of Norway where the impact of long-range transport of air pollutants is most pronounced, a more detailed regional survey of the metals Pb, Cd, Zn and Cu in natural surface soils has recently been carried out (Solberg, W. and Steinnes, E., unpublished). The maximum

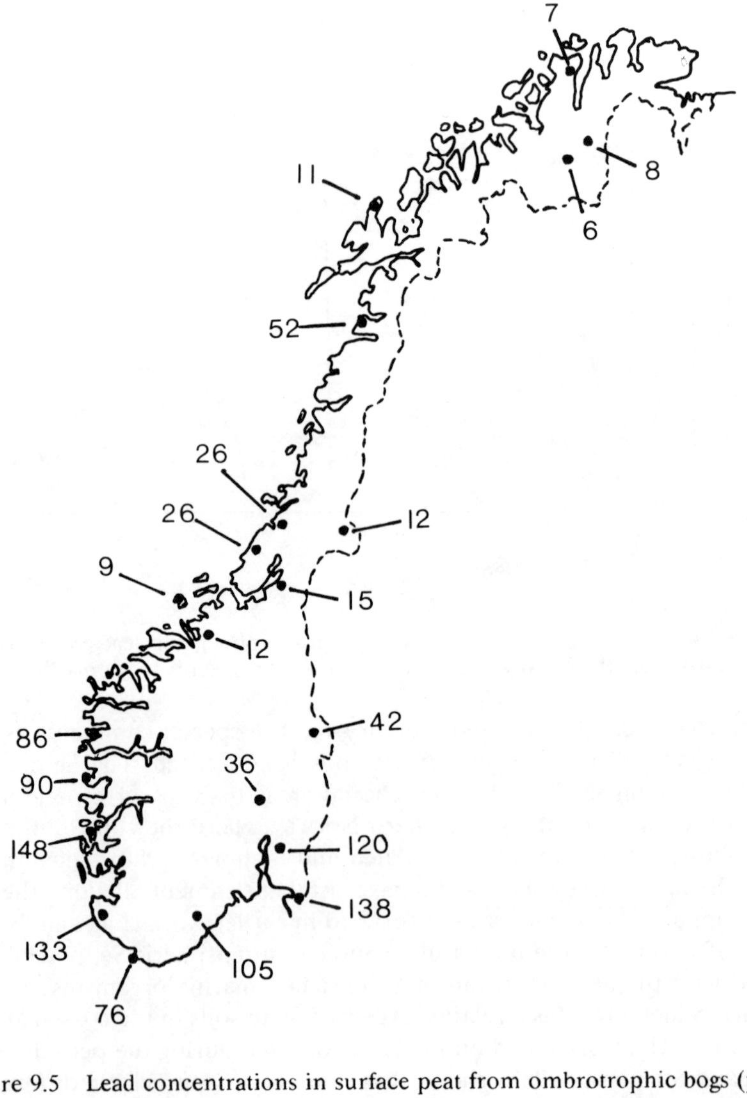

Figure 9.5 Lead concentrations in surface peat from ombrotrophic bogs (ppm)

level of these elements in the soil appears to occur in areas situated about 25–30 km from the coast. These areas have very few local activities that could create any air pollution, but they coincide quite well with the zone of maximum precipitation due to orographic effects. In these local areas Pb levels of 180–200 ppm and Cd levels of 2–3 ppm are typically observed in the humus layer of podzolic soils.

HEAVY METALS IN OMBROTROPHIC BOGS

In an investigation of peat profiles from 21 ombrotrophic bogs in different parts of the country, 11 metals were determined using methodology similar to that indicated for soils (Hvatum *et al.* 1983). The concentration of the relatively volatile elements Pb, Cd, As and Sb appeared to be about ten times higher in surface peat from the southernmost part of Norway as compared with areas in the far north. The data for Pb in surface peat shown in Figure 9.5 are quite similar to the levels observed in surface soils shown in Figure 9.3. Similar but less pronounced differences are shown for Zn, while the concentrations of Cr, Cu, Mn, Fe, Co and Ni do not show any large differences between northern and southern bogs.

In bogs in the far south, the surface levels of Pb, As and Sb are about 20 times higher than those observed at a 50 cm depth, indicating a very substantial increase in the atmospheric supply of these elements in recent times. In Central and Northern Norway where surface levels are 5–10 times lower, there is a less distinct but still significant enrichment in the surface, indicating an impact of air pollution even in these areas.

The data for Se at the surface and at a depth at 50 cm confirm the hypothesis of a natural source of atmospheric Se which also contributed in pre-industrial times; in the coastal districts of Central and Northern Norway, relatively high and constant Se values are apparent at different depths. In continental areas of South-Eastern Norway there is an increase by a factor of 10 from 50 cm depth to the surface, illustrating recent input from long-range atmospheric transport.

HIGHER PLANTS AND ANIMALS

The pattern of distribution of the content of heavy metals in vascular plants in coniferous forest ecosystems is similar to that of the atmospheric deposition of these metals, as illustrated in Figure 9.6 (Solberg and Steinnes, 1983), which is a comparison of metal levels in twigs of some dominant species in Southern and Central Norway. Each bar represents the average of unwashed, randomly collected samples taken at three times during the growing season at four different locations. As is evident from Figure 9.6, Cu shows no significant regional difference in any of the species included.

Zn is significantly higher than Cu, and Pb and Cd are about 5 times higher in Southern Norway than in Central Norway.

The excess heavy metal content of vegetation in Southern Norway may be expected to influence the uptake of these elements in animals feeding on natural vegetation in the areas concerned. The liver Pb content of lambs from different areas of Norway was shown to be closely connected with the Pb deposition (Frøslie *et al.*, 1984) as indicated by a correlation coefficient of 0.94 between lamb liver Pb and moss Pb. Significant positive correlations were also found for Cd, Se, and As.

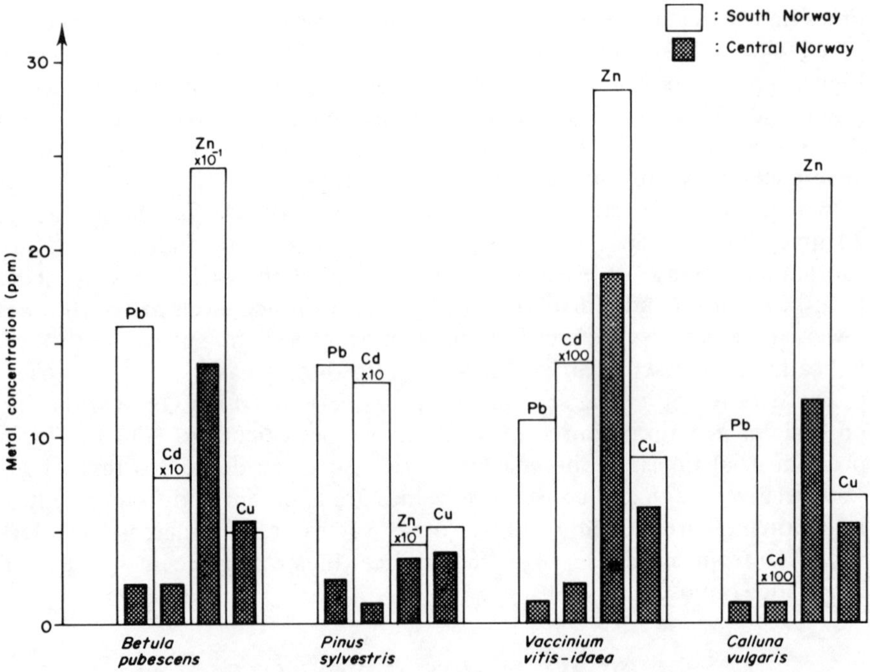

Figure 9.6 Mean concentrations of Pb, Cd, Zn and Cd in twigs of four common plant species from South Norway and Central Norway (ppm). Cd and Zn values are sometimes altered to fit onto the figure as indicated

REFERENCES

Allen, R. A., and Steinnes, E. (1980). Contribution from long-range atmospheric transport to the heavy metal pollution of surface soil. In Drablos, D., and Tollan, A. (Eds), *Ecological Impact of Acid Precipitation*, pp. 102–103, Oslo-Ås.

Frøslie, A., Norheim, G., Rambaek, J. P., and Steinnes, E. (1985). Heavy metals in lamb liver: contribution from atmospheric fallout. *Bull. Environ. Contam. Toxicol*, **34**, 175–182.
Hanssen, J. E., Rambaek, J. P., Semb, A., and Steinnes, E. (1980). Atmospheric deposition of trace elements in Norway. In Drabløs, D., and Tollan, A. (Eds), *Ecological Impact of Acid Precipitation*, pp. 116–117, Oslo-Ås.
Hanssen, J. E., Rambaek, J. P., Semb, A., and Steinnes, E. (1981). Atmospheric deposition of some heavy metals in Norway. In *Heavy Metals in the Environment*, Amsterdam, 1981, pp. 322–325, CEP Consultants, Edinburgh 1981.
Hvatum, O.Ø., Bølviken, B., and Steinnes, E. (1983). Heavy metals in Norwegian ombrotrophic bogs. In Hallberg, R. (Ed.), *Environmental Biogeochemistry. Ecol. Bull. (Stockholm)*, **35**, 351–356.
Låg, J., and Steinnes, E. (1974). Soil selenium in relation to precipitation. *Ambio*, **3**, 237–238.
Låg, J., and Steinnes, E. (1978a). Regional distribution of selenium and arsenic in humus layers of Norwegian forest soils. *Geoderma*, **20**, 3–14.
Låg, J., and Steinnes, E. (1978b). Regional distribution of mercury in humus layers of Norwegian forest soils. *Acta Agr. Scand.*, **28**, 393–396.
Overrein, L. N., Seip, H. M., and Tollan, A. (1981). *Acid Precipitation—Effects on Forest and Fish.* Final report of the SNSF-project 1972–1980. Research report FR 19/80. Oslo-Ås.
Pacyna, J. M., Semb, A., and Hanssen, J. E. (1984). Emission and long range transport of trace elements. *Tellus*, **36B**, 163–178.
Petersen, H. M., and Steinnes, E. (1983). Arsenic in Norwegian surface soils in relation to atmospheric pollution. In *Heavy Metals in the Environment*, Heidelberg 1983, pp. 187–190, CEP Consultants, Edinburgh.
Rambaek, J. P., and Steinnes, E. (1980). *Atmospheric Deposition of Heavy Metals Studied by Analysis of Moss Samples Using Neutron Activation Analysis and Atomic Absorption Spectrometry.* Nuclear Methods in Environmental and Energy Research (CONF-800433) 175–180.
Rühling, Å., and Tyler, G. (1971). Regional differences in the deposition of heavy metals over Scandinavia. *J. Appl. Ecology*, **8**, 497–507.
Solberg, W., and Steinnes, E. (1983). Heavy metal contamination of terrestrial ecosystems from long-distance atmospheric transport. In *Heavy Metals in the Environment, Heidelberg 1983*, pp. 170–173, CEP Consultants, Edinburgh.
Statens forurensningstilsyn (Norwegian State Pollution Control Agency). (1983). *Overvåking av langtransportert forurenset luft og nedbør.* Årsrapport 1982. Report 108/83. 227 pages.
Steinnes, E. (1980). Atmospheric deposition of heavy metals in Norway, studied by analysis of moss samples using neutron activation analysis and atomic absorption spectrometry. *J. Radioanal. Chem.*, **58**, 387–391.

Lead, Mercury, Cadmium and Arsenic in the Environment
Edited by T. C. Hutchinson and K. M. Meema

CHAPTER 10

Cadmium Levels in Soils and Crops in the United States

A. L. PAGE, A. C. CHANG AND MOHAMED EL-AMAMY

Dept. of Soil and Environmental Sciences,
University of California,
Riverside, CA 92521

ABSTRACT

The concentrations of cadmium in agricultural soils depend upon the amounts present in the parent rocks from which the soils form, the amounts added in the form of fertilizers and soil amendments, the amounts deposited onto soils from the atmosphere, and the amounts removed by harvested crops and by leaching. On average, sedimentary rocks contain greater concentrations of Cd than either igneous or metamorphic rocks, and therefore, recent soils derived from sedimentary rock should contain greater concentrations of Cd than those derived from igneous or metamorphic rocks. Among the commercial fertilizers, phosphorus fertilizers contain somewhat elevated levels of Cd. Results show that the long-term use of phosphorus fertilizers will slightly increase the concentration of Cd in surface soils. Due to the combustion of fossil fuels and to the smelting and processing of ores, air in rural regions contains Cd. Results show that the amounts of Cd added to soils via aerial deposition in rural agricultural regions is of the order of a few grams per hectare per year and approximately equal to that added to soil from normal phosphorus fertilization operations. All crops contain Cd, and the harvested portion will serve to remove some Cd from soil. Amounts removed from soil in harvested crops, however, are quite small compared with the amounts present. Cadmium is reasonably immobile in soils, and the available data suggest that the amounts removed by leaching are also small compared with amounts present.

The concentrations of Cd in non-contaminated agricultural soils in the United States range from about 0.1 to 1.0 mg/kg. Except for a limited number of soils derived from parent materials unusually high in Cd, organic soils (Histosols) tend to contain the highest total Cd concentrations, and highly weathered soils (Ultisols and Alfisols) the lowest. Other soil orders contain concentrations of Cd in amounts approximately equal to their geological abundance or from about 0.2 to 0.4 mg Cd/kg. The fact that highly weathered soils are somewhat depleted in Cd suggest that greater quantities of Cd are removed by crops and leaching than are added through fertilization and atmospheric deposition. Surface soils commonly contain higher concentrations of Cd than subsurface horizons. The higher concentrations of Cd in surface horizons are probably due to the cycling of Cd from lower depths to the surface by plants.

Concentrations of Cd in food chain crops vary substantially within and among species. Research conducted during the past has demonstrated that the variations observed are due to differences in the Cd concentrations present in the soil, differences in the growth conditions among the regions where the crops are grown, differences in the cultivars of a particular crop, and differences in soil chemical and physical properties. We have grown a number of vegetable crops in fall and spring seasons and have observed seasonal differences in the amounts of Cd which the crops accumulate. Numerous data from a number of sources have demonstrated that different cultivars of the same plant show differences in their Cd absorption characteristics. Likewise, numerous published data have demonstrated that soil chemical properties, particularly soil pH, influence the amounts of Cd absorbed by crops.

On a dry weight basis the concentration of cadmium in food chain crops grown in non-contaminated soils (except those soils unusually naturally elevated in Cd) range from 0.01 to 1.0 mg/kg. Leafy vegetables (lettuce, spinach) usually show the highest Cd concentrations, grains (wheat, oats, barley) show the lowest concentrations and root vegetables (carrot, radish, onion, potato) are intermediate between the two extremes. Concentrations of Cd in crops grown on soils elevated in Cd, either naturally or from anthropogenic sources, may accumulate substantially greater than 1.0 mg/kg. The amounts accumulated depend upon the level of Cd present in the root zone of the soil, the crop species, and the chemical properties of the soil.

INTRODUCTION

Cadmium (Cd) is regarded as one of the most toxic trace elements in the environment. The increased emissions resulting from its production, use, and disposal, combined with its persistence in the environment, and its relatively rapid uptake and accumulation by food chain crops contribute to its potential environmental hazards. Cadmium may find its way to the human population through food and beverage, drinking water, air, and cigarette smoking. Although acute Cd toxicity caused by food consumption is rare, chronic exposure to high Cd levels in food can significantly increase the accumulation of Cd in certain body organs. When the concentration in the human body reaches levels considered to be harmful (>200 μg/gm wet weight in the kidney cortex according to Kjellström and Nordberg, 1978), cadmium-induced kidney damage, skeletal disorders as well as other diseases may result.

A highly publicized episode of Cd poisoning of humans (itai-itai disease) was reported in Japan in the mid-1950s (Tsuchiya, 1978). The source of excessive Cd to the affected individuals came from rice grown on nearby paddies which had been irrigated with water from a river contaminated by zinc mining operations. The Cd concentrations of the rice, of the river sediments and of the soil in which the rice was grown were considerably greater than those found in uncontaminated regions. However, the actual Cd levels in the rice grain were not unusually high. Since the per capita dietary consumption of rice among the Japanese population is high, the impact of Cd enrichment in the rice on the indigenous population was proportionally magnified. The Japanese incident is the only documented Cd poisoning arising from human consumption of crops grown on Cd-contaminated soils.

For populations not subjected to a Cd-contaminated environment, the main source of Cd human body burden is also from food. Drinking water and ambient air contribute significantly lesser amounts to the daily intake. Concentrations of Cd in domestic water supplies seldom exceed a few μg/litre, and at a consumption of 1–2 litres of water/day the daily intake would be merely a few micrograms (Friberg *et al.*, 1974; Tsuchiya, 1978; Pahren *et al.*, 1979). Cadmium concentrations of ambient air rarely exceed 0.01 $\mu g/m^3$, and at an intake rate of 20 m^3 of air/day the daily Cd intake would not exceed 0.2 μg. Cigarette smoking, however, adds considerably to Cd input via inhalation. Friberg *et al.* (1974) estimated that a daily intake of 2 to 4 μg Cd would be expected from smoking one package of cigarettes per day. They further suggested that the daily ingestion of Cd from all sources for the world's population ranged from 25 to 75 μg/day. Based upon results of market basket surveys over a seven-year period, the US Food and Drug Administration (Mahaffey *et al.*, 1975) showed an average intake of 39 μg Cd/day for the 15 to 20 year old age group males in the United States. If the figure is adjusted according to the recommended daily caloric intake for other age groups, the average daily intake from birth to age 50 for men and women in the US would be 33 and 26 μg/day, respectively (Pahren *et al.*, 1979). These values, however, were derived with the assumption that any market basket food items analyzed contained at least 0.02 mg Cd/kg. More recent data based upon analysis of fecal excretion gave estimated figures of from 18 to 21 μg Cd/day for teenage males (Ryan *et al.*, 1982; Kowal, 1984).

The above-cited figures illustrate the importance of Cd in the food chain crops to the accumulation of Cd in the human body. The concentration of Cd in foods is, in turn, controlled by its concentration in the soil substrate and by the physical and chemical nature of this soil. In the following sections we will review concentrations of Cd in crops and soils in the United States. Sources of Cd in soils, as well as factors influencing its absorption by crops, are also discussed.

NATURAL OCCURRENCE

According to Heinrichs *et al.* (1980), the average concentration of cadmium in the lithosphere is 0.098 mg/kg. It is a rarer element than mercury and about 1/700 as abundant as zinc (whose crustal abundance averages 70 mg/kg). The ranges and mean concentrations of cadmium for some common igneous, sedimentary, and metamorphic rocks are presented in Table 10.1. There are very small differences in the overall abundance of cadmium in

Table 10.1 Abundance of cadmium in common rocks*

Rock type	Cadmium content (mg/kg) Mean	Range
Igneous		
Granite	0.09	0.001–0.60
Basalt	0.13	0.006–0.60
Ultramafic	0.026	0.001–0.03
Metamorphic		
Gneisses	0.04	0.007–0.26
Schists	0.02	0.005–0.87
Ecologite	0.11	0.04–0.26
Sedimentary		
Limestone	0.065	0.001–0.50
Sandstone	0.02	0.01–0.41
Shale, clay	0.03	0.02–11.0
Red clay	0.56	
Organic mud	0.39	
Deep ocean sediments	0.5	0.05–17
Oceanic manganese oxides	8.0	<3.0–21
Phosphorites	25	<10–500
Recent sediments		
Lake sediments	0.91	0.02–6.2
Stream sediments	0.16	0.03–0.40

* Based on compilations by Fleischer *et al.* (1974), Gong *et al.* (1977), Waketa and Schmitt (1970), Horn and Adams (1966), and Marowsky and Wedepohl (1971).

mafic and granitic rocks. It tends to concentrate by sedimentation as evidenced by the high concentrations (0.001–11.0 mg/kg) in sedimentary rocks. Some shales are unusually high in Cd (Wedepohl, 1968). Along the Pacific Coast in the US, the Monterey shale formation in some locations has been found to contain as much as 90 mg Cd/kg (Lund *et al.*, 1981).

Amounts of cadmium in marine sediments in the Atlantic and Pacific Oceans range from 0.1–1.0 mg Cd/kg (Mullin and Riley, 1956). Marine manganese nodules (5.1–8.4 mg Cd/kg) and marine zinc-bearing phosphorite (60–340 mg Cd/kg) are also high in cadmium (Mullin and Riley, 1956; Goldschmidt, 1958; Auer 1977).

CADMIUM IN NON-CONTAMINATED AND CONTAMINATED SOILS

Natural Levels in Soils

Amounts of indigenous cadmium in cultivated and non-cultivated soils are determined by the quantities of Cd in the parent materials together with amounts added through atmospheric deposition, fertilizers, pesticides, and irrigation water, but minus amounts removed by leaching, erosion, and in harvested crops. Based upon Cd concentrations reported for common rocks, one would expect on the average, soils of similar age derived from igneous rocks to be lowest in total Cd, soils derived from metamorphic rocks to be intermediate, and soils derived from sedimentary rocks to contain the largest quantities of Cd.

Recently, an extensive survey was conducted to investigate Cd in surface soils in the major crop producing regions of the continental United States (Holmgren *et al.*, 1986). This survey took great care to ensure that the sites sampled were free from any known source of contamination and to ensure the utmost analytical accuracy. A total of 3305 soil samples were collected from crop-producing areas in 36 states. Concentrations of Cd ranged from 0.005 to 2.4 mg/kg, with mean and median values of 0.27 and 0.20 mg/kg, respectively. A summary of the results by geographical regions (Figure 10.1) is presented in Table 10.2. On average, Cd concentrations in soils from the Western and North Central states are greater than those from the Northeastern and Southern states.

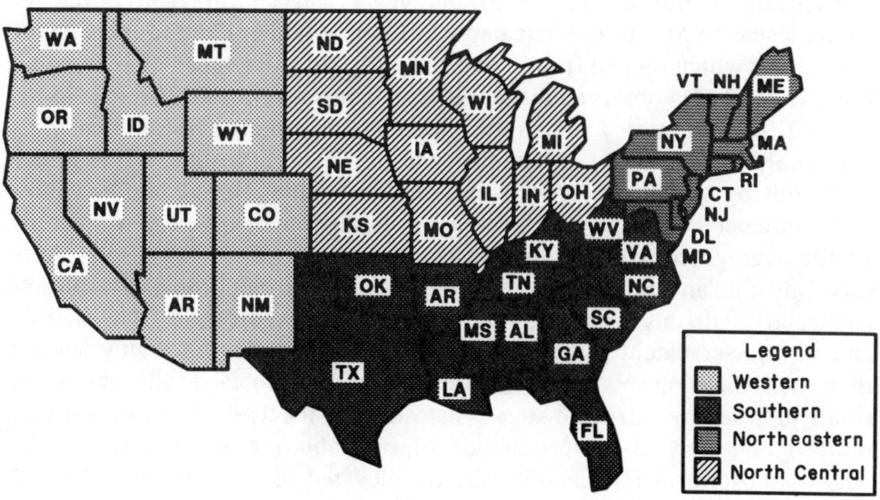

Figure 10.1 Map showing regions and states within the contiguous United States

Data from other sources are in general agreement with those published by Holmgren *et al.* (1986). Logan and Miller (1983) collected surface soils from 237 farms in six counties in Ohio and observed Cd concentrations to range

Table 10.2 Cadmium concentrations of soils from major agricultural regions in the United States

Region*	No. of samples	No. of states sampled	Cd concentration (mg/kg)	
			Range	Mean
Western	742	8	0.20–0.49	0.33
North Central	937	12	0.20–0.94	0.37
Northeast	293	5	0.08–0.21	0.17
Southern	1230	9	0.03–0.44	0.15

Derived from data presented by Holmgren *et al.* (1986), excluding organic soils from New York and high lead samples from West Virgina.
* See Figure 10.1.

from <0.1 to 2.9 mg/kg with a mean concentration of 0.2 mg/kg. Pierce *et al.* (1982) presented data to show the Cd concentration of soils from 16 soil series developed on seven parent materials in Minnesota varied from 0.06 to 0.74 mg/kg. The soil series sampled were considered representative of most of the crop-producing areas in Minnesota. The highest total Cd concentrations occurred in calcareous soils developed in lacustrine sediments and, generally, in soils with free carbonates. In California, soils developed from Monterey shale or alluvium from the shale frequently have elevated Cd concentrations. Burau *et al.* (1973) surveyed 177 agricultural soils developed from or near to Monterey shale deposits (Salinas Valley) and found Cd concentrations which ranged from 0.05 to 10.1 mg/kg with a mean of 1.24 mg/kg. Lund *et al.* (1981) observed highly elevated concentrations of Cd in soils developed from Monterey shale in the Malibu Canyon area in California.

A summary of the survey data for cadmium concentrations in US agricultural soils by taxonomic order is presented in Table 10.3. Excluding those soils influenced by Monterey shale deposits, the data suggest that Histosols on the average contain the highest concentrations of Cd; Mollisols, Vertisols, Entisols, and Inceptisols are intermediate, and Spodosols, Alfisols and particularly Ultisols are the lowest in total Cd concentrations. The Ultisols and to a lesser extent the Spodosols and Alfisols are more highly leached than the remaining soil orders, and less Cd in these surface soils may be the result of leaching. Organic soils (Histosols) in the United States are used rather extensively in the production of vegetable crops which require intensive phosphorus fertilization, and the elevated concentrations in this soil order, at least in part, could be due to phosphorus fertilization. The cycling of Cd from soil to plant and subsequent decomposition in soils combined with retention of Cd by organic matter may also account for elevated levels of Cd in the Histosols. Based on the data summarized in Table 10.3, most uncontaminated agricultural surface soils will contain from about 0.1 to 1.0 mg Cd/kg.

Table 10.3 Cadmium concentrations of agricultural soils in the United States*

Location	No. of samples	Taxonomic Order†	Cadmium concentration (mg/kg) Minimum	Maximum	Mean	(Median)	Reference
Ohio	231	NR (Alfisol)	<0.1	2.9	0.2		Logan and Miller, 1983
	6	NR (Entisol)	0.3	0.4	0.2		
	81	NR	0.08	0.68	0.38	(0.30)	Holmgren *et al.*, 1986
Michigan	91	NR (Alfisol)	NR	NR	0.4	(0.57)	Klein, 1972
	86	NR	0.34	1.54	0.94		Holmgren *et al.*, 1985
Minnesota	21	Mollisol	NR	NR	0.48		Pierce *et al.*, 1982
	18	Alfisol	NR	NR	0.28		
	89	NR	0.12	0.18	0.30		Holmgren *et al.*, 1985
California	283	NR	0.03	0.87	0.31		Holmgren *et al.*, 1985
Soils developed on or in regions near Monterey shale deposits	105	Mollisol	0.077	15	1.54	(0.85)	Burau *et al.*, 1973 and Lund *et al.*, 1981
	39	Alfisol	0.13	6.38	1.12	(0.76)	
	18	Vertisol	0.39	2.10	1.08	(1.10)	
	17	Entisol	0.05	3.70	0.93	(0.69)	
	2	Inceptisol	12	22	17		
Excluding soils influenced by Monterey shale	7	Mollisol	0.01	0.26	0.08	(0.15)	Various sources from authors' file
	10	Alfisol	<0.1	0.6	0.25	(0.2)	
	10	Entisol	0.13	3.5	0.95	(0.75)	
	4	Aridisol	<0.1	0.3		(<0.1)	
Continental US	288	Histosol	0.36	1.44	0.72		Holmgren *et al.*, 1986
	152	Aridisol	0.17	0.71	0.35		
	256	Entisol	0.07	0.70	0.28		
	1076	Mollisol	0.08	0.68	0.28		
	256	Inceptisol	0.05	0.71	0.27		
	91	Vertisol	0.13	0.55	0.27		
	43	Spodosol	0.08	0.47	0.21		
	570	Alfisol	0.03	0.42	0.16		
	571	Ultisol	<0.01	0.30	0.08		

* Range from Holmgren *et al.* (1985) is estimated from the standard deviation (95% confidence interval).
† NR = no record; order in parentheses represent authors' estimate.

Only limited data are available pertaining to concentrations of cadmium in relation to soil depth. Pierce *et al.* (1982) showed that levels of Cd in surface horizons of Mollisols and Alfisols are greater than those found in subsurface horizons. Mean concentrations of Cd in surface horizons were 0.39 mg/kg compared with 0.23 mg/kg for subsurface horizons. The higher concentrations in surface horizons suggest that additions of Cd by atmospheric deposition, fertilizers, and plant cycling exceed losses by leaching and plant removal in the surface horizon.

Sources and Extent of Cadmium Contamination of Soils

Soils may be contaminated with cadmium by fallout from aerial sources, by application of waters, fertilizers, or pesticides which contain cadmium, or by the discharge of cadmium-containing waste materials from industrial, metallurgical or urban activities.

1. Levels Near Mining and Smelting Operations

Processing of industrial metals is an important source of atmospheric emission of trace metals including Cd. Smelting and sintering of non-ferrous metals have resulted in Cd contamination of the nearby environment. The ore-smelting furnaces in which metals enter the flue-gas stream as fine particulates or volatiles are responsible for the majority of discharge from the stack and, eventually, the deposition onto soils and vegetation. Airborne dusts and fumes from charging furnaces, transporting metal ores, and sintering and metal-reducing furnaces are also sources of metals found in and near the metallurgical operations. Soils contaminated by smelting operations exhibited the highest Cd concentrations in the upper horizons near the point of discharge in the direction of the prevailing winds. The Cd concentrations at a depth of about 30–40 cm would return to levels close to the background because of surface retention especially in organic matter and litter.

The extent to which soils in the United States have been contaminated by mining and smelting operations can be found in published reports by Fulkerson and Goeller (1973), Buchauer (1973), Munshower (1977), Ragaini *et al.* (1977), Johnson *et al.* (1975), Lagerwerff and Brower (1975) and US Environmental Protection Agency (1972). Two kilometres northeast of a lead-zinc smelter in Montana, 29 mg Cd/kg in surface soils was observed (Munshower, 1977). The Cd concentration of surface soil decreased with distance and reached near background levels 24 kilometres from the source. Likewise, within one kilometre of a zinc smelter at Palmerton, Pennsylvania, Buchauer (1973) reported concentrations of Cd in the surface organic horizon as great as 750 mg Cd/kg. The Cd concentrations of the soils decreased logarithmically with distance and levels exceeding background were observed up to 21 kilometres from the source. Recently adopted emission control measures should lessen future contamination of soils from mining and smelting. However, those soils already contaminated by past operations will remain high in Cd more or less permanently because Cd is quite immobile in soil.

The extent of contamination of soils resulting from stack emissions of coal-fired power plants varies as a result of differences in emission control devices. The rate of atmospheric discharge of Cd per ton of coal combusted ranges from 3 to 90 mg (Lyon, 1977). With 90% emission control, the annual deposition rate of Cd onto soil adjacent to a 3000 MW coal-fired power plant is 0.00002 mg/kg soil (Ondov, *et al.*, 1976). Assuming a mean Cd concentration of 0.2 mg/kg soil, the enrichment of the soil over the lifetime of the power plant ($\sim$35 years) is approximately 0.35%.

2. *Disposal/Recycling of Municipal Sewage Sludges*

Sewage sludges from municipalities are frequently applied to soils in disposal operations, as soil conditioners, or as a source of plant nutrients. The concentrations of Cd in sewage sludge always exceed those normally found in soils. Municipal sewage sludge contains in the range of approximately 2 to 3500 mg Cd/kg sludge, with medians in the order of 5 to 20 mg/kg (Page, 1974; Sommers, 1977; Logan and Chaney 1984). Application of sludge as a nitrogen fertilizer can increase the cadmium level in surface soils. A 5 tons/ha application of sludge containing 10 to 15 mg Cd/kg may result in an increase of cadmium concentration by as much as 10–15% in the surface soil (assuming background level is 0.22 mg/kg). The applied cadmium may remain available to plants for an extended period of time.

Although the use of municipal sewage sludge as a fertilizer or soil conditioner will result in increased Cd concentrations in soil, the impact of this practice on the Cd concentration of commercial crops which enter the marketing and distribution systems is expected to be minimal. This is so because less than 1.0% of the agricultural land in the United States would be affected if 50% of the sludges generated annually were to be applied to agricultural lands as a source of nitrogen (CAST, 1976). Currently, less than 20% of the sludge generated is applied to lands used to produce commercial crops. In situations where foods grown on sludged lands made up a substantial percentage of the human diet, over a number of years it is possible that consumers could accumulate Cd in amounts considered to be a health hazard.

3. *Phosphorus Fertilizers*

Phosphorus fertilizers frequently contain greater concentrations of cadmium than are typically found in soils. Depending upon the source of phosphorus, commercial fertilizers produced in the United States contain from a few to 200 mg Cd per kilogram with median values from 2 to 20 mg Cd/kg (US Environmental Protection Agency, 1978). The amounts of Cd introduced into the soil through normal phosphorus fertilization, however, are quite small. For example, the application of 50 kg P per hectare from rock phosphate [$Ca_{10}F_2(PO_4)_6$] which contained 20 mg P per kg would introduce 5.4 gm of Cd to the soil. If mixed with the surface 15 cm of a silt loam soil, this application would increase the Cd concentration of a soil containing 0.3 mg Cd/kg by about 1%.

4. Atmospheric Deposition

The extent of contamination of soil arising from deposition from air in agricultural regions of the United States, except near point emission sources, amounts to a few grams per hectare per year or less. For example, data derived from Hunt *et al.* (1971) for 77 cities in the midwestern United States show an annual deposition in residential areas of 1.2 gm Cd per hectare. Likewise, in a rural region of Tennessee, Lindberg *et al.* (1981) observed an annual deposition of 0.9 gm Cd per hectare.

Adjacent to point emission sources, concentrations of Cd in air are considerably greater. Consequently, amounts deposited onto land surfaces are also greater. Data derived from Rupp *et al.* (1978), for example, show annual depositions of Cd of 170, 89, 46, and 23 grams per hectare at distances of 1.6, 3.2, 6.4 and 12.8 kilometres, respectively, downwind from a smelter in Montana.

SOIL FACTORS INFLUENCING THE ACCUMULATION OF CADMIUM BY FOOD CHAIN CROPS

The two most important factors governing the uptake of cadmium by crops are the soil pH and the concentration of Cd in the soil. Factors of less importance include soil temperature, content of hydrous oxides of iron and manganese in soils, redox potential in soil, and interactions with other metals. To delineate Cd accumulation by crops, a wide variety of plants will be referred to by their common names in the remainder of the text. To avoid unnecessary duplication, the scientific names of these crops are summarized in Table 10.4.

Soil pH Effects

If other soil conditions remain unchanged, the cadmium concentration of plant tissue would decrease as the pH of the soil increased. Data selected

Table 10.4 Common and scientific names of listed crops

Common name	Scientific name
Apple	*Malus pumila*
Asparagus	*Asparagus officinalis*
Barley	*Hordeum vulgare*
Beans	*Phaseolus* spp.
Broccoli	*Brassica oleracea botrytis*
Cabbage	*Brassica oleracea capitata*
Cantaloupe	*Cucumis melo*
Carrot	*Daucus carota sativa*
Corn	*Zea mays*
Cucumber	*Cucumis sativus*
Eggplant	*Solanum melongena*
Grape	*Vitis* spp.
Grapefruit	*Citrus paradisi*
Lettuce	*Lactuca sativa*
Oat	*Avena sativa*
Onion	*Allium cepa*
Orange	*Citrus sinensis*
Pea	*Pisum sativum*
Peach	*Prunus persica*
Peanut	*Arachis hypogaea*
Pear	*Pyrus communis*
Pepper	*Capsicum* spp.
Plum	*Prunus americana*
Potato	*Solanum tuberosum*
Radish	*Raphanus sativus*
Rice	*Oryza sativa*
Sorghum	*Sorghum vulgare*
Soybean	*Glycine max*
Spinach	*Spinacea oleracea*
Squash	*Cucurbita* spp.
Swiss chard	*Beta vulgaris* var. *cicla*
Tomato	*Lycopersicon esculentum*
Turnip	*Brassica rapa*
Wheat	*Triticum* spp.

from results published in a report by the Council for Agricultural Science and Technology (1980) illustrate Cd accumulation by Swiss chard and oat grain in relation to soil pH (Table 10.5). The mean concentration of Cd in Swiss chard grown at control plots of four farms with soil pH values ranging from 4.9 to 5.4 was 1.98 mg Cd/kg while that of the limed control plots (soil pH values between 6.1 to 6.4) was 0.57 mg Cd/kg. Liming the same soils which were contaminated with Cd from land application of municipal sewage sludge reduced the mean Cd concentration of the Swiss chard from 40.5 to 3.01 mg/kg. Similar results reported by Chaney *et al.* (1975) showed

Table 10.5 Influence of soil pH and cadmium concentration on the cadmium concentration of Swiss chard and oat grain*

Treatment	Soil Cd conc. (mg/kg)	Soil pH	Cd concentration (mg/kg†) Swiss chard	Oat grain
Control				
Farm 1	0.18	4.9	3.34	0.104
Farm 2	0.10	5.3	2.65	0.076
Farm 3	0.22	5.4	1.48	0.034
Farm 4	0.07	5.9	0.44	0.051
Mean			1.98	0.066
Control plus limestone				
Farm 1	0.15	6.4	0.94	0.052
Farm 2	0.10	6.1	0.37	0.064
Farm 3	0.16	6.4	0.63	0.025
Farm 4	0.07	6.3	0.33	0.044
Mean			0.57	0.046
Sludge-treated				
Farm 1	1.66	4.9	94.80	1.960
Farm 2	9.10	5.5	54.30	2.240
Farm 3	0.98	4.9	3.24	0.209
Farm 4	3.26	5.5	9.51	0.299
Mean			40.50	1.180
Sludge-treated plus limestone				
Farm 1	2.10	6.3	2.20	0.259
Farm 2	7.02	6.2	5.34	0.277
Farm 3	0.94	6.0	0.82	0.065
Farm 4	4.50	6.2	3.68	0.193
Mean			3.01	0.198

* Data from paired plots on sludge utilization farms in four northeastern USA cities. Derived from data presented by Council for Agricultural Science and Technology, 1980.
† Dry weight basis.

a reduction in the Cd concentration of soybean leaves from 33 to 5 mg/kg when the pH of the soil on which the plants were grown was increased by liming from pH 5.3 to pH 7.0. Cadmium content of oat grain also showed a dramatic reduction as the soil pH was raised even when the pH increase was only from 4.9 to 5.3. These data and others demonstrate that one of the most effective means of minimizing the absorption of Cd by plants grown on acid soils is liming.

Under similar chemical, physical, biological and mineralogical conditions in the soil, amounts of Cd absorbed by plants tend to increase as the concentration of Cd in the soil increases (Valadares *et al.*, 1983; Chang *et al.*, 1983). However, the increased Cd concentration in plants may or may not be proportional to the increased concentration in the soil. Also repeated annual applications of fertilizers or soil conditioners which contain Cd (e.g. municipal sewage sludges), even though they increase the concentration of Cd in the soil each subsequent year, may or may not increase contents of Cd in the crop from year to year. Repeated applications alter the soil conditions and in turn the plant availability of Cd. Perhaps the factors which control the availability of Cd to the crop are not necessarily the same each year. Additional information on the availability of Cd to crops following repeated applications of Cd to soil can be obtained from Chang *et al.* (1982) and the Council for Agricultural Science and Technology (1980). These reports demonstrate that the magnitude of increased concentrations of Cd in plants is not always consistent with cumulative amounts applied to soil, and may depend upon a wide variety of environmental factors.

Although only limited data are available, it appears that crops grown on soil naturally elevated in Cd show higher concentrations of Cd than those grown on similar soils whose Cd concentrations are low. Feeney *et al.* (1984) reported results from a national survey of metals content of vegetable species. In the region where soils were naturally elevated in Cd (Salinas Valley, California) mean concentrations of Cd for lettuce, spinach, carrot, and tomato were, without exception, greater than mean concentrations for the same crops grown on soils with more or less normal Cd concentrations (Table 10.6). The Cd concentrations for the soils in the region designated as naturally elevated are not uniformly high throughout, and vary quite substantially from what would be considered normal (Table 10.3). This natural variation among the soils within the region accounts for the wide range of Cd concentrations observed for the vegetable species sampled. Studies by Lund *et al.* (1981) and our laboratory on native vegetation and crops grown on soils naturally high in Cd also show elevated Cd concentrations in plants when compared with the same plants grown on soils with normal levels of Cd. The data demonstrate that background levels for crops will vary with, in addition to pH and possibly other soil properties, the level of Cd in soil.

Both field and greenhouse experiments have been conducted to evaluate the effect of adding Cd to soil on its concentration in plants. Plants grown on Cd-enriched soils in containers in the greenhouse tend to absorb more Cd than the same plants grown on the same soil amended with identical amounts of Cd in the field (Page and Chang, 1978; DeVries and Tiller, 1978). The extension of the field plant roots beyond the contaminated layer results in less Cd absorption by roots and subsequently less accumulation in the plant tissues (Page *et al.*, 1981).

Results derived from field experiments conducted by Giordano *et al.* (1979), illustrate the relationship between the increased Cd added to soils

Table 10.6 Cadmium concentrations of vegetable crops grown in a region naturally elevated in Cd and other regions*

Crop/Region	No. of samples	Conc. in soil (mg/kg)† Range	Conc. in crop (mg/kg)‡ Range	Conc. in crop (mg/kg)‡ Mean
Lettuce				
Naturally elevated	13	0.33–5.98	0.52–24	4.63
Other	12	<0.33–1.38	0.48–1.54	0.88
Spinach				
Naturally elevated	5	0.26–1.26	2.38–3.80	3.09
Other	12	<0.33	0.38–1.98	0.85
Carrot				
Naturally elevated	8	0.16–5.97	0.39–2.79	0.98
Other	15	0.50–1.10	0.04–0.39	0.25
Tomato				
Naturally elevated	10	0.60–8.42	0.46–4.12	2.57
Other	8	NR	0.09–0.78	0.35

* Derived from Feeney *et al.* (1984).
† 0.1 *N* HCl soluble Cd, dry weight basis.
‡ Dry weight basis.
NR = Not reported.

and the corresponding Cd concentration in a variety of vegetables (Table 10.7). Increases in concentration of Cd in foliage and the edible part of all crops tested, except potato, were observed upon addition of Cd to soil at a rate equivalent to 11.2 kg/ha. Even at considerably lower Cd inputs, vegetable crops grown on soils contaminated with Cd experienced significant rises in plant tissue Cd. Four years of data from one of the authors' field experiments are tabulated as an example (Table 10.8). Although the range of plant tissue Cd concentrations for different levels of Cd treatment appeared to overlap one another, statistical analysis showed an increase of Cd in each of the five vegetables tested (carrot, lettuce, radish, Swiss chard and turnip) with increased levels of Cd inputs at $p<0.05$. There are also significant effects due to years of Cd addition and interaction between Cd treatment and years. But Cd in plant tissue did not always increase with years of Cd addition. These effects reflect the influence of plant tissue Cd concentrations by secondary factors (such as climatic conditions, soil pH changes, length of growing season, etc.) which varied from year to year and could not be controlled during the experiment. This set of data demonstrates that Cd content of plant tissue will always be elevated by introducing Cd into the soil. For a given soil, the level of elevation in plant tissue Cd, however, will depend on the level of Cd input. Due to the influence of other factors on plant growth,

Table 10.7 Influence of cadmium applications to soils on the cadmium concentrations of various vegetable crops*

		Cd concentration (μg/gm)† Application rate‡	
Crop	Tissue	0 kg/ha	11.2 kg/ha
Lettuce	Edible part	0.83	4.44
Broccoli	Edible part	0.27	0.89
Eggplant	Foliage	0.81	2.02
	Edible part	0.54	1.64
Tomato	Foliage	1.11	3.61
	Edible part	0.52	1.04
Potato	Foliage	0.80	0.69
	Edible part	0.11	0.10
Corn	Foliage	0.29	16.30
	Edible part	0.10	1.83
Squash	Foliage	0.15	1.40
	Edible part	0.15	0.27
Pepper	Foliage	0.90	7.51
	Edible part	0.24	1.08
Bean	Foliage	0.16	0.72
	Edible part	0.07	0.21
Cabbage	Edible part	0.18	0.27
Carrot	Edible part	0.84	1.77
Cantaloupe	Edible part	0.21	0.63

* Derived from data published by Giordano *et al.* (1979).
† Dry weight basis.
‡ Cadmium was applied in the form of municipal sewage sludge.

the actual Cd content of plants grown in Cd-contaminated soils may vary from crop to crop and from season to season. Additional data on the accumulation of Cd in crops in relation to amounts applied in the form of sewage sludge are presented in a review by Sommers (1982).

The oxidation-reduction potential of soil also has a marked influence on the concentration of Cd absorbed from soil by plants. This is so because under reducing conditions Cd is substantially less soluble in soil solutions than it is under oxidizing conditions. Although the redox potential in soil influences Cd bioavailability, it is only important in the culture of rice because all other agricultural crops will not grow under reducing conditions. Bingham *et al.* (1976) have shown that the concentration of Cd in rice grown under reducing conditions (flooded) is less than that of rice grown under oxidizing

Table 10.8 Cadmium contents (d.w.) of selected vegetables grown on composted sludge-treated soils

	Cadmium contents (μg/gm)				
Treatment	Carrot	Lettuce	Radish	Swiss chard	Turnip
Control					
1976	0.44	0.60	0.13	0.24	0.11
1977	0.10	0.65	0.17	0.20	0.16
1978	0.60	0.40	0.12	0.23	0.10
1979	0.60	0.61	0.25	0.27	0.20
*1.2 kg Cd/ha/yr**					
1976	0.91	1.25	0.16	0.56	0.17
1977	0.91	1.27	0.32	0.55	0.27
1978	0.60	0.90	0.38	0.55	0.20
1979	1.40	1.00	0.68	0.90	0.30
*2.4 kg Cd/ha/yr**					
1976	1.51	2.39	0.27	0.66	0.12
1977	1.64	3.05	0.46	1.35	0.27
1978	0.80	1.25	0.56	1.65	0.30
1979	1.10	2.80	0.78	1.91	0.60

* Cd input from application of composted municipal sludge.

conditions (dry land). Similar observations have been made by Takijima *et al.*, (1973).

Although other soil factors have been reported to influence the absorption of Cd by agricultural crops, their effects are not nearly as consistent and clear-cut as those of soil pH, Cd content of soil, and redox potential. These factors include: (a) cation exchange capacity (Haghiri, 1974; Stenström and Lonsjo, 1974; Mahler *et al.*, 1978); (b) content of oxides of iron and manganese in soil (Forbes *et al.*, 1976; Chaney and Hornick, 1978; Corey *et al.*, 1981); (c) interactive effects of zinc, copper, nickel, manganese, selenium, and phosphorus (Lagerwerff and Biersdorf, 1971; John *et al.*, 1972; Francis and Rush, 1974; Haghiri, 1974; Cunningham *et al.*, 1975; Iwai *et al.*, 1975; McLean, 1976; Williams and David, 1977; Chaney and Hornick, 1978; Mitchell *et al.*, 1978; Street *et al.*, 1978; Valadares *et al.*, 1983); (d) soil temperature (Haghiri 1974; Giordano *et al.*, 1979). Effects of these soil factors are beyond the scope of this chapter. Details, however, can be obtained from the references cited.

CONCENTRATIONS OF CADMIUM IN FOOD CHAIN CROPS

Cadmium is a naturally occurring element present in all soils in at least trace quantities. For this reason, all food chain crops contain at least trace amounts of Cd. The concentrations of Cd in plants vary among species and cultivars. Different plant parts (leaves, stems, fruit, roots) accumulate different amounts of Cd (Table 10.9). The concentration of cadmium in the leaves and seeds of pea, wheat and corn (Table 10.10) are representative of results reported in the literature. The concentration of Cd in a particular plant part is also influenced by its physiological state of development (Chang *et al.*, 1984).

In the past decade numerous reports on the concentrations of Cd in agricultural crops grown in the United States have appeared in the literature. Most of these reports, however, deal with a limited number of crops grown at only one location. The recent reports of Shacklette (1980), Wolnick *et al.* (1983, 1985) and Feeney *et al.* (1984), however, are exceptions in that they deal with a survey of a variety of crops grown in a number of locations throughout the continental United States. Data reported in these surveys along with those obtained from control plots of various field experiments are presented in Table 10.9. The data pertain to situations where the soils on which the crops were grown were not known to be subject to any major external sources of Cd contamination. They represent more than one cultivar of the same plant species and a wide variety of soil types and climatic conditions. The concentrations of Cd reported for the various crops, therefore, should be representative of natural background levels.

The mean concentrations of the various crops as reported by a number of investigators vary by a factor of about 2 to 10. For example, the range of means among the various investigators for tomatoes, carrots, lettuce, barley, and field corn are, respectively, 0.08 to 0.52; 0.15 to 0.71; 0.42 to 0.88; 0.027 to 0.14 and <0.02 to 0.15 mg/kg (Table 10.9). The variations are most likely due to a combination of differences in substrate Cd concentration, soil properties such as soil pH, and different cultivars of the crop grown. It is worthwhile to single out the data reported by Wolnik *et al.* (1983, 1985). These data are derived from a national survey of the main crop production areas in the United States. The survey was jointly conducted by the US Department of Agriculture, Food and Drug Administration, and the Environmental Protection Agency. Special care was taken to insure that all sites selected were remote from sources of contamination, and care was taken in the handling, packaging and shipping of samples. All samples were analyzed in a specially equipped Food and Drug Administration laboratory. The senior author of this review (A.L. Page) served on the advisory committee for the study, and in his opinion the data are among the most reliable in existence. A summary showing maximum, minimum, mean, and median concentrations for the various crops surveyed is presented in Table 10.11. Generally, the results of others fall within the range reported by Wolnik *et al.* (1983, 1985).

The data presented in Table 10.12 are condensed by crop class (fruits,

Table 10.9 Mean concentrations of cadmium in various crops from locations throughout the United States

Crop	No. of samples	Areas represented (by state(s))	Cd conc. (mg/kg)*	Reference
Fruits				
Grapes				
American	21	MI,NY,WA	0.0110	Shacklette, 1980
European	12	CA,WA	0.0043	Shacklette, 1980
Oranges	20	AZ,CA,FL,TX	0.0050	Shacklette, 1980
Grapefruit	23	AZ,CA,FL,TX	0.0061	Shacklette, 1980
Apples	36	MI,NY,NJ,WA,CO	0.0340	Shacklette, 1980
Peaches	24	CA,CO,NY,WA	0.0110	Shacklette, 1980
Pears	38	CA,CO,MI,NY,WA	0.0057	Shacklette, 1980
Plums	16	CO,MI,NY,WA	0.0058	Shacklette, 1980
Vegetable fruits				
Peppers	2	MI	0.021	Shacklette, 1980
	–	AL	0.250	Giordano *et al.*, 1979
Cucumber	22	CA,MI,NY	0.093	Shacklette, 1980
Tomatoes	48	CA,FL,MI,WA	0.110	Shacklette, 1980
	8	CA†,FL,GA,IL,IN,MI,TN	0.350	Feeney *et al.*, 1984
	3	MN	0.080	Dowdy and Larson, 1975
	–	AL	0.520	Giordano *et al.*, 1979
	231	CA,FL,NY,OH,PA,TX	0.270	Wolnik, *et al.*, 1985
Squash	1	AL	0.150	Giordano *et al.*, 1979
Cantaloupe	2	AL	0.210	Giordano *et al.*, 1979
Eggplant	1	AL	0.540	Giordano *et al.*, 1979
	2	MI	0.380	Shacklette, 1980
Seed vegetables				
Dry beans	35	CA,CO,ID,NY	0.110	Shacklette, 1980
Snap beans	42	ID,FL,MI,NY,NJ	0.024	Shacklette, 1980
Green beans	9	CA,FL,IL,IN,LA,NC,TX	0.130	Feeney *et al.*, 1984
Peas	1	MN	<0.030	Dowdy and Larson, 1975

Sweet corn	40	ID,FL,MI,NJ	0.026	Shacklette, 1980
	268	MD,MN,NY,OR,WA,WI	0.016	Wolnik *et al.*, 1983
Root and bulb crops				
Carrots	20	CA,TX	0.150	Shacklette, 1980
	2	AL	0.710	Giordano *et al.*, 1979
	207	FL,MI,NY,TX,WA,WI	0.250	Wolnik *et al.*, 1985
	15	CA†,AZ,CO,FL,MI,TX	0.250	Feeney *et al.*, 1984
	6	CA	0.460	Authors' unpublished data
Radish	17	CA	0.310	Authors' unpublished data
Potatoes	40	ID,NY,NJ,WA	0.029	Shacklette, 1980
	1	AL	0.110	Giordano *et al.*, 1979
	297	AL,CA,CO,ID,ME,NC,NY, OR,TX,WA,WI,FL	0.170	Wolnik, *et al.*, 1983
Turnip	5	CA	0.170	Author's unpublished data
Onion	10	TX	0.050	Shacklette, 1980
	228	CA,CO,MI,NM,NY,OR,TX	0.100	Wolnik *et al.*, 1983
Peanuts	320	AL,GA,NC,TX	0.090	Wolnik *et al.*, 1983
Asparagus	10	CA	0.032	Shacklette, 1980
Leaf vegetables				
Cabbage	24	AR,MI,NJ,TX	0.093	Shacklette, 1980
	2	AL	0.160	Giordano *et al.*, 1979
Lettuce	40	CA,FL,NJ,TX	0.420	Shacklette, 1980
	9	MD	0.830	CAST, 1980
	8	AL	0.830	Giordano *et al.*, 1979
	150	AZ,CA,FL,NY,TX,WI	0.620	Wolnik *et al.*, 1983
	12	CA†,IL,MO,NJ,OR,PA,TX	0.880	Feeney *et al.*, 1984
	8	CA	0.570	Author's unpublished data
Spinach	104	CA,DE,MD,NJ,TX	0.800	Wolnik *et al.*, 1985
	12	CA†,IL,MO,NJ,OR,PA,TX	0.850	Feeney *et al.*, 1984
Swiss chard	16	CA	0.470	Authors' unpublished data
	6	MD	0.580	CAST, 1980
	7	CA,IL,NJ	0.350	Feeney *et al.*, 1984
Broccoli	1	AL	0.270	Giordano *et al.*, 1979

Table 10.9 Continued

Field crops				
Wheat	9	CA	0.055	Hyde *et al.*, 1979
	288	CO,ID,IL,KS,MT,ND, NE,OK,SD,TX,WA	0.048	Wolnik *et al.*, 1983
Barley	37	CA	0.055	Vlamis *et al.*, 1978
	29	CA	0.027	Chang *et al.*, 1979
	3	MT	0.080	Munshower, 1977
	75	AL,AZ,CA,CO,FL,IL,IN, MD,MI,MN,OH,OR,WI	0.140	‡CSRS Regional Report, 1984
Rice	166	AZ,CA,LA,TX	0.013	Wolnik *et al.*, 1985
Field corn	9	CA	0.020	Hyde *et al.*, 1979
	16	IL	0.080	CAST, 1980
	7	IL	0.150	Hinesly *et al.*, 1979
	3	MN	<0.020	Dowdy and Larson, 1975
	10	MD	0.090	CAST, 1980
	277	GA,IA,IL,IN,MN,MO, NC,NE,OH,WI	0.014	Wolnik *et al.*, 1985
Oat	6	MD	0.090	Chaney and Hornick, 1977
	6	MD	0.060	CAST, 1980
Sorghum	36	CA	0.033	Chang *et al.*, 1979
Soybeans	6	MD	0.021	Chaney and Hornick, 1977
	322	AR,GA,IA,IL,IN,LA, MN,NC	0.064	Wolnik *et al.*, 1983

AL=Alabama; AR=Arkansas; AZ=Arizona; CA=California; CO=Colorado; DE=Delaware; FL=Florida; GA=Georgia; IA=Iowa; ID=Idaho; IL=Illinois; IN=Indiana; KS=Kansas; LA=Louisiana; MD=Maryland; ME=Maine; MI=Michigan; MN=Minnesota; MO=Missouri; MT=Montana; NC=North Carolina; ND=North Dakota; NE=Nebraska; NJ=New Jersey; NM=New Mexico; NY=New York; OH=Ohio; OK=Oklahoma; OR=Oregon; PA=Pennsylvania; SD=South Dakota; TN=Tennessee; TX=Texas; WA=Washington; WI=Wisconsin.

* Oven dry basis (~70° C).
† Excluding samples from the Salinas Valley.
‡ Unpublished report of the CSRS Regional Project W-124. A.L.Page, University of California Riverside, and Lee Sommers, Purdue University, West Lafayette, Indiana, co-chairpersons.

Table 10.10 Cadmium concentrations of vegetative and reproductive tissues from plants grown on sewage sludge-amended soils in relation to species, application rate and soil pH

Cd added from sludge (kg/ha)	Soil pH	Cadmium concentrations (mg/kg)*					
		Pea		Wheat		Corn	
		Leaves	Seeds	Leaves	Grain	Leaves	Grain
0	4.4	0.94	0.20	0.19	0.06	0.11	0.03
2.2	4.6	4.40	0.64	0.82	0.25	1.44	0.06
4.5	5.0	3.80	1.11	1.10	0.32	1.83	0.08
9.0	5.2	6.05	1.45	0.89	0.58	3.37	0.09
0	7.8	0.43	0.14	0.14	0.04	0.07	0.01
2.2	7.6	0.43	0.17	0.18	0.12	0.14	0.02
4.5	7.4	0.42	0.18	0.25	0.23	0.41	0.03
9.0	7.3	0.48	0.30	0.33	0.34	0.92	0.03

* Dry weight basis.

Table 10.11 Cadmium concentrations of various crops from major producing regions throughout the United States*

Crop	No. of samples	Regions sampled†	Cadmium concentration (mg/kg)‡			
			Minimum	Maximum	Median	Mean
Wheat	288	NC,S,W	<0.001	0.22	0.036	0.048
Rice	166	S,W	0.001	0.25	0.005	0.013
Field corn	277	NC,S	<0.001	0.35	0.004	0.014
Sweet corn	268	NC,NE,W	<0.001	0.23	0.008	0.016
Soybeans	322	NC,S	0.002	1.20	0.045	0.064
Peanuts	320	S	0.011	0.66	0.068	0.090
Potatoes	297	NC,NE,S,W	0.009	1.00	0.140	0.170
Onions	228	NC,NE,S,W	0.011	0.34	0.090	0.100
Carrots	207	NC,NE,S,W	0.015	1.20	0.160	0.250
Tomatoes	231	NC,NE,S,W	0.045	0.79	0.220	0.270
Spinach	104	NE,S,W	0.160	1.90	0.800	0.830
Lettuce	150	NC,NE,S,W	0.034	3.80	0.440	0.620

* Derived from Wolnik *et al.* (1983, 1985).
† NC=North Central; NE=Northeast; S=Southern; W=Western; (see Figure 10.1).
‡ Dry weight basis.

vegetables, field crops). This data show that on a dry weight basis fruits contain the lowest ($\bar{x}$ = 0.005 mg/kg) and leafy vegetables contain the highest concentrations of cadmium ($\bar{x}$ = 0.560 mg/kg). Seed crops (bean, sweet corn) and grains on the average, show mean concentrations of cadmium of approximately 0.03 mg/kg; root/tuber (carrots, potatoes, etc.) and fruity

Table 10.12 Concentrations of cadmium in crops and range of mean concentration by crop class*

Crop class	No. of samples	Cadmium concentration (mg/kg) Minimum	Maximum	Mean
Fruits	190	0.0043	0.012	0.005
Vegetables				
Seed	394	0.016	0.13	0.028
Root/Bulb†	878	0.029	0.71	0.208
Fruity	322	0.021	0.54	0.237
Leafy‡	297	0.093	0.88	0.560
Field Crops				
Grains	1302	0.014	0.21	0.047

* Derived from data presented in Table 10.9.
† Excluding peanuts.
‡ Excluding asparagus.

(tomatoes, cucumbers, etc.) vegetables show mean concentrations of 0.21 and 0.24 mg/kg, respectively.

SUMMARY

Concentrations of cadmium in soils from major agricultural regions throughout the United States not subject to external sources of contamination vary from about 0.1 to 1.0 mg/kg, with a median concentration of about 0.3 mg/kg. Soils from the Northeastern and Southeastern United States, on average, contain less total cadmium than do soils from the North Central and Western regions. In terms of taxonomic order classification, Histosols show the highest total cadmium concentrations; Aridisols, Entisols, Mollisols, Inceptisols, and Vertisols are intermediate; while Spodosols, Alfisols, and particularly Ultisols contain the lowest total cadmium concentrations. The lower concentrations of cadmium in the Spodosols, Alfisols and Ultisols are probably due to the more intense leaching these soil orders experience. Exceptions to the above generalizations are soils developed from parent materials unusually high in cadmium. Concentrations as high as 22 mg Cd/kg have been observed for soils developed from shales along the Pacific Coast.

Mining, smelting, and sintering of non-ferrous metals have resulted in cadmium contamination of nearby soils. In regions near these types of industrial operations, surface soils with concentrations in excess of 24 mg Cd/kg have been frequently reported. The contamination is restricted to the surface horizons; subsurface soils usually show concentrations representative of background levels for the region.

The major soil factors which influence the accumulation of cadmium by plants are soil pH and cadmium concentration. All other soil factors being equal, the accumulation of cadmium by crops increases as the pH of the soil decreases and as the concentration of cadmium in soil increases. Although fertilization and atmospheric deposition are sources of cadmium, amounts added to rural agricultural soils from these sources are small and amount to a few grams per hectare per year or less.

Concentrations of cadmium in crops vary quite substantially among species and even among cultivars of the same species. Different parts of the plant (leaves, stems, fruit, roots) accumulate different amounts of cadmium. For the above-ground parts, leaves usually contain greater concentrations of cadmium than the flowering and fruiting parts of the plant. On a dry weight basis, fruits (e.g. grapes, citrus, apples, peaches, etc.) contain the lowest concentrations of cadmium ($\bar{x} = 0.005$), and leafy vegetables (lettuce, spinach, etc.) the highest total concentration ($\bar{x} = 0.560$). Seed crops (bean, sweet corn) and grains on the average, show mean concentrations of cadmium of approximately 0.03 mg/kg; root/tuber (carrots, potatoes, etc.) and vegetable fruits (tomatoes, cucumbers, etc.) show mean concentrations of 0.21 and 0.24 mg/kg, respectively. The efficacy with which leafy crops take up Cd and transport it to the foliage suggests caution and monitoring of sewage sludge additions to land on which these crops are or might be grown.

The concentrations of crops grown in the United States vary within and among crop species. Within any particular plant species the variation in concentration of crops grown at one location to those of the same crop given another location can be as much as a factor of 600; factors ranging from 50 to 300 are common. For example, the concentration of cadmium in wheat was observed to vary from a minimum of 0.001 mg Cd/kg when grown at one location to a maximum of 0.22 mg Cd/kg when grown at another. Similarly, from one location to another the concentrations of cadmium for carrots and potatoes varied from 0.015 to 1.2 and 0.009 to 1.0 mg/kg, respectively.

Third World countries, as compared with the Western world, are less likely to be subjected to a Cd-contaminated environment. The main source of Cd human burden for those populations is food. Since the main source of Cd human burden for those populations is food, and the per capita dietary consumption of grains is high, the impact of Cd enrichment may be proportionately magnified. Food crops imported from or grown in areas high in Cd may result in a greater impact. Other factors such as the less stringent emission control and waste disposal measures for the developing industries contribute to the potential environmental hazards of Cd.

REFERENCES

Auer, C. (1977). *Cadmium in Phosphate Fertilizer Production.* US Environmental Protection Agency, Midwest Research Institute Report, Washington, DC.

Bingham, F. T., Page, A. L., Mahler, R. J., and Ganje, T. J. (1976). Cadmium availability to rice in sludge-amended soil under 'flood' and 'nonflood' culture. *Soil Sci. Soc. Am. J.*, **40**, 715–719.

Buchauer, M. J. (1973). Contamination of soil and vegetation near a zinc smelter by zinc, cadmium, copper and lead. *Environ. Sci. Technol.*, **7**, 131–135.

Burau, R. G., Kaita, K. Y., Inouye, T. S., and Miller, M. (1973). *Chemical Analysis of Soil Samples from the Salinis Valley, California for Cadmium, Zinc, and Phosphate.* Report to State Water Resources Control Board, University of California, Davis.

CAST (Council for Agricultural Science and Technology) (1976). *Application of Sewage Sludge to Cropland: Appraisal of Potential Hazards of the Heavy Metals to Plants and Animals.* Council for Agricultural Science and Technology, No. 64, Ames, Iowa, 63 pages.

CAST (Council for Agricultural Science and Technology) (1980). *Effects of Sewage Sludge on the Cadmium and Zinc Content of Crops.* Council for Agricultural Science and Technology, No. 83, Ames, Iowa, 77 pages.

Chaney, R. L., White, M. C., and Simon, P. W. (1975). Plant uptake of heavy metals from sewage sludge applied to land. In *Proceedings of Second National Conference on Municipal Sludge Management.* p. 169, Information Transfer, Inc., Rockville, MD.

Chaney, R. L., and Hornick, S. B. (1978). Accumulation and effects of cadmium on crops. In *Proceedings of First International Cadmium Conference*, p. 125, San Francisco, Metal Bulletin Ltd, London.

Chang, A. C., Page, A. L., Lund, L. J., Pratt, P. F., and Bradford, G. R. (1979). Land application of sewage sludges—a field demonstration. Final report, Regional Wastewater Solids Management Program, Los Angeles/Orange County Metropolitan Area (LA/OMA Project), 311 pages.

Chang A. C., Page, A. L., Warneke, J. E., and Johanson, J. B. (1982). Effects of sludge application on the Cd, Pb and Zn levels of selected vegetable plants. *Hilgardia*, **50**, 1–14.

Chang, A. C., Page, A. L., Warneke, J. E., Resketo, M. R., and Jones T. E. (1983). Accumulation of cadmium and zinc in barley grown in sludge-treated soils: A long-term field study. *J. Environ. Qual.* **12**, 391–397.

Chang, A. C., Warneke, J. E., Page A. L., and Lund L. J. (1984). Accumulation of heavy metals in sewage sludge-treated soils. *J. Environ. Qual.* **13**, 87–91.

Corey, R. B., Fujii, R., and Hendrickson, L. L. (1981). Bioavailability of heavy metals in soil-sludge. In *Conference on Application Research Practices of Municipal and Industrial Waste*, p. 449, University of Wisconsin Extension, Madison.

Cunningham, J. D., Keeney, D. R., and Ryan, J. A. (1975). Yield and metal composition of rye grown on sewage sludge-amended soil. *J. Environ. Qual.*, **4**, 448–454.

DeVries, M. P. C., and Tiller, K. G. (1978). Sewage sludge as a soil amendment with special reference to Cd, Mn, Ni, Pb and Zn comparisons of results from experiments conducted inside and outside a glasshouse. *Environ. Pollut.*, **16**, 231–240.

Dowdy, R. H., and Larson, W. E. (1975). The availability of sludge-borne metals to various vegetable crops. *J. Environ. Qual.*, **4**, 278–282.

Feeney, S., Peterson, J. R., Zenz, D. R., and Lue-Hing, C. (1984). *National Survey of the Metals Content of Seven Vegetable Species.* Dept. of Research and Development—Report No. 84-4 Metropolitan Sanitary District of Greater Chicago, Chicago, Illinois.

Fleischer, M., Sarofim, A. F., Fassett, D. W., Hammond, P., Shacklette, H. T., Nisbet,

I. C., and Epstein, S. (1974). Environmental impact of cadmium; A review by the panel on hazardous trace substances. *Environ. Health Persp.*, **7**, 253–323.

Forbes, E. A., Posner, A. M., and Quirk, J. P. (1976). The specific adsorption of divalent Cd, Co, Cu, Pb and Zn on geothite. *Soil Sci.*, **27**, 154–160.

Francis, C. W. and Rush, S. G. (1974). Factors affecting uptake and distribution of cadmium in plants. In Hemphill, D. D. (Ed.), *Trace Substances in Environmental Health VII*, p. 75, University of Missouri Press, Columbia.

Friberg, L., Piscator, M., Nordberg, G., and Kjellström, T. (1974). *Cadmium in the Environment.* 2nd ed. CRC Press, Cleveland, Ohio, 248 pages.

Fulkerson, W., and Goeller, H. E. (Eds) (1973). *Cadmium, the Dissipated Element.* Oak Ridge National Laboratory, Oak Ridge, Tennessee, ORNL/NSF/EP-21.

Giordano, P. M., Mays, D. A., and Behel, A. D-Jr. (1979). Soil temperature effects on the uptake of cadmium and zinc by vegetables grown on sludge-amended soil. *J. Environ. Qual.*, **8**, 232–236.

Goldschmidt, V. M. (1958). *Geochemistry.* Oxford University Press, London, 730 pages.

Gong, H., Rose, A. W., and Suhr, N. H. (1977). The geochemistry of cadmium in some sedimentary rocks. *Geochim. Cosmochim. Acta.*, **41**, 1687.

Haghiri, F. (1974). Plant uptake of cadmium as influenced by cation exchange capacity, organic matter, zinc and soil temperature. *J. Environ. Qual.*, **3**, 180–183.

Heinrichs, H., Schulz-Dobrick, B., and Wedepohl, K. H. (1980). Terrestrial Geochemistry of Cd, Bi, Ti, Pb, Zn and Rb. *Geochim. Cosmochim. Acta.*, **44**, 1519–1532.

Hinesly, T. D., Ziegler, E. L., and Barrett, G. L. (1979). Residual effects of irrigating corn with digested sewage sludge. *J. Environ. Qual.*, **8**, 35–38.

Holmgren, G. G. S., Meyer, M. W., Daniels, R. B., Kubota, J., and Chaney, R. L. (1986). Cadmium, lead, zinc, copper, and nickel in agricultural soils in the United States. *J. Environ. Qual.*, **16** (in press).

Horn, M. K., and Adams, J. A. S. (1966). Computer-derived geochemical balances and element abundances. *Geochim. Cosmochim. Acta.*, **30**, 279.

Hunt, W. F., Pinkerton, C., McNulty, O., and Creason, J. A. (1971). A study of trace element pollution of air in 77 midwestern cities. In Hemphill, D. D. (Ed.), *Trace Substances in Environmental Health IV*, p. 56, University of Missouri, Columbia.

Hyde, H. C., Page, A. L., Bingham, F. T., and Mahler, R. J. (1979). Effect of heavy metals in sludge on agricultural crops. *J. Water Pollut. Control Fed.*, **51**, 2475–2486.

Iwai, I., Hara, T., and Sonoda, Y. (1975). Factors affecting cadmium uptake by corn plant. *Soil Sci. Plant Nutr.*, **21**, 37–46.

John, M. K., Van Laerhoven, C. J., and Chuah, H. H. (1972). Factors affecting plant uptake and phytotoxicity of cadmium added to soils. *Environ. Sci. Technol.*, **6**, 1005–1009.

Johnson, R. D., Miller, R. J., Williams, R. E., Wai, C. M., Wies, A. C., and Mitchell, J. E. (1975). The heavy metal problem of Silver Valley, Northern Idaho. In Hutchinson, T. C. (Ed.), *Proceedings of International Conference on Heavy Metals in the Environment.* Vol. II, Part 2, p. 465, University of Toronto, Toronto, Canada.

Kjellström, T., and Nordberg, G. F. (1978). A kinetic model of cadmium metabolism in human beings. *Environ. Res.*, **16**, 248–269.

Klein, D. H. (1972). Mercury and other metals in urban soils. *Env. Sci. Technol.*, **6**, 558–559.

Kowal, N. E. (1984). An overview of public health effects. In Page, A. L., Gleason, T. L., Smith, J. E., Iskandar, I. K., and Sommers, L. E. (Eds), *Utilization of Municipal Wastewater and Sludge on Land*, p. 329. University of California Press, Riverside, California.

Lagerwerff, J. V., and Biersdorf, G. T. (1971). Interactions of zinc with uptake and translocation of cadmium in radish. In Hemphill, D. D. (Ed.), *Trace Substances in Environmental Health V*, p. 515, University of Missouri Press, Columbia, Missouri.

Lagerwerff, J. V., and Brower, D. L. (1975). Source determination of heavy metal contaminants in the soil of a mine and smelter area. In Hemphill, D. D. (Ed.), *Trace Substances in Environmental Health IX*, p. 207, University of Missouri Press, Columbia, Missouri.

Lindberg, S. E., Turner, R. R., Shriner, D. S., and Huff, D. D. (1981). Atmospheric deposition of heavy metals and their interaction with acid precipitation in a North American deciduous forest. In *Proceedings of International Conference on Heavy Metals in the Environment*, p. 306, CEP Consultants Ltd, Edinburgh, U. K.

Logan, T. J., and Miller, R. H. (1983). *Background Levels of Heavy Metals in Ohio Farm Soils*. Research Circular 275. The Ohio State University Agricultural Research and Development, Wooster, Ohio.

Logan, T. J., and Chaney, R. (1984). Metals. In Page, A. L., Gleason, T. L., Smith, J. E., Iskandar, I. K., and Sommers, L. E. (Eds), *Utilization of Municipal Wastewater and Sludge on Land*, p. 235, University of California Press, Riverside, California.

Lund, L. J., Betty, E. E., Page, A. L., and Elliott, R. A. (1981). Occurrence of high Cd levels in soil and its accumulation by vegetation. *J. Environ. Qual.*, **10**, 551–556.

Lyon, W. S. (1977). *Trace Element Measurements at the Coal-fired Power Plant*. CRC Press, Cleveland, OH, 136 pages.

Mahaffey, K. R., Corneliussen, P. E., Jelinek, C. F., and Fiorino, J. A. (1975). Heavy metal exposure from foods. *Environ. Health Perspec.*, **12**, 63–69.

Mahler, R. J., Bingham, F. T., and Page, A. L. (1978). Cadmium-enriched sewage sludge application to acid and calcareous soils: I. Effect on yield and cadmium uptake by lettuce and Swiss chard. *J. Environ. Qual.*, **7**, 274–281.

Marowsky, G., and Wedepohl, K. H. (1971). General trends in the behavior of Cd, Hg, Ti, and Bi in some major rock forming processes. *Geochim. Cosmochim. Acta.*, **35**, 1255.

McLean, A. J. (1976). Cadmium in different plant species and its availability in soils as influenced by organic matter and additions of lime, P, Cd and Zn. *Canadian J. Soil Sci.*, **56**, 129–138.

Mitchell, G. A., Bingham, F. T., Page, A. L., and Nash, P. (1978). Yield and metal composition of lettuce and wheat grown on soils amended with sewage sludge enriched with Cd, Cu, Ni and Zn. *J. Environ. Qual.*, **7**, 165–171.

Mullin, J. B., and Riley, J. P. (1956). The occurrence of cadmium in sea water and in marine organisms and sediments. *J. Marine Res.*, **15**, 103.

Munshower, F. F. (1977). Cadmium accumulation in plants and animals of polluted and nonpolluted grasslands. *J. Environ. Qual.*, **6**, 411–413.

Ondov, J. M., Regaini, R. C., Heft, R. E., Fisher, G. L., Silberman, D., and Prentice, B. A. (1976). Interlaboratory comparison of neutron activation and atomic absorption analyses of size-classified stack fly ash. *Eighth Materials Research Symposium on Methods and Standards for Environmental Measurement*. Gaithersberg, MD.

Page, A. L. (1974). *Fate and Effects of Trace Elements in Sewage Sludge When Applied to Agricultural Lands.* U. S. Environmental Protection Agency Special Publication EPA 670/2-74-005.

Page, A. L., and Chang, A. C. (1978). Trace elements impact on plants during cropland disposal of sewage sludges. In *Proceedings of Fifth National Conference on Acceptable Sludge Disposal Techniques*, p. 91, Information Transfer Inc., Rockville, Maryland.

Page, A. L., Bingham, F. T., and Chang, A. C. (1981). Cadmium in terrestrial plants. In Lepp, N. W., (Ed.) *Effect of Heavy Metal Pollution on Plants.* Vol. 1. *Effects of Trace Metals on Plant Function*, p. 77, Applied Science Publishers, London, U. K.

Pahren, H. R., Lucas, J. B., Ryan, J. A., and Dotson, G. R. (1979). Health risks associated with land application of municipal sludge. *J. Water Pollut. Control Fed.*, **51**, 2588–2601.

Pierce, F. J., Dowdy, R. H., and Grigal, D. F. (1982). Concentrations of six trace metals in some major Minnesota soil series. *J. Environ. Qual.*, **11**, 416–422.

Ragaini, R. C., Ralston, H. R., and Roberts, N. (1977). Environmental trace metal contamination in Kellogg, Idaho near a lead smelting complex. *Environ. Sci. Technol.*, **11**, 773–781.

Rupp, E. M., Parzyck, D. C., Walsh, P. J., Booth, R. S., Raridon R. J., and Whitfield, B. L. (1978). Composite hazard index for assessing limiting exposures to environmental pollutants: Application through a case study. *Environ. Sci. Technol.*, **12**, 802–807.

Ryan, J. A., Pahren, H. R., and Lucas, J. B. (1982). Controlling cadmium in the human food chain: A review and rationale based on health effects. *Environ. Res.*, **28**, 251–302.

Shacklette, H. T. (1980). *Elements in Fruits and Vegetables from Areas of Commercial Production in the Conterminous United States.* Geological Survey Professional Paper No.1178. US Government Printing Office, Washington, DC. 149 pages.

Sommers, L. E. (1977). Chemical composition of sewage sludges and analysis of their potential use as fertilizers. *J. Environ. Qual.*, **6**, 225–232.

Sommers, L. E. (1982). Toxic metals in agricultural crops. In Bitton, G., Damron, B. L., Edds, G. T., and Davidson, J. M. (Eds), *Sludge-Health Risks of Land Application*, p. 105, Ann Arbor Science Publishers/The Butterworth Group, Ann Arbor, Michigan.

Stenström, T., and Lonsjo, H. (1974). Cadmium availability to wheat: a study with radioactive tracers under field conditions. *Ambio*, **3**, 87–90.

Street, J. J., Sabey, B. R., and Lindsay, W. L. (1978). Influence of pH, phosphorus, cadmium, sewage sludge, and incubation time on the solubility and plant uptake of cadmium. *J. Environ. Qual.*, **7**, 286–290.

Takijima, Y., Katsumi, F., and Takezawa, K. (1973). Cadmium contamination of soils and rice plants caused by zinc mining. II. Soil conditions of contaminated paddy fields which influence heavy metal contents of rice. *Soil Sci. Plant Nutr.*, **19**, 173–182.

Tsuchiya, K. (Ed). (1978). *Cadmium Studies in Japan–A Review.* Elsevier/North Holland Biomedical Press, Amsterdam.

US Environmental Protection Agency (1972). *Helena Valley, Montana Area Environmental Pollution Study.* EPA Office of Air Programs Publ. AP-91 Research Triangle Park, North Carolina.

US Environmental Protection Agency. (1978). *Cadmium additions to Agricultural Lands Via Commercial Phosphate Fertilizers—A Preliminary Assessment*, p. 35, Office of Water and Waste Management, Washington DC. Soil Waste Management Series No. SW-718.

Valadares, J. M. A. S., Gal, M., Mingelgrin, U., and Page. A. L. (1983). Some heavy metals in soils treated with sewage sludge, their effects on yield, and their uptake by plants. *J. Environ. Qual.*, **12**, 49–57.
Vlamis, J., Williams, T. E., Fong, K., and Corey, J. E. (1978). Metal uptake by barley from field plots fertilized with sludge. *Soil Sci.*, **126**, 49–55.
Waketa, H., and Schmitt, R. A. (1970). Cadmium. In Wedepohl, K. H. (Ed.), *Handbook of Geochemistry*. Springer-Verlag, New York, Chapter 48.
Wedepohl, K. H. (1968). Origin and distribution of elements. In Ahrens, L. H. (Ed). *International Symposium on the Origin and Distribution of the Elements*, p. 999, Pergamon Press, New York.
Williams, C. H., and David, D. J. (1977). Some effects of the distribution of cadmium and phosphate in the root zone on the cadmium content of plants. *Australian J. Soil Res.*, **15**, 59–68.
Wolnik, K. A., Fricke, F. L., Capar, S. G., Braude, G. L., Meyer, M. W., Satzger, R. D., and Bonnin, E. (1983). Elements in major raw agricultural crops in the United States. I. Cadmium and lead in lettuce, peanuts, potatoes, soybeans, sweet corn and wheat. *J. Agric. Food Chem.*, **31**, 1240–1244.
Wolnik, K. A., Fricke, F. L., Capar, S. G., Meyer, M. W., Satzger, R. D., Bonnin, E. and Gaston, C. M. (1985). Elements in major raw agricultural crops in the United States. III. Cadmium, lead and other elements in carrots, field corn, onion, rice, spinach and tomatoes. *J. Agric. Food Chem.*, **33**, 807–811.

Lead, Mercury, Cadmium and Arsenic in the Environment
Edited by T. C. Hutchinson and K. M. Meema

CHAPTER 11

Metal Transfer Across the Air–Sea Interface: Myths and Mysteries

PATRICK BUAT-MÉNARD

*Centre des Faibles Radioactivites,
Laboratoire Mixte CNRS-CEA, PB 1,
91190, Gif sur Yvette, France*

and

ROBERT A. DUCE

*Center for Atmospheric Chemistry Studies,
Graduate School of Oceanography,
University of Rhode Island, Kingston,
RI 02881, USA*

ABSTRACT

Our increasing awareness of the global atmospheric mobilization of heavy metals such as As, Cd, Hg and Pb by human activities has raised the question of the perturbation of the oceanic cycles of these elements. Such a perturbation is now evident for Pb and probably for Sn, and the level of perturbation appears closely related to the atmospheric input rate for these metals. However, there is growing evidence of significant spatial and temporal variability of atmospheric concentrations of heavy metals and their fluxes to the ocean. This variability is a major obstacle to obtaining representative field data.

A number of misconceptions, or 'myths', have developed over the years concerning heavy metals in the atmosphere and their exchange with the oceans. Several of these are evaluated critically in this chapter. For example, the concept of constant background concentrations of metals in air, rain, and dry deposition is certainly erroneous, and recent investigations indicate clearly that collection of precipitation in remote regions for trace metal analysis is not a routine task. Problems include sample contamination during collection and the representativeness of the rain samples collected. Problems with the use of scavenging ratios for estimating wet deposition and pitfalls in field studies of heavy metal dry deposition are also discussed.

Due to the recycling of metals from the ocean into the atmosphere, problems with determining net air/sea exchange rates arise, even if accurate total deposition measurements in remote marine areas can be made.

Several 'mysteries' still prevent us from attaining a comprehensive understanding of air/sea metal exchange, and several of these are discussed as well. For example, are there significant unknown natural or anthropogenic sources for atmospheric metals? What are the physical and chemical forms of metals in the marine atmosphere? What is the fate of atmospheric metals when they enter the ocean?

INTRODUCTION

Over the last 15 years there has been an increasing concern about the levels of toxic metals such as lead (Pb), cadmium (Cd), mercury (Hg) and arsenic (As) in remote marine air and rain. There is more and more evidence that, at least in the northern hemisphere, the atmospheric cycles of such elements are now strongly perturbed by human activities, primarily high-temperature combustion processes (Andreae *et al.*, 1984). How such a geochemical perturbation in the atmosphere interacts with the oceanic cycles of these elements is a matter of crucial interest, since many trace metals are both essential for the development of marine life and potentially toxic for marine life above certain concentration levels. The assessment of the human factor in the present cycling of metals through the atmosphere and oceans is, however, a very difficult exercise since natural sources are still poorly known. Nevertheless, recent years have seen a tremendous increase in our knowledge of the global atmospheric chemistry of trace metals. The use of improved sampling and analytical techniques has enabled us to document reasonably well latitudinal and longitudinal distributions of the atmospheric concentrations of trace metals. The combined use of geochemical and meteorological tracers has enabled considerable insight to be made into the relative contributions of the various natural and man-made sources (Buat-Ménard, 1983).

One major conclusion which now emerges is that the magnitude of the anthropogenic input of metals from the atmosphere to the ocean exhibits a strong spatial and temporal variability. This is primarily because of (a) meteorological factors (air mass circulation patterns) and (b) the fact that, with the exception of mercury, trace metals are found primarily associated with atmospheric particulate matter and, as a consequence, have very short atmospheric residence times (~ one week). Such regional variability is also a characteristic of the oceanic cycle of some of these elements. Elements such as manganese (Mn) and lead have a mean oceanic residence time of the order of 100 years (Bruland, 1983), much shorter than the mixing time of the ocean (1000 years). Thus any reliable assessment of the impact of the atmospheric deposition of trace metals on oceanic chemistry cannot be obtained simply by a consideration of global mass balance. Together with the obvious requirement for data on the chemical speciation of metals in the marine atmosphere (which addresses the problem of their biological availability once they enter the ocean), the need for accurate data on net air-to-sea fluxes is evident for any oceanic region. These data are extremely difficult to obtain because of theoretical, and especially experimental, difficulties.

The theoretical aspects of dry and wet deposition of aerosol particles to the ocean surface will not be reviewed here since this topic has already been treated extensively (e.g. Slinn, 1983). In principle, knowledge of aerosol particle deposition processes combined with the use of atmospheric transport models should allow an evaluation to be made of anthropogenic air-to-sea fluxes to a given oceanic region, provided accurate source-emission inventory data are available. At the present time such an approach is not possible for trace metals and we must rely on experimental field measurements for such an assessment.

In this chapter we will evaluate some of the commonly held myths relative to air-sea exchange of metals, for both wet and dry deposition, and outline some of the current mysteries which make the accurate quantitative determination of air/sea fluxes so difficult.

MYTHS

There are 'Background Concentrations' for Metals in Air, Rain and Dry Deposition

Ten to fifteen years ago it became fashionable to discuss a fairly uniform 'background aerosol', a particle population that was believed to have a rather uniform chemical composition and size distribution and that was relatively constant in space and time in the free troposphere. This led to the concept of background air and rain concentrations for those substances found primarily on particles in the atmosphere. During the last few years a number of rather intensive studies of the chemical composition and size distribution of aerosol particles in the remote marine atmosphere have given us a much more accurate description of the geographical and temporal variability of trace substances in the global troposphere. They have shown clearly that the background marine atmosphere is anything but constant relative to its trace element, and particularly metal, content. For example, over the North Pacific mean weekly differences in the aluminosilicate aerosol concentration can range up to a factor of 100 or more (Uematsu *et al.*, 1983), with similar differences in the rain concentrations. These differences are much larger than the variability due to analytical precision and/or accuracy of the chemical analysis of aerosol particles for trace metals over the ocean. Analytical variability ranges from 10 to perhaps 50%, depending on the metal. Geographical differences in the atmospheric concentration distribution of the metals of interest in this workshop can also span several orders of magnitude over the world's oceans, as is shown for lead in Table 11.1. While there have been relatively few careful studies of the metal content of precipitation over the ocean, the few data that are available (see Table 11.2) confirm that there is no 'background' concentration of metals in rain.

Table 11.1 Lead in rain carefully collected from islands and ships

Location	Pb conc. (ng/kg)	Reference
North Atlantic		
Tower–Florida Keys, 25 °N, 81 °W	770–960	Settle *et al.*, 1982
Ship's bow 34 °N, 66 °W	550	Settle *et al.*, 1982
North Pacific trades		
Tower–Enewetak, 11 °N, 162 °E	6–63	Settle *et al.*, 1982
Ship's bow, 11 °N, 155 °W	43	Settle and Patterson, 1982
South Pacific trades		
Tower–Samoa, 14 °S, 170 °W	7	Settle *et al.*, 1982
Ship's bow, 15 °S, 150 °W	9	Settle *et al.*, 1982
South Pacific westerlies		
Tower–New Zealand, 34 °S, 173 °E	17	Patterson and Settle, unpublished data
Ship's bow, 35 °S, 161 °E	41	Patterson and Settle, unpublished data

Table 11.2 Metal concentrations in marine rains (μg/kg)

Metal	Shetland Islands	Bay of Bengal	Bermuda	Enewetak	Samoa
Pb	11	—	0.77	0.023	0.007
Cd	—	—	0.06	0.0041	~0.002
Hg	—	—	—	0.002	—
As*	<8	—	—	—	—
Cu	15	—	0.66	0.03	~0.01
Zn	29	100	1.2	0.15	~0.1
V	1.8	—	—	0.03	—
Fe	170	30	4.8	1.5	~0.4
Mn	6	—	0.27	0.08	~0.01

* Hawaii: 0.019; California coast: 0.019 (Andreae, 1980).
Sources: Shetland Is. from Cambray *et al.*, 1975; Bay of Bengal from Mahadevan *et al.*, 1982; Bermuda from Jickells *et al.*, 1984; Enewetak from Arimoto *et al.*, 1985; Fitzgerald *et al.*, 1983; Settle *et al.*, 1982; Samoa from Duce *et al.*, unpub. data; Settle *et al.*, 1982.

Concentrations of metals in air and rain are, of course, related to the source strengths for the metals and the wind flow patterns over the ocean. For metals in aerosol particles derived primarily from pollution sources, we would expect the highest concentrations to be observed over the North Atlantic, downwind of industrial and population centers of North America. Lower concentrations would be expected over the North Pacific and northern Indian Oceans, while the lowest concentrations would be expected over the South Pacific and South Atlantic. The data in Table 11.3 confirm this

Table 11.3 Typical atmospheric metal aerosol concentrations in marine regions (ng/m^3)

Location	Pb	Cd	As particulate	As vapor	Hg particulate	Hg vapor	Reference
Shetland Islands	30	<9	1.3	–	<0.05	–	Peirson *et al.*, 1974
Western Mediterranean	50	0.5	1.8	–	0.17	–	Arnold *et al.*, 1982
Bermuda area	3	0.2	0.16	–	0.0007	1.6	Duce *et al.*, 1976; Fitzgerald *et al.*, 1983; Walsh *et al.*, 1979a
North Atlantic	10	–	–	–	<0.07	1.6	Buat-Ménard and Chesselet, 1979; Slemr *et al.*, 1981
Hawaii	2	0.02	0.06	–	–	–	Hoffman *et al.*, 1972; Walsh *et al.*, 1979a
Enewetak	0.12	0.003	–	–	0.0006	1.6	Duce *et al.*, 1983; Fitzgerald *et al.*, 1983; Settle and Patterson, 1982
South Atlantic	–	–	–	–	–	1.05*	Slemr *et al.*, 1981
Samoa Area	~0.02	<0.005	0.017	0.002	–	1.05*	Walsh *et al.*, 1979a; Fitzgerald *et al.*, 1983; Duce, unpub. data; Patterson unpub. data

* 10°–20° S.

general picture. For metals such as mercury, which may have significant natural sources, such as marine and terrestrial biological activity and volcanoes as well as pollution sources, the distribution pattern might not be expected to be so variable. Indeed, a significant marine biological source for atmospheric Hg has been suggested by Fitzgerald *et al.* (1984) on the basis of measurements during a crossing of the productive equatorial upwelling region in the Pacific Ocean.

There are very few data for arsenic in marine rain. Andreae (1980) has provided the most extensive data set (Table 11.2), but samples have only been analyzed from the west coast of the United States and Hawaii. In both locations the As concentration in rain is approximately the same. Walsh *et al.* (1979b) have calculated, and Andreae (1980) concurs, that approximately 75% of the global emissions of As to the atmosphere each year are from pollution sources. Measurements of particulate As (generally less than 10% of the As in the marine atmosphere is in vapor form) confirm that concentrations decrease from the North Atlantic to the North Pacific to the South Pacific (Walsh *et al.*, 1979a—see Table 10.3), supporting the importance of anthropogenic sources for As in the northern hemisphere.

Collection of Precipitation for Metal Analysis is Simple

Obviously the first step toward evaluation of the wet deposition of heavy metals to the surface of the ocean is collection of precipitation samples. This would appear to be a most trivial task, but it is perhaps here that most research studies of metal wet deposition in remote marine (or even continental) regions run aground. We will restrict the discussion to the collection of rain, where a number of problems quickly arise—some unique to rain collection over the sea and some applicable to rain collection anywhere. We will also consider only true wet deposition measured by continuously open collectors.

Representativeness of Rain Samples

All rain and rain samples are not the same. Rain over the ocean is associated with a variety of precipitation systems, ranging from simple trade wind showers, with a typical horizontal cross-section of 500 to 1000 meters and vertical extent of as little as 1500 to 2000 meters, to mesoscale storm systems in mid-latitudes, with horizontal dimensions of 1000 to 1500 km or more and a vertical extent up to or through the tropopause. The chemical composition of rain from these precipitation systems may be quite dissimilar. This can arise from several factors, ranging from differences in the vertical distribution of the rain-forming portions of the cloud systems to differences in duration, intensity, and droplet size of the precipitation.

As an example, over the North Pacific region most of the mineral aerosol particles are derived from Asian deserts. Particle production in these desert regions is very seasonally dependent, dropping by over an order of magnitude from the windy high dust period in the late winter and spring to the very low dust period in the late summer and fall (Duce *et al.*, 1980; Uematsu *et al.*, 1983; Parrington *et al.*, 1983; Parrington and Zoller 1984). In addition, upper level wind patterns (700 mb and above) result in strong transport to the east over the North Pacific during the late winter and spring, with weaker and more variable winds during the rest of the year. The temporal changes in the atmospheric mineral aerosol concentration are reflected in the rain concentrations. Measurements of metals derived from Asia in rain over the North Pacific during only October, for example, could give very misleading information relative to the annual flux of these metals to the North Pacific.

Mineral particles and associated metals from Asia enter the marine boundary layer from above as a result of gravitational settling, turbulent mixing, and/or precipitation, and thus are relatively uniformly distributed vertically. On the other hand, most metals whose source in the marine atmosphere is the ocean are found concentrated in the marine boundary layer, generally in the lower 1–2 km. In addition, metals present on the smallest particles (e.g. $< 1\mu m$), whether derived from continents or the ocean, have longer atmospheric residence times and are more uniformly distributed vertically in the atmosphere. Obviously rain from mesoscale storm systems, with their greater vertical extent, would more effectively remove the metals transported in the upper troposphere and those on smaller size particles than the more vertically limited trade wind showers.

While there are essentially no data on the metals of interest here as a function of altitude over the ocean, and really not even sufficient data to distinguish concentration differences among different precipitation system types, we can illustrate the vertical variability in concentration using mineral aerosol and sea salt. For example, in Table 11.4 the mean and range of the aluminum (Al) concentration (an indicator of mineral aerosol from continental regions) and the sodium (Na) concentration (an indicator of sea salt) are presented for near sea level on Oahu and 3400 m on Hawaii, approximately 300 km apart horizontally. Note the very different Al concentrations during the high dust and clean periods at both sites. The standard deviation of the mean Al concentration during these two time periods is rather high because the dust storm activity in Asia results in pulses of dust being transported over the North Pacific. The analytical uncertainty in measurement of the Al and Na concentrations is 10% or less. Note also that there is relatively little difference in the mean Al concentrations at the two altitudes during the individual time periods, while the Na (or sea salt) concentration decreases from 2 to 3 orders of magnitude from sea level to 3400 m.

Table 11.4 Atmospheric concentrations at sea level and 3400 m in Hawaii* (ng/m^3)

Location	Aluminum		Sodium	
	Dust season	Clean season	Dust season	Clean Season
Oahu, 15 m (1981–84)				
Mean	95 ± 82	17 ± 20	4300 ± 1400	4300 ± 2200
Range	12 – 390	1.2 – 120	2400 – 6600	400 – 11,400
No. of samples	66	47	67	54
Hawaii, 3400 m (1979–82)				
Mean	71 ± 51	6.7 ± 2.3	25 ± 13	11 ± 9
No. of samples	88	89	88	89

* From Uematsu *et al.*, 1983; Parrington *et al.*, 1983; Uematsu (personal communication).

Contamination of Rain Samples

Contamination of rain samples collected in remote open ocean and continental regions is a fundamental problem that is often given relatively little attention. Contamination can occur as a result of (1) unsatisfactory collection of material, (2) inadequate cleaning of the collection system, (3) improper storage of the system between rain events, and (4) collection of locally produced material in the sampling system which is not representative of the marine rain being investigated.

The collection system must be constructed of material which will not itself contaminate the sample, for example, through leaching processes, etc. For most heavy metals, one of the cleanest collection materials is conventional (low density) polyethylene, which should be cleaned carefully and regularly with ultra-pure acid and water (see Settle *et al.*, 1982). The collector must be carefully covered between rain events and washed anew before each collection. Clean room techniques are required during the actual collection and processing of the precipitation samples.

Collection of a rain sample representative of rain falling to the ocean surface, rather than locally derived material in the air, is particularly difficult over the ocean. Collection of rain samples on land away from the coast is useless for obtaining information on metal transport from the atmosphere to the ocean. Collection along windward coastlines and/or from ships is necessary. In both cases, however, the sampling is fraught with pitfalls! To avoid local contamination from sea spray, sand and soil along a coastline, it is preferable to collect the rain from a tower several meters above ground level. Care must be taken to sample only when the wind is off the ocean and has been for some time. This is particularly difficult in convective rains, when the wind circulation in and around the shower itself is quite complex and unpredictable.

As an example, consider rain collected from an 18 meter high tower on the eastern tip of the island of Tutuila, American Samoa, as part of the Sea/Air Exchange (SEAREX) Program. Table 11.5 presents data on the atmospheric particulate concentration of lead and cadmium collected when the wind was coming directly off the ocean (O) and off the land (L). Note that even at this very remote and windward location on this relatively uninhabited island, the atmospheric lead concentrations were approximately an order of magnitude higher when the air flow was off the island, with the difference for Cd being somewhat less. The opportunity for contamination of rain with locally derived Pb and Cd during the passage of a tradewind shower is obvious.

Table 11.5 Atmospheric lead and cadmium in Samoa, 1981* (pg/m^3)

Sample no.	Pb	Cd
1-O†	25 ± 5	2 ± 1
1-L‡	120 ± 20	10 ± 3
2-O	22 ± 5	4 ± 2
2-L	130 ± 20	14 ± 4
3-O	1.5 ± 0.8	8 ± 2
3-L	210 ± 40	8 ± 2
4-O	5 ± 2	<1
4-L	290 ± 50	33 ± 8

† Collected only when the air flow was off the ocean and condensation nuclei counts were $<600/cm^3$.
‡ Collected only when the air flow was off the island or the CN counts were $>600/cm^3$.
* Duce, unpublished data.

Collection of precipitation samples from ships also provides ample opportunity for contamination as a result of eddying of the wind around ship structures. However, the maneuverability of a ship often provides the opportunity to avoid these problems. The collection of uncontaminated rain samples at sea requires close coordination between the bridge and scientific teams. Rain areas should be carefully tracked by radar to establish the development, direction of movement, and speed of the rain system relative to the ship. If the rain system appears suitable for sampling, the ship should intercept the rainstorm from a direction that will cause the ship's exhaust plume to be carried downwind of the bow. When these precautions are taken and small towers, platforms, or poles are used for collection on the bow, clean rain samples can be obtained. Table 11.1 presents data for Pb in rains collected in a very careful and contamination-free manner in several open ocean regions from both island coastal towers and on the bows of ships. Similar concentrations of lead were observed from the different collection platforms in the individual regions, indicating that both types of rain collection can be made satisfactorily with proper care and planning.

Washout Factors are a Panacea for Estimating Rain Fluxes

Washout factors, or scavenging ratios, are often used to relate atmospheric concentrations of trace metals to their concentrations in rain. Washout factors have been defined in several ways, but all involve a ratio of the rain concentration to the atmospheric concentration of the substance of interest. A detailed review of washout factors and precipitation scavenging is presented by Slinn (1983) and Scott (1981), and these processes will not be discussed in depth here. We will define the washout factor, W, for any metal, M, as $(W)_M$, given by:

$$(W)_M = [(C)_{M,\mathrm{rain}} \times \rho]/(C)_{M,\mathrm{air}} \quad (1)$$

where $(C)_{M,\mathrm{rain}}$ is the concentration of metal M in rain in gm/kg, $(C)_{M,\mathrm{air}}$ is the concentration of metal M in the air in gm/m^3, and ρ is the density of air (1.20 kg/m^3) at standard conditions.

In this form washout factors are dimensionless,* but the density term is not used by all authors. Once the relationship between metal concentration in the air and in the rain has been established, one can utilize atmospheric metal concentrations to predict the concentrations in rain.

The use of washout factors to calculate metal deposition rates implicitly assumes that the concentrations of the metals in rain and air are linearly related. However, it is very difficult, if not impossible, to rigorously test the

* Washout is usually defined as

$$W = \frac{C(M, \mathrm{rain})}{C(M, \mathrm{air})}$$

with dimensions

$$\frac{\text{gm metal/kg rain}}{\text{gm metal/m}^3\text{ air}}$$

leading to m^3 air washed per kg rain. The Buat-Ménard and Duce definitions above lead to the dimensions

$$\frac{\text{gm metal/kg rain}}{\text{gm metal/m}^3\text{ air}} \times \frac{\text{kg air}}{\text{m}^3\text{ air}} = \frac{\text{kg air}}{\text{kg rain}} = \text{amount of air cleaned of all its metal by 1 kg rain}$$

significance of this relationship. Problems arise because of possible differences in the removal efficiency of particles as a function of chemical composition and rain droplet size. Rain concentrations represent an integrated removal throughout the atmospheric column through which the rain is falling, while the air concentration is determined only at the surface. Thus, metals with different vertical concentration profiles but similar surface air concentrations might have very different washout factors. An additional problem faced in truly remote marine regions is that it may take many hours, occasionally even days, to collect aerosol particle samples large enough to analyze, whereas rain samples may be collected over a period of minutes to hours. Thus, the true atmospheric concentration, even at the surface, appropriate to the washout factor calculation may not really be known. Scott (1981) also points out that smaller washout ratios would be expected in 'cold' clouds, that is, those where a much greater fraction of the precipitation occurs as a result of vapor growth rather than droplet coalescence processes. The latter apparently dominated in 'warm' rains, that is, those not containing ice particles or super-cooled droplets. Typical washout ratios for material present on particles in cold rains might be ~100, while in warm rains they might be ~1000.

One apparent advance in attempting to understand some of the factors leading to the variability in washout factors was the discovery that in urban or near-continental marine areas there was a relationship between the washout factor and the mass median diameter (MMD) of the particles containing the metals. For example, Scott (1981) has shown the relationship between $(W)_M$ and the MMD for metals collected in St Louis (Gatz, 1975) and Chilton, UK (Cawse, 1974) (Figure 11.1), and Duce *et al.* (1979) have found similar results in the Florida Keys ($r = 0.76$, significant at $p = 0.01$) (Figure 11.2). In general, the washout factor increased with increasing particle size at all three locations. Scott (1981) and Slinn (1983) suggest that these results can be explained by the lower collection efficiency of the raindrops for the smaller particles (down to perhaps a few tenths of a micrometer radius), with the chemical composition, particularly solubility, of the different size particles also playing a role. The latter would be particularly important in marine areas for the sea salt aerosol particles, the major source for such metals as Na, K, Mg and Ca. However, for rains collected at Enewetak Atoll, over 5000 km from the nearest continent, there is strong evidence that there is no relationship between washout factor and particle size, as can be seen in Figure 11.3 ($r = 0.39$, not significant at $p = 0.1$). Corroboration of this result was obtained at Enewetak by Buat-Ménard *et al.* (1983). They measured the size distribution of aluminosilicate particles in the atmosphere and filtered from rain samples and were thus able to calculate washout factors for the various size particles. The results, presented in Table 11.6, show no statistical difference (1σ) in the values for $(W)_{\text{mineral aerosol}}$ for the different size

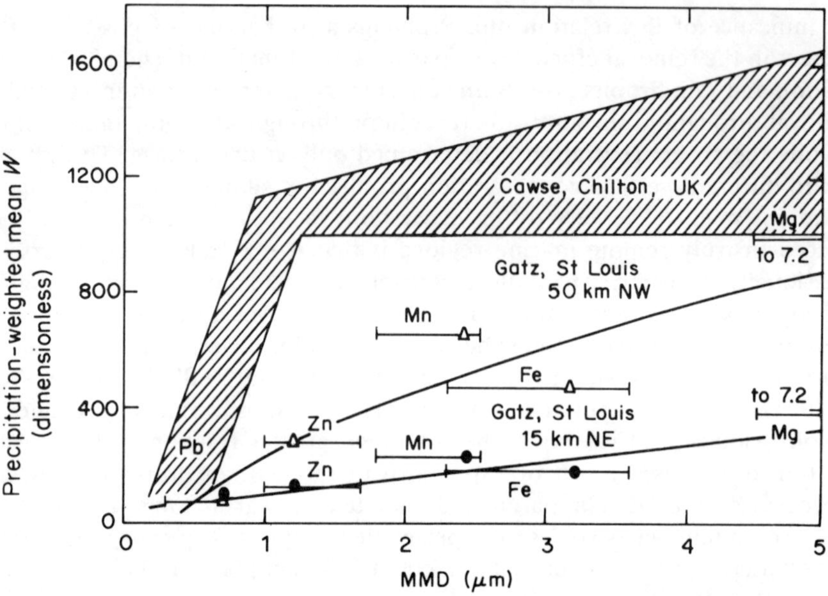

Figure 11.1 Variation of washout factor or scavenging ratio, *W*, as a function of particle size for various trace elements in urban areas (from Scott, 1981)

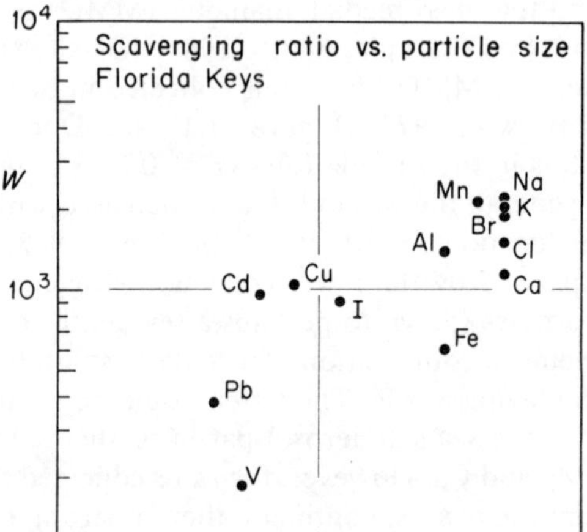

Figure 11.2 Variation of washout factor or scavenging ratio, *W*, as a function of particle size for various trace elements in the Florida Keys (from Duce *et al.*, 1979)

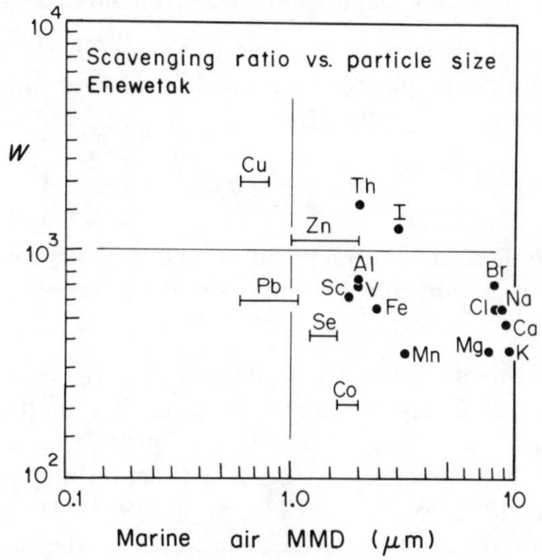

Figure 11.3 Variation of washout factor or scavenging ratio, *W*, as a function of particle size for various trace elements at Enewetak Atoll

Table 11.6 Washout factors for mineral aerosols, Enewetak*

Particle diameter (μm)	$(W)_{\text{mineral aerosol}}$ †
<0.5	940±1100
0.5–1	800±620
1.0–1.5	500‡
1.5–3	310±250
3.0–6	270±170
>6	320±120

* From Buat-Ménard *et al.*, 1983.
† One standard deviation, based on five air samples and three rain samples.
‡ Assumed value.

fractions. The cause of these differences between continental and remote marine regions is uncertain, although Slinn (1983) has suggested that in truly remote marine regions, where there are very few cloud condensation nuclei, essentially all particles are activated regardless of size and chemical composition. It is clear that any relationships developed between washout factors and particle size distribution in urban or continental regions may not apply over open ocean regions far from land.

A Field Approach to Dry Deposition Measurements is Straightforward

The dry deposition of aerosol particles to the ocean surface refers to all deposition processes except precipitation. It is often estimated by utilizing the deposition velocity, v_d, given by:

$$v_d = F/M \quad (2)$$

where v_d is the deposition velocity in cm/sec, F is the flux of particles to the surface in gm/cm^2 sec, and M is the atmospheric concentration in gm/cm^3.

Direct measurements of the dry deposition of trace elements to the ocean surface are not technically feasible at present. One method for obtaining reasonably accurate estimates of dry deposition is the use of available dry deposition models to a water surface as a function of particle size (Slinn and Slinn, 1980, 1981; Williams, 1982; Slinn, 1983). However, the use of such models requires an accurate measurement of the mass-size distribution of the trace metals. In addition, deposition velocities depend markedly upon particle size, relative humidity and wind speed, as shown for sea-salt deposition to the ocean in Figure 11.4.

Model calculations of dry deposition have been rather successful for sea-salt aerosol particles, provided extreme care was taken to obtain air samples that represent the true sea-salt particle mass and size distribution (McDonald *et al.*, 1982). Acceptably accurate size distribution for sea-salt aerosol particles can be deduced from the use of high-volume cascade impactors, when proper corrections are made for particle loss in the largest size range. However, the use of such sampling devices may cause serious problems for other types of aerosol particles to which trace metals may often be attached. For example, it appears that particle 'bounce' effects, especially for dry particles such as aluminosilicate minerals, can result in an apparent shift of the size distribution to smaller sizes (Buat-Ménard *et al.*, 1983). The situation is even worse for trace elements such as Pb, As and Cd because a major fraction of the mass of these elements occurs on particles less than $\simeq 0.25$ μm radius, and cascade impactors currently in use do not separate particles below that size range.

Since direct measurements of dry deposition of metals such as lead, cadmium and arsenic to the ocean are not possible, and model calculations suffer numerous pitfalls as well, most presently published dry deposition data are based on the use of surrogate surfaces, such as open vessels, plates or filters, for deposition measurements. The use of such surfaces has been seriously criticized because they do not accurately mimic the characteristics (microstructure, roughness, etc.) of natural water surfaces (Hicks *et al.*, 1980). Moreover, such measurements must be performed 10 to 20 m above

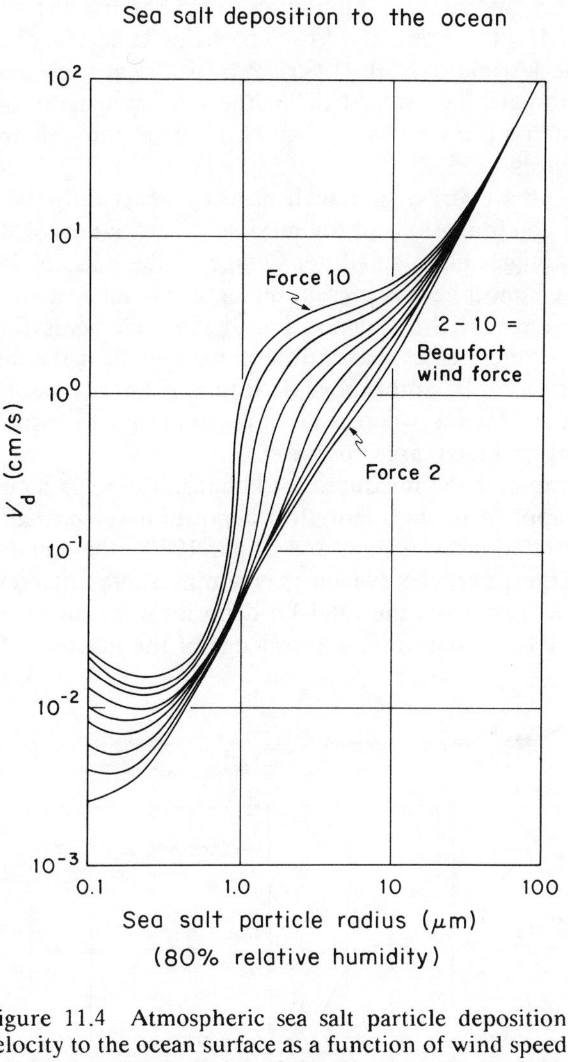

Figure 11.4 Atmospheric sea salt particle deposition velocity to the ocean surface as a function of wind speed and particle size, as derived from the theoretical considerations of Slinn and Slinn (1980)

sea-level (ships, islands), where the relative humidity, which influences the size of certain types of particles, is quite different from that close to the sea surface.

However, although the biases introduced by the use of a surrogate surface are real, model calculations appear to indicate that dry deposition of aerosol particles to the sea surface and to a surrogate surface such as a rimless plastic plate should agree to within a factor of two to three over a broad range of wind speeds and relative humidities. This has been shown for sea-salt aerosol particles and mineral aerosol particles collected over the tropical North Pacific (Arimoto *et al.*, 1985). Provided that such agreement holds for sub-micrometer aerosol particles, the correct magnitude of total dry deposition of some trace metals may be inferred through the use of such sampling devices.

In any event, we stress that such measurements must be performed in parallel with determinations of the mass-size distribution of the elements of interest. Although only a small percentage of the mass of Pb, Cd, and As in the marine atmosphere is present on large aerosol particles, that fraction may dominate the total dry deposition of these elements. The large particle fraction, in most cases, has a different origin than the sub-micrometer fraction. For example, anthropogenic lead is primarily associated with the sub-micrometer fraction whereas Pb of crustal origin or associated with sea-salt is present on much larger particles.

As an example, if the measured lead concentration as a function of particle size is applied to the theoretical deposition velocity as a function of particle size relationship (Slinn and Slinn, 1980), the small percentage of Pb on the largest particles (which may simply represent recycled Pb from the sea surface) controls the total Pb deposition, as shown in Figure 11.5 from Duce (1982). Therefore, a knowledge of the mass-size function of Pb

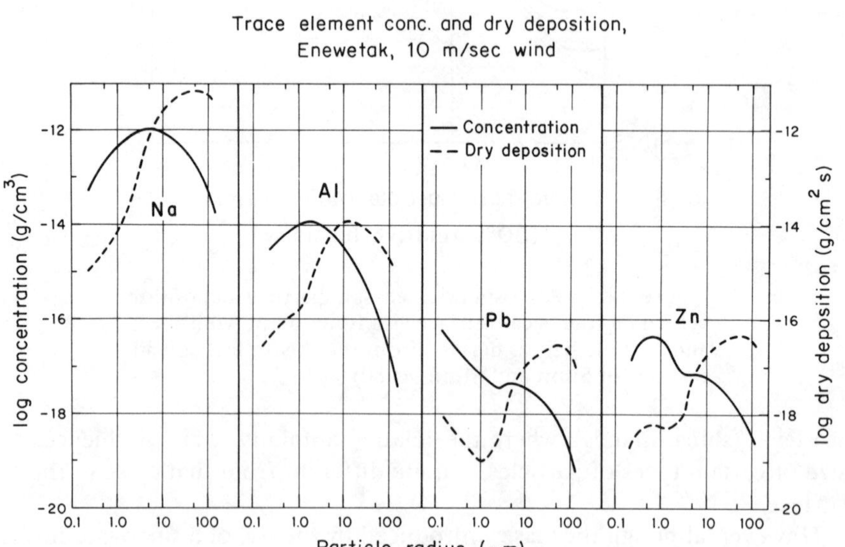

Figure 11.5 Atmospheric concentration and modeled dry deposition of trace elements at Enewetak Atoll as a function of particle size (from Duce, 1982)

is necessary to assess the relative contributions of the various Pb sources to the total dry deposition of this element.

Moreover, the atmospheric concentrations of particles derived from different sources vary temporally, so the relative contributions of the different sources to the total dry deposition will also vary temporally. Sampling must be undertaken at various times of the year to obtain representative annual dry deposition data at a given location. However, individual dry deposition samples should be collected over short enough time intervals to indicate the natural temporal variability in the deposition.

We have only considered the case of metals attached to aerosol particles. More than 90% of atmospheric mercury is present as gaseous elemental mercury. Fitzgerald *et al.* (1983) estimated the gas phase transfer of elemental Hg from the atmosphere to the ocean using the exchange model of Liss and Slater (1974) and assuming that the diffusivity of gaseous Hg is similar to radon and the sea surface is a perfect sink. Clearly, for metallic vapor phase compounds in the atmosphere, a completely different approach to that described here is needed to assess the significance of their dry deposition to the ocean (e.g. see Liss, 1983).

Accurate Deposition Measurements Guarantee Accurate Net Air/Sea Exchange Rates

While an accurate measurement of the concentration of metals in rain and dry deposition is a necessary first step in evaluating fluxes of these substances to the ocean, it is not sufficient in itself. The total amount of rainfall in the period of interest, per year for example, must be known. Potential problems arise concerning variability of rainfall amount and intensity with season and how this will affect the metal concentrations, as well as how seasonal changes in wind flow patterns or particle production processes will affect atmospheric metal concentrations at the marine location. Even taking these factors into consideration, significant problems remain. In the marine environment the gross deposition of a metal to the ocean is composed of a net input as well as a component associated with recycled sea spray. The importance of the atmosphere as a transport path for material from the continents to the ocean can only be assessed accurately if the relative contributions of the net and recycled components can be distinguished (Settle and Patterson, 1982; Jickells *et al.*, 1984; Arimoto *et al.*, 1985). There is strong evidence that atmospheric sea salt particles produced by bubbles bursting at the sea surface contain many metals in concentrations considerably higher than would be expected on the basis of the metal-to-sodium ratio of near-surface water. It is apparent that some fraction of these metals is associated with surface active organic material and is scavenged by the rising bubbles and concentrated on the sea-salt particles produced from a very thin layer of the air/sea interface when the bubbles burst (Weisel *et al.*, 1984). If this fractionation is not taken into account, the calculated net deposition to the ocean will be anomalously high.

As one approach to this problem, Arimoto *et al.* (1985) determined the metal/Na ratio on the largest atmospheric particles at Enewetak Atoll in the Marshall Islands, using only particles collected on the first stage of a cascade impactor. This represented, in general, particles with radii greater than about 3.5 μm. Most particles in this size range are sea salt, so this size range gave a good estimate of the metal/Na ratio in particles produced by bursting bubbles. These ratios also agreed well with direct measurements of the metal/Na ratios on sea-salt particles produced and collected artificially using the Bubble Interfacial Microlayer Sampler in the North Atlantic by Weisel *et al.* (1984) (see also Buat-Ménard, 1983 for an extensive discussion of this problem). The recycled component for any metal in the rain was calculated by

$$(\mathrm{M})_{\mathrm{recycled}} = (\mathrm{M/Na})_{\mathrm{Stage\ 1}} \times (\mathrm{Na})_{\mathrm{rain}} \quad (3)$$

where $(\mathrm{M})_{\mathrm{recycled}}$ is the concentration of metal M recycled from the sea in gm/kg, $(\mathrm{M/Na})_{\mathrm{Stage\ 1}}$ is the metal/Na ratio on stage 1 of the impactor and $(\mathrm{Na})_{\mathrm{rain}}$ is the sodium concentration in rain in gm/kg.

Recycled components calculated in this way for several metals ranged from 15% to about 50%, as shown in Table 11.7, and are comparable with the recycled fraction calculated using a similar approach by Jickells et al. (1984) at Bermuda. For Pb, ~30% was recycled in Enewetak rain according to Arimoto *et al.* (1985), compared with 10% at Enewetak calculated by Settle and Patterson (1982) and 17% at Bermuda determined by Jickells *et al.* (1984).

Table 11.7 Recycled metals in marine rains

Metal	Rain location	Percent metal Recycled	Reference
Pb	Bermuda	17	Jickells *et al.*, 1984
	Enewetak	30	Arimoto *et al.*, 1985
	Enewetak	10	Settle and Patterson, 1982
Cu	Bermuda	4	Jickells *et al.*, 1984
	Enewetak	48	Arimoto *et al.*, 1985
Zn	Bermuda	11	Jickells *et al.*, 1984
	Enewetak	15	Arimoto *et al.*, 1985
Fe	Bermuda	21	Jickells *et al.*, 1984
V	Enewetak	33	Arimoto *et al.*, 1985

Recycled components calculated in the same way for dry deposition measurements to a surrogate surface are shown in Table 11.8. They range from 12 to 100%. For lead, the percentage recycled varied considerably between samples. In two of the samples, approximately 50 to 60% of the gross Pb deposition was due to recycled material, but in a third sample, essentially all of the Pb could be attributed to sea spray.

Table 11.8 Dry deposition (gm/cm^2s) of lead, zinc and copper to a plastic plate*

Element	DD†	Total measured deposition	Deposition due to recycled sea salt	Net deposition from crustal weathering	Net deposition from other source(s)	Percent recycled
Pb	1	2.6×10^{-16}	1.4×10^{-16}	2.9×10^{-18}	1.2×10^{-16}	54
Pb	2‡	1.9×10^{-16}	2.1×10^{-17}	1.2×10^{-17}	0	100
Pb	3	$<8.5 \times 10^{-16}$	2.1×10^{-16}	3.6×10^{-18}	$<6.4 \times 10^{-16}$	>25
Pb	4‡	1.3×10^{-16}	7.4×10^{-17}	1.1×10^{-19}	5.6×10^{-17}	57
Zn	2	1.9×10^{-15}	2.3×10^{-16}	6.8×10^{-17}	1.6×10^{-15}	12
Cu	1	$\sim 5 \times 10^{-16}$	1.8×10^{-16}	1.3×10^{-17}	3.1×10^{-16}	~36
Cu	3	$\sim 3 \times 10^{-16}$	2.7×10^{-16}	1.6×10^{-17}	1.4×10^{-17}	~90
Cu	4	$\sim 2 \times 10^{-16}$	1.1×10^{-16}	4.9×10^{-19}	$\sim 9 \times 10^{-17}$	~55

* From Arimoto *et al.* (1985).
† Dry deposition sample number.
‡ Pb dry deposition data from Settle and Patterson (1982).

These examples give an indication of the state of the art in this area. Clearly, more sophisticated techniques for accurately evaluating this recycled fraction must be developed. Future work should focus in two areas (a) the use of adequate tracers (stable or radioactive) during field measurements, and (b) an improvement of our knowledge of metal/Na ratios as a function of sea-salt particle size through carefully designed *in situ* or laboratory experiments.

MYSTERIES

Are there Significant Unknown Natural or Anthropogenic Sources for Atmospheric Metals?

One of the major mysteries in this field of research is our lack of knowledge of all the sources for metals in the atmosphere. As emphasized in other chapters of this book, a number of anthropogenic sources are still not well documented. Among these are waste incineration and other high temperature sources. Of particular concern is the need for accurate emission inventory data for these metals, especially for Third World countries. However, even less is known about natural sources. In recent years some information has emerged on the importance of volcanic activity, and we now have approximate values for the global volcanic source strength, especially for rather volatile metals such as Cd, Hg, Pb and As (Arnold *et al.*, 1981; Settle and Patterson, 1982; Fitzgerald *et al.*, 1983; Zoller, 1984). On the other hand, there is a dearth of data for such sources as crustal degassing and the terrestrial biosphere. Metal release from plants during their growth has been suggested as a source of zinc in the atmosphere (Beauford *et al.*, 1977). The importance of terrestrial biological sources for trace elements has yet to be established on a global basis. Little is known about mobilization of trace elements by forest fires or other land-burning practices. Also, the role of the oceans in the atmospheric chemistry of metals cannot be simply restricted to the ejection of sea-salt particles and their associated trace metals into the atmosphere. Micro-organisms in the marine environment can produce methylated forms of As, Hg, Se, Sn and other metals which may be released into the air. Evidence for this has been shown for Hg by Fitzgerald *et al.* (1984) and for Se by Mosher and Duce (1983).

What are the Physical and Chemical Forms of Metals in the Marine Atmosphere?

Before accurate estimates of the air-to-sea flux of metals can be made, we must have better information on the physical and chemical forms of the metals. For the metals of interest here, there is no evidence to date of any vapor phase for cadmium in the atmosphere—it is apparently all present on particles, although a careful search for a vapor phase has yet to be made. In the case of lead, there is some evidence that a small fraction is present in the vapor phase, at least in urban areas, very likely as tetraalkyl lead compounds (Harrison and Perry, 1977; Harrison and Laxen, 1978; Rohbock *et al.*, 1980), although there is no evidence for this as yet in the remote marine atmosphere. Evidence suggests that ~10% or less of the arsenic in the marine atmosphere is present either in the gas phase or on particles considerably smaller than 0.2 μm (Walsh *et al.*, 1979b). The chemical form of the vapor phase As is unknown.

In the case of mercury, well over 99% is in the vapor phase in the atmosphere, with only a very small fraction present on particles (Table 11.3) (Fitzgerald *et al.*, 1983). For Hg particularly, it is critical to know the physical and chemical form to evaluate the wet and dry deposition. From Hg measurements at Enewetak Atoll, Fitzgerald *et al.* (1983) have pointed out that all the Hg in rain can be accounted for by the very small particulate Hg concentration if the washout factor for particulate Hg is about 1200, a reasonable figure. It thus appears that only a small fraction of the wet deposition of Hg to the ocean results from removal of the primary physical form of Hg in the atmosphere, gaseous Hg, while removal of the 0.1% particulate Hg may account for essentially all the Hg wet deposition. This conclusion is complicated, however, by the fact that relatively little is known about the chemical species in the vapor phase. Fitzgerald *et al.* (1983) have found that from 27 to 82% of the gaseous Hg over coastal Long Island Sound is organically associated and suggested that dimethyl mercury and/or methyl chloride are possible species. However, Slemr *et al.* (1981) found that <10% of the gaseous Hg over the open North and South Atlantic was dimethyl mercury while Fitzgerald *et al.* (1983) found <20% of the gaseous mercury was methylated at Enewetak. Thus, it appears that the primary gaseous form is elemental Hg, although considerable additional research is needed in this area. Removal of gaseous forms of Hg by precipitation will be very species dependent, and adequate modelling of the wet deposition of Hg to the ocean cannot be undertaken until the chemical forms of Hg can be accurately quantified and their geographical and temporal variability determined.

What is the Fate of Atmospheric Metals when they Enter the Ocean?

We know virtually nothing about what happens to metals in air and rain after they enter the ocean. To date there is virtually no information from remote marine regions on the fraction of metals in rain which are in the particulate or dissolved form. We know nothing yet about reactions that occur during the mixing of rain and sea water. Solubilization, sorption, or flocculation processes involving metals may occur, and such processes would change the physical and chemical forms of the metals in the ocean and may have a significant impact on the cycling of the metals in surface waters. Similar concerns are obvious relative to dry deposition of aerosol particles containing metals. Reactions of these kinds would affect the residence times of the metals in the ocean and their interactions with biological systems.

It is now well established that the atmosphere is the primary transport path for lead found in the surface waters of the North Atlantic and North Pacific (Schaule and Patterson, 1983). Figure 11.6 shows the vertical profile of dissolved Pb at open ocean sites in both oceans. The strong concentration gradient near the surface identifies the atmospheric source for this Pb, and indicates that a considerable fraction of the atmospheric Pb entering the surface waters is soluble. Note that similar seawater profiles and conclusions relative to atmospheric input have been made for Sn by Byrd and Andreae (1982).

Walsh and Duce (1976) showed that a major fraction of the anthropogenically derived vanadium present on coastal North Atlantic aerosol particles is soluble in seawater. Similar results were found for several metals

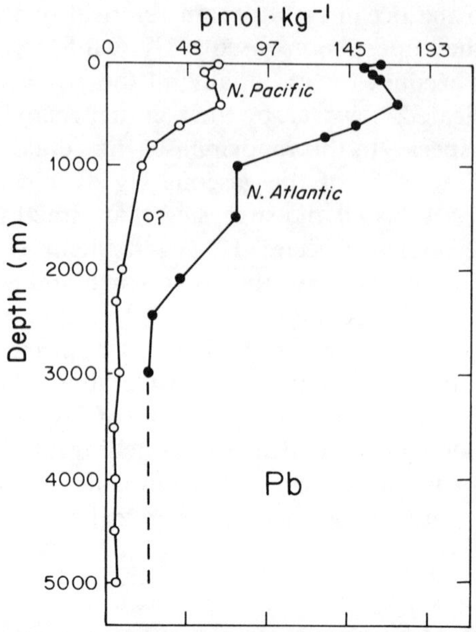

Figure 11.6 Depth profile of lead in the North Atlantic and North Pacific Oceans (from Schaule and Patterson, 1983)

during aerosol particle studies along the southern California coast (Hodge *et al.*, 1978), coastal Washington (Crecelius, 1980), and the southeastern US (Mullins, 1978). Results of some of these seawater solubility studies are presented in Table 11.9.

To determine whether particulate material atmospherically supplied to the ocean surface may be significantly leached, not only by direct interaction with sea water but also as a result of zooplankton grazing, Moore *et al.* (1984) have leached samples of atmospheric dust at pH values between 5.4 and 8.0. The elements Cu, As and V showed significant increases in the proportion leached as the pH was lowered. Approximately 10% of the iron was leached, independent of pH. An estimate of the atmospheric flux of iron in a leachable form supports the idea that this source may be the primary contributor to the Fe requirements of the biota. Since the efficiencies with which phytoplankton can grow at low Fe concentrations vary among species, it follows that the meteorological factors that govern the rate of transport of atmospheric dust to ocean areas could affect plankton ecology and productivity.

Indeed, numerous studies, based principally on the behavior of natural and artificial radionuclides, now support the conclusion that scavenging by sinking biogenic particles is the major removal process for many trace

Table 11.9 Seawater soluble fraction of metals on coastal aerosols

Metal	Percent soluble	Metal	Percent soluble
Quillayute, WA (1 hr)*		La Jolla, CA (3 hr)†	
As	50	Pb	39
Zn	65	Cd	84
		Zn	68
Quillayute, WA (24 hr)*		Ensenada, Mexico (3 hr)†	
As	49	Pb	13
Zn	72	Cd	80
		Zn	24
Narragansett, RI (1 hr)‡			
Excess V	72		

* Crecelius, 1980;
† Hodge *et al.*, 1978;
‡ Walsh and Duce, 1976.

elements from surface waters reservoir to deep waters. The short residence times of many trace elements (particulate and dissolved) in surface waters imply that relatively fast (of the order of a few months to a few years) geochemical coupling exists between the atmospheric input of trace metals (and other reactive substances) and their transport to the deep ocean (Buat-Ménard, 1984). This rapid particulate transport 'short-circuits' water mixing processes and has major geochemical implications which are of great importance to the health of the oceans. Induced chemical changes in the deep water column will be first observed in the particulate flux.

Using a steady state approach and indirect flux estimates, it has been calculated that atmospheric deposition is a significant source for some particulate trace metals in North Atlantic deep waters (Buat-Ménard and Chesselet, 1979). Such a source appears dominant in the case of lead, for which it can be inferred that the particulate flux in the ocean is, at present, primarily anthropogenic. For other trace elements anomalously enriched in marine aerosols possibly because of an anthropogenic component (Cu, Zn, Cd), the present anthropogenic air-sea flux still appears to be small relative to natural fluxes in the ocean. However, one can speculate that if the atmospheric input was to be one order of magnitude higher than at present over the North Atlantic for some metals, the detectable effect would be recorded in oceanic suspended matter and would alter the oceanic cycle of such elements. Such effects are extremely difficult to predict at present. Indeed, changes in the rate of input from the atmosphere will be biologically cycled at the same rate only if such changes do not affect the biological cycle itself. For example, increased levels of some trace metals in surface waters could either reduce or enhance the biogeochemical cycling.

CONCLUSION

Clearly much more work is needed before we can properly evaluate the impact of atmospheric input on the biogeochemical cycling of metals in the ocean. There is a need for carefully coordinated field programs to estimate more accurately net atmospheric fluxes to the ocean, as well as vertical particulate fluxes in the ocean at the mesoscale (coastal regions, enclosed seas) and at the global scale. Data from the southern hemisphere, where the atmospheric flux, both natural and anthropogenic, is a minimum, should be obtained with high priority to assess natural processes and fluxes in the ocean. Also, the observed temporal variabilities for both the atmospheric flux and the biogenic particulate flux clearly indicate that steady-state approaches are not valid on time scales of the order of one month, or when we consider gradually increasing inputs of pollutants. Sampling programs over several years are probably needed at a given location. Together with atmospheric flux assessments, the sampling strategy in the water column should consider concurrent sampling with sediment traps, large volume water in situ filtration units and conventional, small-volume water sampling. Finally, continued intercalibration of collection and analytical methods for trace elements in the marine atmosphere and ocean is mandatory.

REFERENCES

Andreae, M. O. (1980). Arsenic in rain and the atmospheric mass balance of arsenic. *J. Geophys. Res.*, **85**, 4512–4518.

Andreae, M. O., Asami, T., Bertine, K. K., Buat-Ménard, P. E., Duce, R. A., Filip, Z., Forstner, U., Goldberg, E. D., Heinrichs, H., Jernelov, A. B., Pacyna, J. M., Thornton, I., Tobschall, H. J., and Zoller, W. H., BUA(1984). Changing Biogeochemical Cycles. In Nriagu, J. O., (Ed.), *Changing Metal Cycles and Human Health,* pp. 359–374, Dahlem Konferenzen, Berlin, Springer-Verlag.

Arimoto, R., Duce, R. A., Ray, B. J., and Unni, C. K. (1985). Atmospheric trace elements at Enewetak Atoll: 2. Transport to the ocean by wet and dry deposition. *J. Geophys. Res.*, **90**, 7391–2408.

Arnold, M., Buat-Ménard, P., and Chesselet, R. (1981). *An Estimate of the Input of Trace Metals to the Global Atmosphere by Volcanic Activity,* IAMAP Third Scientific Assembly, Hamburg, FRG (abstract)

Arnold, M. A., Seghaier, D., Martin, D., Buat-Ménard, P. and Chesselet, R. (1982). *Geochimie de l'aerosol marin au-dessus de la Mediterranée Occidentale,* VI, Journées Etud. Pollutions, Cannes, CIESM, 27–37.

Beauford, W. T., Barber, T., and Barringer, A. R. (1977). Release of particles containing metals from vegetation to the atmosphere. *Science,* **195**, 571–573.

Bruland, K. W. (1983). Trace elements in sea water. In Riley, J. P., and Chester, R. (Eds.), *Chemical Oceanography,* pp. 157–220. Vol 8, Academic Press, New York.

Buat-Ménard, P. (1983). Particle geochemistry in the atmosphere and oceans. In Liss, P. S. and Slinn, W. G. N. (Eds), *Air–Sea Exchange of Gases and Particles.* pp. 455–532, Reidel, Dordrecht.

Buat-Ménard, P. (1984). Fluxes of metals through the atmosphere and oceans. In Nriagu, J. O. (Ed.)., *Changing Metal Cycles and Human Health,* pp. 43–69, Dahlem Konferenzen, Springer-Verlag, Berlin.

Buat-Ménard, P. and Chesselet, R. (1979). Variable influence of the atmospheric flux on the trace metal chemistry of oceanic suspended matter. *Earth Planet. Sci. Lett.*, **42**, 399–411.

Buat-Ménard, P., Ezat, U. and Gaudichet, A. (1983). Size distribution and mineralogy of aluminosilicate dust particles in tropical Pacific air and rain. In *Precipitation Scavenging, Dry Deposition, and Resuspension*, Proc. of the Fourth Intern. Conf., Santa Monica, California, Vol. 2, pp. 1259–1270, Elsevier, New York.

Byrd, J. T. and Andreae, M. O. (1982). Tin and methyltin species in seawater: concentrations and fluxes. *Science*, **218**, 565–569.

Cambray, R. S., Jeffries, D. F., and Topping, G. (1975). *An Estimate of the Input of Atmospheric Trace Elements into the North Sea and the Clyde Sea.* (1972–73), AERE-R7733, United Kingdom Atomic Energy Authority, AERE Harwell, HMSO, London, 30 pages.

Cawse, P. A. (1974). A survey of atmospheric trace elements in the United Kingdom. AERE Harwell Report R7669, UKAEA, HMSO, London.

Crecelius, E. (1980). The solubility of coal fly ash and marine aerosols in water, *Mar. Chem.*, **8**, 245-250.

Duce, R. A. (1982). Sea salt and trace element transfer across the sea/air interface. Presented at the Session on Ocean/Atmosphere Material Exchange, Joint Oceanographic Assembly, Halifax, Nova Scotia, Canada, 4 August.

Duce, R. A., Hoffman, G. L., Ray, B. J., Fletcher, I. S., Walsh, P. R., Fasching, J. L., Piotrowicz, S. R., Hoffman, E. J., Miller, J. M., and Heffter, J. L. (1976). Trace metals in the marine atmosphere: sources and fluxes. In Windam, H. L. and Duce, R. A. (Eds), *Marine Pollutant Transfer*, pp. 77–119. D. C. Heath and Co., Lexington, MA.

Duce, R. A., Unni, C. K., Harder, P. J., Ray, B. J., Patterson, C. C., Settle, D. M. and Fitzgerald, W. F., (1979). *Wet and Dry Deposition of Trace Metals and Halogens in the Marine Environment.* Presented at the IAMAP/CACGP Symposium on the Budget and Cycles of Trace Gases and Aerosols in the Atmosphere, University of Colorado, Boulder, CO, 12–18 August.

Duce, R. A., Unni, C. K., Ray, B. J., Prospero, J. M., and Merrill, J. T. (1980). Long-range atmospheric transport of soil dust from Asia to the tropical North Pacific: temporal variability. *Science*, **209**, 1522–1524.

Duce, R. A., Arimoto, R., Ray, B. J., Unni, C. K., and Harder, P. J. (1983). Atmospheric trace elements at Enewetak Atoll: 1. Concentrations, sources, and temporal variability. *J. Geophys. Res.*, **88**, 5321–5342.

Fitzgerald, W. F., Gill, G. A., and Hewitt, A. D. (1983). Air–sea exchange of mercury. In Wong, C. S., Boyle, E., Bruland, K. W., Burton, J. D. and Goldberg, E. D. (Eds.). *Trace Metals in Sea Water*, pp. 297–315, Plenum Press, New York.

Fitzgerald, W. F., Gill, G. A., and Kim, J. P., (1984). An equatorial Pacific Ocean source of atmospheric mercury. *Science*, **224**, 597–599.

Gatz, D. F. (1975). Scavenging ratio measurements in Metromex. In Beadle, R. W. and Semonin, R. G. (Eds.), *Precipitation Scavenging—1974*, pp. 71–87, ERDA Symposium Series, CONF-741014.

Harrison, R. M. and Perry, R. (1977). The analysis of tetraalkyl lead components and their significance as urban air pollutants. *Atmos. Environ.*, **11**, 847–852.

Harrison, R. M. and Laxen, D. H. P. (1978). Sink processes for tetraalkyl lead compounds in the atmosphere. *Environ. Sci. and Technol.*, **12**, 1384–1392.

Hicks, B. B., Wesely, M. L., and Durham, J. L., (1980). *Critique of Methods to Measure Dry Deposition.* Workshop Summary, EPA 1600.9-80-050, 70 pages.

Hodge, V., Johnson, S. R., and Goldberg, E. D., (1978). Influence of atmospherically transported aerosols on surface ocean water composition. *Geochem. J.*, **12**, 7–20.

Hoffman, G. L., Duce, R. A., and Hoffman, E. J., (1972). Trace metals in the Hawaiian marine atmosphere. *J. Geophys Res.*, **77**, 5322–5329.

Jickells, T. D., Knap, A. H., and Church, T. M. (1984). Trace metals in Bermuda rainwater. *J. Geophys. Res.*, **89**, 1423–1428.

Liss, P. S. (1983). Gas transfer: experiments and geochemical implications. In Liss, P. S. and Slinn, W. G. N. (Eds), *Air–Sea Exchange of Gases and Particles*, pp. 261–298, Reidel, Dordrecht.

Liss, P. S., and Slater, P. G. (1974). Fluxes of gases across the air-sea interface. *Nature*, **247**, 181–186.

McDonald, R. L., Unni, C. K., and Duce, R. A. (1982). Estimation of atmospheric sea salt dry deposition: wind speed and particle size dependence. *J. Geophys. Res.*, **87**, 1246–1250.

Mahadevan, T. N., Sadasivan, S., and Mishra, U. C. (1982). Chemical composition of precipitation in tropical marine atmosphere around Indian subcontinent. *Science of the Total Environment*, **24**, 275–285.

Moore, R. M., Milley, J. E. and Chatt, A. (1984). The potential for biological mobilization of trace elements from aeolian dust in the ocean and its importance in the case of iron. *Oceanol. Acta*, **7**, 221–228.

Mosher, B. W., and Duce, R. A., (1983). Vapor-phase and particulate selenium in the marine atmosphere. *J. Geophys. Res.*, **88**, 6761–6768.

Mullins, B. M. (1978). Geochemical aspects of atmospherically transported trace metals over the Georgia Bight. MS Thesis, Georgia Institute of Technology, Atlanta, GA, 36 pages.

Parrington, J. R., Zoller, W. H., and Aras, N. K. (1983). Asian dust: seasonal transport to the Hawaiian Islands. *Science*, **220**, 195–197.

Parrington, J. R. and Zoller, W. H. (1984). Diurnal and longer-term temporal changes in the composition of atmospheric particles at Mauna Loa, Hawaii. *J. Geophys. Res.*, **89**, 2522–2534.

Peirson, D. H., Cawse, P. A., and Cambray, R. S. (1974). Chemical uniformity of airborne particulate material and a maritime effect. *Nature*, **251**, 675–679.

Rohbock, E., Georgii, H. W., and Muller, J. (1980). Measurements of gaseous lead alkyls in polluted atmospheres. *Atmos. Environ.*, **14**, 89–98.

Schaule, B. K., and Patterson, C. C. (1983). Perturbations of the natural lead depth profile in the Sargasso Sea by industrial lead. In Wong, C. S., Boyle, E., Bruland, K. W., Burton, J. D. and Goldberg, E. D. (Eds), *Trace Metals in Sea Water*, pp. 487–503, Plenum Press, New York.

Scott, B. C., (1981). Modeling of atmospheric wet deposition. In Eisenreich, S. J. (Ed.), *Atmospheric Pollutants in Natural Waters*, pp.3–21, Ann Arbor Science, Ann Arbor, MI.

Settle, D. M., and Patterson, C. C. (1982). Magnitudes and sources of precipitation and dry deposition fluxes of industrial and natural leads to the North Pacific at Enewetak. *J. Geophys. Res.*, **87**, 8857–8869.

Settle, D. M., Patterson, C. C., Türekian, K. K., and Cochran, J. K. (1982). Lead precipitation fluxes at tropical oceanic sites determined from ^{210}Pb measurements. *J. Geophys. Res.*, **87**, 1239–1245.

Slemr, F., Seiler, W., and Schuster, G. (1981). Latitudinal distribution of mercury over the Atlantic Ocean. *J. Geophys. Res.*, **86**, 1159–1166.

Slinn, W. G. N. (1983). Air-to-sea transfer of particles. In Liss, P. S. and Slinn, W. G. N., (Eds), *Air–Sea Exchange of Gases and Particles.* pp. 299–405, Reidel, Dordrecht.

Slinn, S. A., and Slinn, W. G. N. (1980). Predictions for particle deposition on natural waters. *Atmos. Environ.*, **14**, 1013–1016.

Slinn, S. A., and Slinn, W. G. N. (1981). Modeling of atmospheric particulate deposition to natural waters. In Eisenreich, S. J. (Ed.), *Atmospheric Pollutants in Natural Waters*, pp. 23–53. Ann Arbor Science, Michigan.

Uematsu, M., Duce R. A., Prospero, J. M., Chen, L., Merrill, J. T., and McDonald, R. L. (1983). Transport of mineral aerosol from Asia over the North Pacific Ocean. *J. Geophys. Res.*, **88**, 5343–5352.

Walsh, P. R., and Duce, R. A. (1976). The solubilization of anthropogenic atmospheric vanadium in sea water. *Geophys. Res. Lett.*, **3**, 375–378.

Walsh, P. R., Duce, R. A., and Fasching, J. L. (1979a). Tropospheric arsenic over marine and continental regions. *J. Geophys. Res.* **84**, 1710–1718.

Walsh, P. R., Duce, R. A., and Fasching, J. L. (1979b). Considerations of the enrichment, sources, and flux of arsenic in the troposphere. *J. Geophys. Res.*, **84**, 1719–1726.

Weisel, C. P., Duce, R. A., Fasching, J. L., and Heaton, R. W. (1984). Estimates of the transport of trace metals from the ocean to the atmosphere. *J. Geophys. Res.*, **89**, 11,607–11,618.

Williams, W. M. (1982). A model for the dry deposition of particles to natural water surfaces. *Atmos. Environ.*, **16**, 1933–1938.

Zoller, W. H. (1984). Anthropogenic perturbation of metal fluxes into the atmosphere. In Nriagu, J. O. (Ed.). *Changing Metal Cycles and Human Health*, pp. 27–41. Dahlem Konferenzen, Springer-Verlag, Berlin.

Lead, Mercury, Cadmium and Arsenic in the Environment
Edited by T. C. Hutchinson and K. M. Meema

CHAPTER 12

Speciation and Cycling of Arsenic, Cadmium, Lead and Mercury in Natural Waters

DAVID R. TURNER

Marine Biological Association of the UK, Citadel Hill,
Plymouth PL1 2PB,
England

ABSTRACT

The cycling of the elements in natural waters is driven by a combination of chemical and biological processes. This paper will focus on two key features of the natural water chemistry of As, Cd, Pb and Hg, chemical speciation and mean oceanic residence time.

Calculation of equilibrium speciation in solution is heavily dependent on the accuracy of both the complex stability constants used and of the procedure used for correcting them to the required ionic strength. Some progress has been made in this respect following the publication of critical data-bases, but there remains considerable uncertainty for lead in particular as to the relative significance of different complexes. Treatment of organic complexation is further complicated by our lack of understanding of the structure of natural organic matter. Calculations and measurements based on thermodynamic equilibrium are, however, only the first stage in the study of chemical speciation. Biological processes can give rise to metastable chemical species with sufficiently long lifetimes to be important components in an element's speciation. Arsenic, mercury and lead alkyls, for instance, fall into this category, and lead alkyls also have an anthropogenic source. Redox speciation must also be considered in the case of arsenic; arsenic (V), which is the stable oxidation state in oxic waters, can be reduced to arsenic (III) in anoxic systems. Arsenic (III) released to oxic waters can then be reoxidized to arsenic (V) within a timescale of some days.

The mean oceanic residence time gives a measure of the tendency of an element in seawater to be incorporated into particulate phases: a very short residence time (<1000 years) indicates that the element is removed very rapidly to the sediments while a very long residence time (10^8 years; Na, Br, C1) indicates that the element is very unreactive towards particulate matter. Residence times for the elements considered here range from about 10^5 years (As) to 10^3 years or less (Pb).

Each of the processes considered above relates to the chemical properties of the element concerned, and hence to its fundamental properties and to its position in the periodic table. Recent developments in modelling chemical processes in natural systems in terms of fundamental elemental parameters will be used to place the cycling of the four elements chosen in a more general chemical context.

INTRODUCTION

The differing chemistries of the four elements, arsenic, cadmium, lead and mercury, are considered in terms of their chemical speciation and timescales of cycling through natural waters. Each of the four elements has a different range of possible oxidation states, and different chemical properties within each oxidation state. We will concentrate first on the equilibrium oxidation state of each element in oxygenated waters and its complexation reactions, and then consider metastable oxidation states. Finally, consideration of the interaction of the dissolved forms with particulate matter can be related to the residence times of the elements in seawater.

Equilibrium Speciation Calculations

In oxygenated waters lead, cadmium and mercury are present at equilibrium in the +2 oxidation state, whose chemistry can be considered in terms of complexation of the M^{2+} cation; these elements have featured in a large number of speciation models in recent years. Arsenic, in contrast, is present at equilibrium in the +5 oxidation state which interacts so strongly with water that its chemistry is best considered in terms of the acid H_3AsO_4 and its anions; this element has received rather less attention in speciation models.

SELECTION OF STABILITY CONSTANTS

Equilibrium speciation models are only as good as the data (in the form of stability constants) put into them. Many trace element speciation models have been published in recent years, with differences arising from the selection of stability constants from the literature and the correction of these constants to the appropriate ionic strength. These differences will remain so long as there is room for dispute as to the constants or the correction procedures to be used. The means for resolving these problems lie in the careful investigation of the data in the literature, together with new measurments where necessary, resulting in the publication of authoritative critical compilations of relevant stability constants. The year 1976 saw the publication of two such critical compilations (Smith and Martell, 1976; Baes and Mesmer, 1976), the latter of which also contained procedures for correcting the constants to ionic strengths in the range 0–3 M. These two critical compilations clearly provide the most reliable basis for speciation modelling.

However problems still arise, firstly, because some important interactions were not included because no data of sufficient quality were available when the compilations were prepared, and secondly because in the compilation of Smith and Martell (1976) critical stability constants for the same complex at different ionic strengths (I) are in some cases incompatible: thus, for example, correcting constants from 0.5 M or 1 M to the ionic strength of seawater (0.7 M) may result in different values. Moreover, these compilations are now almost a decade old and many new determinations of stability constants have been made in the intervening period. Updating and extending such compilations while maintaining their critical standards would represent a major undertaking, yet is essential if we are to see a convergence of different workers' chemical speciation models together with an improvement in their accuracy.

RESULTS OF SPECIATION CALCULATIONS

We will restrict the discussion to two recent speciation models based on critical compilations (Dyrssen and Wedborg, 1980; Turner *et al.*, 1981) since these will suffice to illustrate the major features of the chemistry of the four elements under consideration. Calculations have been made for seawater, low alkalinity river water (LARW) and high alkalinity river water (HARW); the two models differ significantly in the alkalinity and pH values assigned to these river waters (Table 12.1). The results of the speciation calculations are summarized in Tables 12.2 and 12.3, and are quoted only to the nearest percent since the accuracy of the models themselves does not justify further detail. Values of less than 0.5% are therefore shown as zero.

Table 12.1 pH and alkalinity of model systems

	Turner *et al.* (1981)		Dyrssen and Wedborg (1980)	
Water	pH	Alkalinity/mM	pH	Alkalinity/mM
Seawater	8.2	2.24	8.20	2.4
LARW	6.0	0.21	7.31	0.1
HARW	9.0	1.41	8.44	1.4

It can be seen that the major features are in accord (Table 12.2); however each model differs in the exact percentages calculated for complexes with individual ligands, sometimes by large margins (Table 12.3). These differences arise from different ionic strength correction procedures, different stability constants for the lead carbonate complexes (not included in the critical compilation owing to lack of data), and the inclusion or otherwise

Table 12.2 Major species of Cd, Hg, Pb and As in oxygenated waters*

Element	Seawater	LARW†	HARW†	Remarks
Cd(II)	Cl complexes	Free cation	CO_3 complex	
Hg(II)	Cl complexes	Cl complexes	OH complexes	
Pb(II)	Cl, CO_3, OH complexes	Free cation	Cl, CO_3,OH complexes	
As(V)	$HAsO_4^{2-}$	$H_2AsO_4^-$	$HAsO_4^{2-}$	No data for association with major cations

* Dyrssen and Wedborg (1980), Turner *et al.* (1981).
† Table 12.1.

Table 12.3 Seawater speciation models based on critical compilations

	Cd		Hg		Pb		
Ligand	T*	D&W†	T*	D&W†	T*	D&W†	T&S‡
Free metal	3	2	0	0	3	4	3
OH^-	0	0	0	0	9	9	8
Cl^-	97	97	100	82	47	75	46
(ClBr)	–§	0	–§	18	–§	0	–§
CO_3^{2-}	0	0	0	0	41	11	42
SO_4^{2-}	0	0	0	0	1	0	1
% + charged	37	43	0	0	26	47	25
% uncharged	43	37	3	5	57	53	58
% – charged	21	20	97	95	18	0	17

* Turner *et al.* (1981).
† Dyrssen and Wedborg (1980).
‡ As * but using carbonate data from Sipos *et al.* (1980).
§ Mixed ligand complexes not included in this model.

of mixed ligand complexes. In the case of lead-carbonate complexation the most reliable constants are probably those of Sipos *et al.* (1980) determined at seawater ionic strength; these constants however make little difference to the model of Turner *et al.* (1981) (Table 12.3). The inclusion of mixed ligand C1Br complexes of mercury by Dyrssen and Wedborg (1980) is based on statistical grounds in the absence of experimental evidence for their existence: this step is therefore to some extent arbitrary but with a significant effect on the calculated speciation (Table 12.3). Similarly for lead, which is bound by a variety of ligands (Table 12.2 and 12.3), the introduction of mixed ligand complexes will effect a change in the speciation: Whitfield and

Turner (1980) estimated 14% of mixed ligand complexes of lead in seawater. Cadmium, which is weakly bound by only one ligand would not be affected by the inclusion of mixed ligand complexes.

Speciation calculations for arsenic (V) are limited to the dissociation of the tribasic acid H_3AsO_4 as a function of pH owing to the lack of other data (Turner *et al.*, 1981). The analogous phosphate system has been well studied, and in seawater 65% of the total phosphate is calculated to be complexed with the major cations Na, K, Ca and Mg (Turner *et al.*, 1981) so that we can expect similar results for arsenic when the data becomes available.

General Trends

The most important conclusions to be drawn from speciation models are not usually the percentages of individual complexes but the trends in strength of complexation and in the properties which directly affect the biological and geochemical reactivity of the element concerned. The most important of these are the proportion of free metal ion, which is the biologically available fraction for weakly complexed metals such as Cd and Pb (Turner, 1984), and the charge distribution of the complexes which will affect their surface interactions. Both these properties are shown in Table 12.3, and it can be seen that agreement between the models is good, with the exception of anionic lead complexes (with chloride) which are not included in the model of Dyrssen and Wedborg. Mercury in seawater in almost entirely negatively charged while lead and cadmium are distributed between the different charge types. Arsenic is also largely negatively charged in all waters (Table 12.2).

The proportion of free metal ion (Table 12.3) also gives an indication of the strength of complexing of the cations which increases in the order Cd $\sim$ Pb $\ll$ Hg in seawater. Two important trends in complex stability constants contribute to these sequences; the chloride affinity of the cations which increases in the order Pb $<$Cd $\ll$Hg and the degree of hydrolysis which increases in the order Cd $<$Pb $\ll$Hg. The reversal of Pb and Cd in these two sequences is important in comparing the two elements; cadmium speciation depends mainly on salinity while lead speciation is significantly pH-dependent. Thus in river water the strength of complexing follows the hydrolysis sequence Cd $<$Pb $\ll$Hg; compare this with the seawater sequence above. Such sequences can be rationalized in terms of ionic and covalent bonding in complex formation. Ionic bonding depends largely on the sizes and charges of the ions concerned, and thus varies little between these three cations which have the same charge (+2) and similar ionic radii. Sulphate complexation is typical of this type, with the metal-sulphate stability constants being almost identical for all three cations. The chloride ion, which prefers to bind covalently, shows a very definite sequence of affinities as noted above; this is part of a more general sequence of metals from class (a) which form essentially ionic complexes (e.g. Al) to class (b) which form largely covalent complexes (e.g. Hg) (Ahrland, 1975; Turner *et al.*, 1981). This classification has a similar basis to the hard and soft acid and base (HSAB) concept (Pearson, 1969). The hydroxide ion, in common with many other oxygen-binding ligands of high p*K*, shows somewhat ambivalent behaviour in being able to bond strongly to class (b) metals as a covalently bonded ligand, and also to class (a) metals by ionic bonding. The hydrolysis sequence shown above is thus less easy to interpret in terms of the properties of the cations. The same sequence is shown by carbonate and by fulvic acids (below).

Organic Complex Formation

In addition to inorganic complexing ligands, natural waters contain significant quantities of ill-defined organic matter (humic and fulvic acids) which can complex trace metals and thus modify the speciation patterns considered above. Table 12.4 shows the results of adding 10^{-6} M fulvic acid to the speciation model of Turner *et al.* (1981) using various metal-FA stability constants from the literature. The concentration of 10^{-6} M used in Table 12.4 is roughly equivalent to 1 mg/litre DOC since fulvic acids typically show metal binding capacities of the order of 1 meq/gm. No corrections have been made for ionic strength, so the calculations are necessarily very approximate. Furthermore several of the measurements have been made at pH values lower than those of seawater or HARW so that complexation will tend to be underestimated in those waters; at higher pH values the metal binding will be enhanced owing to increased proton dissociation of the binding sites (see e.g. Saar and Weber, 1979). Despite these limitations we can draw some useful general conclusions from Table 12.4. It can be seen that the strength of complexation by FA increases in the order Cd $<$Pb $<$Hg which parallels the hydrolysis sequence noted above. For all three metals the importance of FA complexation also increases in the order seawater $<$HARW $<$LARW. This sequence arises firstly from competition by the major ions Ca^{2+} and Mg^{2+} in seawater which decreases FA binding of trace metals relative to freshwaters, and also from increased hydrolysis and carbonate complexation of the trace metals at higher pH which decreases the importance of FA binding in HARW relative to LARW.

Redox Speciation

The above calculations all depend on the assumption of equilibrium conditions. In practice, non-equilibrium behaviour is frequently found, generally in the form of metastable oxidation states, including metal alkyls. Such

Table 12.4 Metal-organic interactions in natural waters

				% organic bound metal*		
Metal	log *K*	I	pH	Seawater	LARW†	HARW†
Hg‡	18.0	0.02	8.0	0	100	62
Cd‡	4.69	0.02	8.0	0	4	2
Cd§	4.08	0.1	8.0	0	1	1
Pb¶,‖	$K_1=5.1$ $K_2=10.1$	0.1	6.0	0	11	0
Pb‖,**	$K_1=5.5$ $K_2=10.4$	0.1	6.8	0	23	0
Pb††	6.78 (5%) 4.19 (95%)	0.1	6.4	0	22	0

* Calculated for 10^{-6}M organic matter; in seawater the ligand concentration is reduced to 3.4×10^{-8}M due to association with Ca^{2+} and Mg^{2+}. No corrections have been made for pH or ionic strength changes.
† Table 12.1.
‡ Mantoura *et al.* (1978).
§ Saar and Weber (1979).
¶ Saar and Weber (1980).
‖ K_2 refers to bidentate binding, i.e. Pb $(FA)_2$.
** Buffle *et al.* (1977).
†† Varney (1982).

metastable states are usually produced either within organisms or in anoxic environments. Figure 12.1 shows the changes in redox states predicted at thermodynamic equilibrium as seawater is reduced. The pE scale on the horizontal axis describes the electron free energy level and is analogous to pH (Stumm and Morgan, 1981); low pE indicates reducing conditions. The reductions of bicarbonate to methane and sulphate to sulphide mark the limits of pE attainable by microbial redox processes. It can be seen that this is sufficiently low to reduce As(V) to As(III), Hg(II) to Hg(0), and to form sulphides of all four elements. Arsenic(III) has in fact been observed in both estuarine waters (Knox *et al.*, 1984) and seawater (Andreae 1978, 1979). The proportion of As(III) is, however, low in these waters, and even in anoxic interstitial waters complete reduction of arsenic has not been observed (Andreae, 1979; Knox *et al.*, 1984). It can be seen from Figure 12.1 that at the limit of very low pE arsenic is predicted to be present as a solid phase (As_2S_3) while cadmium, lead and mercury are predicted to remain in the dissolved state, as solid sulphides of cadmium and lead occur only in a narrow pE range. However, the pE limit marked in Figure 12.1 runs very close to two boundaries between solid and dissolved phases, those for $CdS/Cd(HS)_2$ and $As(OH)_3/As_2S_3$. The occurrence or otherwise of solid phases has an important bearing on the ultimate fate of arsenic and cadmium discharged into estuarine and coastal waters, since precipitated As_2S_3

or CdS would be immobilized in anoxic systems such as sediment interstitial waters, while soluble As(III) or Cd(II) could be remobilized by diffusion and other transport processes. There are, of course, uncertainties in the thermodynamic data used in these calculations, and also wide variations in the sulphide concentrations occurring in anoxic systems. The limit marked in Figure 12.1 assumes 10% of the total sulphur to be present as dissolved sulphide as an extreme case. Although almost 50% dissolved sulphide has been observed in a Norwegian fjord (Skei, 1983), a more common figure would be 0.1–1%, and less still where iron(II) precipitates much of the sulphide. Such variations, however, cause only minor changes in Figure 12.1. Since the reduction of sulphate to sulphide requires 8 electrons, 5 pE units (the horizontal scale marking in Figure 12.1) corresponds to 40 orders of

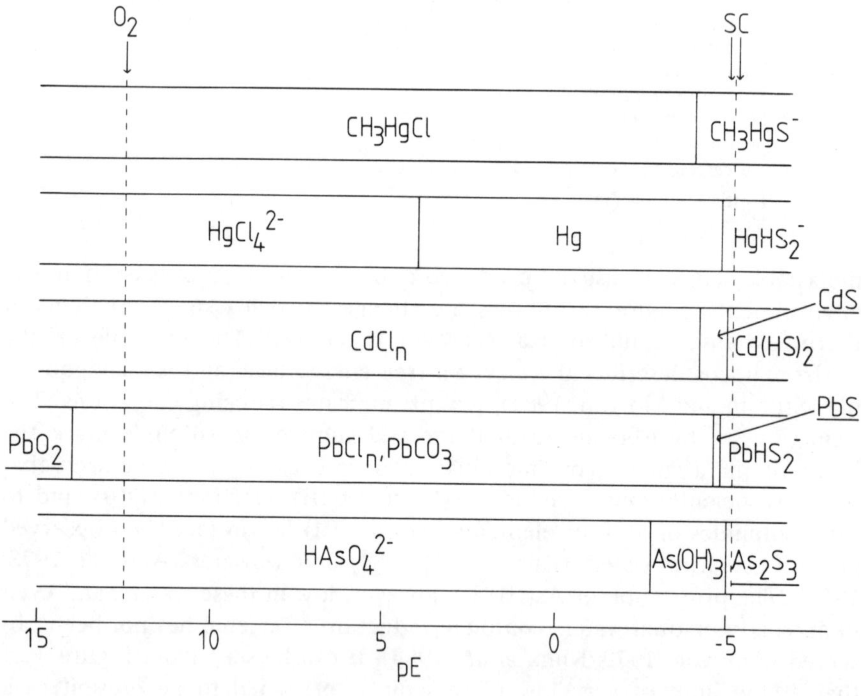

Figure 12.1 Predominance diagram for As, Pb, Cd, Hg and CH_3Hg^+ in seawater at pH 8.2; solid phases are underlined. The pE of oxygenated seawater is indicated by the arrow marked 'O_2', and the pE of anoxic systems by the arrows marked 'S' and 'C' correspond to 10% reduction of sulphate and bicarbonate, respectively. The calculations are based on the speciation model of Turner *et al.* (1981) with additional data from Dobos (1975), Dryssen and Wedborg (1980), Smith and Martell (1976) and Stumm and Morgan (1981), and using oceanic concentrations given by Bruland (1983)

magnitude in the sulphate/sulphide ratio, so that changes of only a few orders of magnitude correspond to very small shifts in the lines drawn. Thus while firm predictions cannot be made at this stage, the possible formation of these solid phases merits further investigation.

Data for a range of methylmercury species (CH_3Hg^+, CH_3HgOH, CH_3HgCl, $(CH_3)_2Hg$, CH_3HgS^-, $(CH_3Hg)_2S$) were included in the calculations for Hg for Figure 12.1, but even when bicarbonate is fully reduced to methane all these species are thermodynamically unstable with respect to Hg(O) and to Hg-sulphide complexes. This strongly suggests that methylmercury species cannot be formed in anoxic systems by chemical processes alone, and that therefore all methylmercury must be formed within living organisms.* Once formed, the CH_3Hg^+ unit is kinetically very stable and can persist for a long time despite its thermodynamic instability: it is therefore included as a separate species in Figure 12.1. Alkyl species of lead and arsenic have not been included in these calculations owing to a lack of data, which leaves an important gap in these thermodynamic models, particularly in the case of arsenic. Monomethyl and dimethyl arsenic compounds can make significant contributions to the speciation of arsenic in natural waters (Andreae, 1978, 1979), so that their thermodynamic stability as a function of pE is of considerable interest.

Adsorption and Residence Time

The discussion has thus far been limited to homogeneous reactions in the dissolved phase, except for some precipitation processes noted in Figure 12.1. A more important heterogeneous process for the four elements considered here, and for the trace elements generally, is adsorption at the surface of particulate or colloidal matter. This process can be considered as a further contribution to the speciation of the element concerned, and is also the key factor determining the residence time of the element in seawater, the major removal process being the sedimentation of particulate matter with its adsorbed trace elements. The study of the adsorption of trace elements on particle surfaces is thus important not only in understanding the distribution of these elements over short time and space scales in natural water systems, but also in understanding the residence times of the elements in the global ocean.

An important development here is the surface complexation model which allows adsorption to be modelled as a chemical process analogous to complexation in the dissolved phase (Davis *et al.*, 1978, Davis and Leckie, 1978). This model treats the particle surface as an array of immobilized ionizable

---

* This opinion is not one that is generally accepted. Many feel that methylation can take place without the presence of living biota. (*Editors*)

hydroxyl groups which can exchange protons and bind anions and cations of both the major and trace elements of the dissolved phase, thus developing the surface charge of the particles. The surface complexation constants are potential-dependent and thus depend in turn on the surface charge developed. The intrinsic surface complexation constants (i.e. the constants at zero surface charge) are found to be directly proportional on a logarithmic scale to the corresponding hydrolysis constants of the cations concerned (Balistrieri *et al.*, 1981), thus providing further encouragement to the development of the model. This approach has proved successful in modelling adsorption data in simple media (Davis and Leckie, 1978, 1979) and is now being extended to more complex systems which more closely resemble those encountered in natural waters, that is seawater instead of simple ionic media and natural particulate matter instead of synthetic, homogeneous solid phases (Balistrieri and Murray, 1982, 1983).

The application of such models to natural systems is thus at an early stage of development, but as a first approximation we might expect the intensity of adsorption of the three cationic metals to follow the hydrolysis sequence Cd $<$Pb $\ll$Hg. In practice this simple picture will be modified firstly by the complexation of the cation in the dissolved phase as discussed above, and secondly by the tendency of charged complexes, particularly hydrolysis products, to be adsorbed in addition to the free metal ion (Davis and Leckie, 1978, 1979). The overall importance of adsorption in the marine environment can, however, be assessed in order of magnitude terms by studying the sequence of oceanic residence times.

The mean oceanic residence time of an element is defined as the total quantity present in the oceans divided by the input rate (from rivers) or the output rate (to the sediment). It is thus a rather approximate quantity since it implicitly assumes the whole ocean to be a well-stirred system, and ignores several important input and output terms (exchanges with the atmosphere and with ridge spreading centres, hydrothermal vents etc.). Nevertheless, residence times give a useful overall picture of the geochemical reactivities of the elements in the oceans. Mean oceanic residence times of the four elements are shown in Table 12.5, together with scavenging residence times in the deep ocean which have been derived from particle and elemental flux measurements. The latter residence times can be expected to be more closely related to adsorption processes, and indeed Balistrieri *et al.* (1981) have shown that the scavenging residence times of Pb, Cd and five other elements can be successfully correlated with a speciation model which incorporates the surfaces of sinking particles. It will be noted from Table 12.5 that the mean oceanic residence time of Hg is greater than that of Pb despite its far greater tendency to hydrolysis. Although a range of factors will contribute to this difference, the strong complexing of Hg by chloride in seawater will undoubtedly tend to inhibit adsorption. The mean oceanic residence times shown in

Table 12.5 also give some idea of the timescale of response of oceanic concentrations to perturbations in the input terms, and are thus important in assessing the effects of anthropogenic inputs to the marine system. The residence time concept is strictly only valid for times significantly greater than the mixing time of the ocean waters themselves (~1500 years). Lead with its very short residence time can thus be seen to be qualitatively as well as quantitatively different from the other three elements, as shown in their concentration profiles in the oceans: recent measurements have shown strongly enhanced concentrations of Pb in ocean surface waters attributable to anthropogenic inputs, while Cd shows a stable nutrient-type depth profile which appears to result from long-term cycling within the oceans (Bruland, 1983).

Table 12.5 Residence times in seawater (years)

Element	Mean oceanic*	Scavenging†
As	63 000	–
Cd	92 000–500 000‡	177 800
Pb	340	47
Hg	14 500	–

* Whitfield (1979), calculated from river input data.
† Balistrieri *et al.* (1981), calculated for the deep ocean.
‡ Range of values calculated from sedimentation rates.

REFERENCES

Ahrland, S. (1975). Metal complexes present in seawater. In Goldberg, E. D. (Ed.), *The Nature of Seawater*, pp. 219–244, Dahlem Konferenzen, Berlin.

Andreae, M. O. (1978). Distribution and speciation of arsenic in natural waters and some marine algae. *Deep-Sea Res.*, **25**, 391–402.

Andreae, M. O. (1979). Arsenic speciation in seawater and interstitial waters: the influence of biological-chemical interactions on the chemistry of a trace element. *Limnol. Oceanogr.*, **24**, 440–452.

Baes, C. F., and Mesmer, R. E. (1976). *The Hydrolysis of Cations*, Wiley, New York, 489 pages.

Balistrieri, L., Brewer, P. G., and Murray, J. W. (1981). Scavenging residence times of trace metals and surface chemistry of sinking particles in the deep ocean. *Deep-Sea Res.*, **28**A, 101–121.

Balistrieri, L. S., and Murray, J. W., (1982). The adsorption of Cu, Pb, Zn and Cd on goethite from major ion seawater. *Geochim. Cosmochim. Acta*, **46**, 1253–1265.

Balistrieri, L. S., and Murray, J. W. (1983) Metal–solid interactions in the marine environment: estimating apparent equilibrium binding constants. *Geochim. Cosmochim. Acta*, **47**, 1091–1098.

Bruland, K. W., (1983). Trace elements in seawater. In Riley, J. P. and Chester, R. (Eds), *Chemical Oceanography*, Volume 8, pp. 157–220, Academic Press, London.

Buffle, J., Greter, F. L. and Haerdi, W. (1977). Measurement of complexation properties of humic and fulvic acids in natural waters with lead and copper ion-selective electrodes. *Anal. Chem.*, **49**, 216–222.

Davis, J. A., James, R. O. and Leckie, J. O. (1978). Surface ionisation and complexation at the oxide/water interface. I. Computation of electrical double layer properties in simple electrolytes. *J. Coll. Interfac. Sci.*, **63**, 480–499.

Davis, J. A., and Leckie, J. O. (1978). Surface ionisation and complexation at the oxide/water interface. II. Surface properties of amorphous iron oxide and adsorption of metal ions. *J. Coll. Interfac. Sci.*, **67**, 90–107.

Davis, J. A., and Leckie, J. O. (1979). Speciation of adsorbed ions at the oxide/water interface. In Jenne E. A. (Ed.) *Chemical Modelling in Aqueous Systems*, pp. 299–317, Amer. Chem. Soc., Washington, DC.

Dobos, D. (1975). *Electrochemical Data.* Elsevier, Amsterdam, 339 pages.

Dyrssen, D. and Wedborg, M. (1980). Major and minor elements, chemical speciation in estuarine waters. In Olausson E. and Cato I. (Eds), *Chemistry and Biogeochemistry of Estuaries*, pp. 71–119, Wiley, Chichester.

Knox, S., Langston, W. J., Whitfield, M., Turner, D. R. and Liddicoat, M. I. (1984). Statistical analysis of estuarine profiles: II. Application to arsenic in the Tamar Estuary (S. W. England), *Est. Cstl. Shelf Sci.*, **18**, 623–638.

Mantoura, R. F. C., Dickson, A. G. and Riley, J. P., (1978). The complexation of metals with humic materials in natural waters, *Est. Cstl. Mar. Sci.*, **6**, 387–408.

Pearson, R. G. (1969). Hard and soft acids and bases, *Surv. Prog. Chem.*, **5**, 1–52.

Saar, R. A., and Weber, J. H. (1979). Complexation of cadmium (II) with water- and soil-derived fulvic acids: effect of pH and fulvic acid concentration. *Can. J. Chem.*, **57**, 1263–1268.

Saar, R. A., and Weber, J. H. (1980). Lead (II)-fulvic acid complexes. Conditional stability constants, solubility and implications for lead (II) mobility. *Environ. Sci. Technol.*, **14**, 877–880.

Sipos, L., Valenta, P., Nurnberg, H. W. and Branica, M. (1980). Voltammetric determination of the stability constants of the predominant labile lead complexes in seawater. In Branica M. and Konrad Z. (Eds.), *Lead in the Marine Environment*, pp. 61–76, Pergamon, Oxford.

Skei, J. M. (1983). Permanently anoxic marine basins—exchange of substances across boundaries. In Hallberg, R. O. (Ed.) *Environmental Biogeochemistry*, pp. 419–429, **Ecol. Bull.** (Stockholm), 35.

Smith, R. M. and Martell, A. E. (1976). *Critical Stability Constants, Volume 4 Inorganic complexes*, Plenum, New York, 257 pages.

Stumm, W. and Morgan, J. J. (1981). *Aquatic Chemistry*, 2nd edition, Wiley, New York, 780 pages.

Turner, D. R. (1984). Relationships between biological availability and chemical measurements. In Sigel H. (Ed.), *Metal Ions in Biological Systems*, Volume 18, pp. 137–164, Marcel Dekker, New York.

Turner, D. R., Whitfield, M. and Dickson, A. G. (1981). The equilibrium speciation of dissolved components in freshwater and seawater at 25° C and 1 atm. pressure. *Geochim. Cosmochim. Acta*, **45**, 855–881.

Varney, M. S. (1982). Organometallic interactions of fulvic acid extracted from natural waters. PhD Thesis, University of Liverpool.

Whitfield, M. (1979). The mean oceanic residence time (MORT) concept—a rationalisation. *Mar. Chem.*, **8**, 101–123.

Whitfield, M. and Turner, D. R. (1980). Theoretical studies of the chemical speciation of lead in seawater. In Branica M. and Konrad Z. (Eds), *Lead in the Marine Environment*, pp. 109–148, Pergamon, New York.

Lead, Mercury, Cadmium and Arsenic in the Environment
Edited by T. C. Hutchinson and K. M. Meema

CHAPTER 13

Studies of Heavy Metal Pollution in Australia with Particular Emphasis on Aquatic Systems

BARRY T. HART

Water Studies Centre,
Chisholm Institute of Technology,
Caulfield East, Australia

and

P. S. LAKE

Zoology Dept.
Monash University
Clayton, Australia

ABSTRACT

The reported work on heavy metal contamination in Australia has been reviewed. The great bulk of this work has been concerned with obtaining information on heavy metal concentrations in waters, sediments and biota, with the emphasis on the latter being often to allay public health concerns. There are fewer data on the effects of heavy metals on Australian biota and that available mostly relates to laboratory-based toxicity studies. Very few data are available which relate to the processes controlling the behaviour of contaminating heavy metals within the Australian environment, that is their speciation, transportation, transformation from one compartment to another and their ultimate fate.

This lack of any adequate process understanding of contaminating heavy metals is of particular concern in Australia because of the quite distinctive features of many Australian aquatic ecosystems (certainly the freshwater ones). This distinctiveness makes the relevance of overseas data relating to the effects of heavy metals on many Australian aquatic ecosystems questionable.

INTRODUCTION

Many heavy metals, including copper, zinc, cobalt and nickel, are essential trace elements. Australia has been actively involved in research on trace elements, especially their agricultural implications (Underwood, 1971). In this chapter, however, we will be concerned with environmental contamination by heavy metals.

Australia is not without examples of serious heavy metal pollution. The most notable have been associated with the mining and metal processing industry. On a smaller scale there have also been examples of localized pollution by electroplating wastes, metal-based pigments and a number of other industrial wastes. This has resulted in considerable activity over recent years, particularly by Government departments, in the study of heavy metals. Most of this activity, however, has resulted simply in the measurement of concentrations in waters, sediments and biota, leaving considerable gaps in the understanding of the processes influencing the behaviour of heavy metals in the environment and of their effects on Australian biota.

This paper reviews the current state of knowledge relating to heavy metals in the Australian environment, with particular emphasis on the aquatic environment. This emphasis arises for two reasons. First, both authors' areas of research relate to aquatic systems, and second there is a paucity of data on heavy metal contamination of Australian terrestrial ecosystems. Most attention has been given to the four metals of primary interest to this Workshop—arsenic, cadmium, lead and mercury. The chapter is structured with reviews on three aspects:

(a) the sources of heavy metals entering the Australian aquatic environment and their levels in various compartments (particularly biota, sediments and waters);

(b) the biogeochemical cycling of heavy metals (taking into consideration speciation, transportation, transformation, mobilization and eventual fate);

(c) the effects of heavy metals on Australian ecosystems (taking into account laboratory toxicity studies and *in situ* studies at population, community and ecosystem levels).

SOURCES

Natural processes, such as weathering and volcanic activity, continually add heavy metals to the environment (Forstner and Wittmann, 1981; Salomons and Forstner, 1984). In addition, there are now increasingly large quantities being added by man's activities. For many of the heavy metals, the amounts contributed globally from anthropogenic sources now exceed those from natural resources. This is particularly so in the case of lead, copper and zinc. Of course, a proportion of the metals mobilized by man will remain bound-up, for example, in buildings and rubbish dumps. Despite this, the amount eventually discharged into the atmosphere or the hydrosphere can be large.

A distinction needs to be made between *point* and *non-point* sources of heavy metal pollution. Point sources are those in which the pollution can be easily identified as coming from a particular factory or treatment works. Non-point sources are more diffuse; examples would include heavy metals in rainfall, urban runoff and agricultural runoff. Most of the Australian examples where serious trace metal pollution has occurred have been very localized and due mainly to one point source.

Little has been done in Australia to assess the importance of anthropogenic emissions of heavy metals to the atmosphere. Salomons and Forstner (1984) have collected together data relating to the natural and anthropogenic inputs of heavy metals to the atmosphere. These data suggest that for the metals of interest to the Workshop, anthropogenic sources exceed natural sources in the case of cadmium and lead.

The only estimates of atmospheric emissions within Australia are for lead. The Australian Academy of Science (AAS, 1981) estimated that 7000 to 8000 tonnes of lead are emitted annually to the Australian atmosphere, with motor cars contributing over 5000 tonnes annually. Considerable work has also been done on pollution from the world's largest lead smelter at Port Pirie, South Australia, where it has been shown that atmospheric emission was the largest source of lead to the environment (Cartwright *et al.*, 1977; Merry and Tiller, 1978; Ward and Young, 1981).

Wind-blown dust can also contribute large quantities of heavy metals to the atmosphere. This can be particularly severe with certain industrial complexes as was shown in Tasmania where fine contaminated dust from a zinc producing company was a significant source of zinc to the Derwent estuary and residential areas of Hobart around Risdon (Ayling and Bloom, 1976; Bloom and Ayling, 1977).

Mining operations, particularly those established a number of years ago, have resulted in serious heavy metal pollution in a number of areas in Australia. The most notable of these have been reviewed recently by Hart (1982a). These include the King River in Tasmania (mainly Cu, also Zn, Pb and Cd); South Esk River, Tasmania (Zn, Cd); Derwent River estuary, Tasmania (mainly Zn and Cd, also Hg); Captains Flat and Molongolo River in New South Wales (mainly Zn, also Cu, Cd, Pb and As); Rum Jungle and Finniss River in the Northern Territory (Cu, Zn). Also well studied have been two rivers in Victoria (Lerderderg River, Goulburn River) which contain elevated mercury levels as a result of old gold mining areas in their headwaters (Bycroft *et al.*, 1982; McCredie, 1982; Ealey *et al.*, 1983).

Although there are a considerable number of possible industrial sources of heavy metals in Australia, most States now require industrial waste discharges to a stream or the sewer to be licensed. This has resulted in two major improvements. First, considerably more attention is paid to the adequate treatment of wastes and to in-plant housekeeping. Second, there has been a searching reappraisal of many of the uses of heavy metals, with the result that some compounds have been totally banned (e.g. phenylmercury compounds used as slimicides in the paper industry), and others have been severely restricted in use.

Sewage effluents and sludges also represent a potentially large source of heavy metals to the environment. We are not aware of any reports in which the concentrations of heavy metals in Australian wastewaters or sludges have been systematically collated. Nor indeed has there been any National assessment of the amounts of heavy metals emitted to the Australian environment from these sources. The major proportion of the sewage and sludge disposal in Australia is to estuarine and coastal marine waters.

A recent study of urban stormwater quality in Melbourne (a city of about 3.5 million people) has shown that this can be an important diffuse source of heavy metals (AWRC, 1981). Table 13.1 summarizes the concentrations of heavy metals found in stormwater from four of the catchments studied.

Table 13.1 Concentrations of heavy metals in μg/litre |mean(range)| in unfiltered urban stormwater samples taken from four stormwater drains in Melbourne*

	Catchment			
Metal	A	B	C	D
Cd	11(10–20)	10(2–10)	9(9–10)	10(9–10)
Cu	480(10–6800)	91(9–310)	38(9–320)	26(9–100)
Cr	580(20–4200)	24(7–30)	20(20–30)	23(20–50)
Pb	490(50–1950)	530(70–2800)	260(30-1370)	160(30–1020)
Ni	20(10–50)	22(9–360)	26(10–40)	22(10–50)
Hg	0.1(<0.1–0.1)	0.1(<0.1–0.1)	—	0.1(0.1–0.2)
Zn	5800(420–19 400)	1230(190–4400)	550(10–2750)	210(9–1050)

Source: AWRC, 1981.
A: Mainly light industrial; B: residential, with two large commercial shopping complexes; C: residential, less than 10 years old; D: mainly undeveloped, orchards and grazing.
* Numbers of samples for Cd, Cu, Cr, Pb, Ni, Zn: A-17, B-55, C-59, D-51; for Hg: A-3, B-15, C-0, D-7.

In summary, the main sources of heavy metals entering the Australian aquatic environment have been identified. In very few cases, however, have we been able to quantify with any degree of confidence the total quantities coming from any of these sources. Such information is, of course, essential if one is to place any priorities upon reducing the amounts.

LEVELS

As noted in the introduction, there has been considerable activity in recent years, particularly by Government departments, in the study of heavy metal contamination in Australia. However, most of this activity has resulted simply in the measurement of concentrations of heavy metals in biota, waters and sediments. Table 13.2 summarizes the number of published papers devoted to the measurement of heavy metal concentrations in various polluted and unpolluted aquatic systems in Australia. Table 13.3 shows more specifically the studies relating to the four metals of most interest to this Workshop.

Table 13.2 Studies on aspects of the biogeochemical cycling of heavy metals through Australian aquatic ecosystems (×= a published study)

System	Polluted/ unpolluted	Sources, levels	Information on process understanding	Effects
Molongolo river (NSW/ACT)	Polluted	××	×	××××××××
King river (Tas.)	Polluted	××	××	×
South Esk river (Tas.)	Polluted	××	×	×××××
Finniss river (NT)	Polluted	××	×	××
Lerderderg river (Vic.)	Polluted	××	×	
Goulburn river (Vic.)	Polluted	××××	×	×
Derwent estuary (Tas.)	Polluted	×××××××××	×	×
Port Pirie/ Spencer's Gulf (SA)	Polluted	××××××××××	××	××××
Port Phillip Bay/ Yarra river (Vic.)	Part-polluted	×××××××××××	××××	×××
Magela creek (NT)	Unpolluted	××××	×××××	××××
Westernport bay (Vic.)	Unpolluted	×××	×××	×××××××××
Great Barrier Reef (Qld.)	Unpolluted	×××××		××
Cockburn Sound (WA)	Unpolluted	××××××		×

Table 13.3 Information on As, Cd, Pb and Hg in the Australian aquatic environment

	Levels, source	Process understanding		Effects	
Arsenic					
Use in Australia waters	BMR, 1979	As forms in:	Edmonds and Francesconi, 1977		
	DOE, 1973	rock lobster	Edmonds *et al.*, 1977		
	Harrison and Dix, 1973		Cannon *et al.*, 1979		
	Bloom and Ayling, 1977	school whiting	Edmonds and Francesconi, 1981		
	Smith, 1984	giant clams	Benson and Summons, 1981		
Sediments	DOE, 1973				
	Harrison and Dix, 1973				
	Davies, 1974				
	Bos, 1984				
Biota (seafoods)	WGMF, 1980				
	Mackay *et al.*, 1975				
	Bebbington *et al.*, 1977				
Cadmium					
Sewage effluent	Hart and Davies, 1979	Speciation	Florence, 1977, 1982a	Toxocity:	
Urban stormwater	AWRC, 1981		Hart and Davies, 1981a,b	invertebrates	Lake *et al.*, 1979
Polluted waters	Doolan and Smythe, 1973		Blutstein and Smith, 1978		Thorp and Lake, 1974
	Norris *et al.*, 1981		Batley and Gardner, 1978		Skidmore and Firth, 1983
	HEC, 1979		Fisher and Frood, 1980		Ahsanullah, 1976
	Ferguson, 1983		Mackey, 1984		Ahsanullah and Arnott, 1978
Unpolluted waters	Hart *et al.*, 1982a				Arnott and Ahsanullah, 1979
	Mackey, 1984	Uptake mechanisms:			Ahsanullah *et al.*, 1981a,b
	Burdon-Jones and Denton, 1984	oysters	Ward, 1982a,b		Sullivan, 1977
Sediments	Bloom, 1975	algae	Gipps and Coller, 1980, 1982	fish	Negilski, 1976
	Norris *et al.*, 1981	sea grass	Fabris *et al.*, 1982		Cassidy and Lake, 1975
	Smith *at al.*, 1981	mussels(FW)	Jones and Walker, 1979		Ehl, 1975
	Ellaway *et al.*, 1982	mussels (MW)	Marshall and Talbot, 1979	field study	Norris *et al.*, 1982
	Ferguson, 1983	invertebrates	Lake *et al.*, 1979	seagrass fauna	Ward, 1984
Biota	Hart, 1982a (summary)				
	Talbot and Chegwidden, 1982	Transport	Hart *et al.*, 1982a		
	Norris and Lake, 1984				
	Talbot, 1985				
	Burdon-Jones and Denton, 1984				
Food intake	Miller *et al.*, 1976				
Soil	Beavington, 1973				

Lead					
Sources	Noller and Smythe, 1974	Speciation	Batley and Florence, 1976	Field study:	
	AAS, 1981		Batley and Gardner, 1978	seagrass fauna	Ward and Young, 1982
	Ferguson, 1983		Blutstein and Smith, 1978		Ward, 1984
Stormwater	AWRC, 1981		Florence, 1977	epifauna	Ward and Young, 1983, 1984
Polluted waters	Anon., 1974; HEC, 1979		Hart and Davies, 1981a,b		Ward *et al.*, 1984
Unpolluted waters	Burdon-Jones and Denton, 1984				
Biota	Hart, 1982a, (summary)				
	Talbot, 1985	Uptake:			
Sediments	Smith *et al.*, 1981	marine mussels	Marshall and Talbot, 1979		
	Ellaway *et al.*, 1982				
	Ward and Young, 1981				
	Ferguson, 1983				
	Ward *et al.*, 1984				
Soils	Cartwright *et al.*, 1977				
	Dossis and Warren, 1980				
Mercury					
Sources	WGMF, 1980	Speciation in biota	WGMF, 1980	Toxicity:	
	Bycroft *et al.*, 1982		Walker, 1982	invertebrate	Ahsanullah, 1982
	Coller, 1983		Walker *et al.*, 1982		Wisely and Blick, 1967
Stormwater	AWRC, 1981			algae	DeFilippis and Pallaghy, 1976a–c
Biota	WGMF, 1980 (summary)	Biota uptake	Denton and Burdon-Jones, 1981	invertebrate	Wisely and Blick, 1967
	Hart, 1982a (summary)			Field study	Beumer and Bacher, 1982
	Bycroft *et al.*, 1982				
	McCredie, 1982				
	Denton and Breck, 1981				
	Walker, 1976, 1981, 1982				
	Walker *et al.*, 1982				
	Lyle, 1984				
	Burdon-Jones and Denton, 1984				
Sediments	Bloom, 1975				
	Bycroft *et al.*, 1982				
	McCredie, 1982				
	Thomas *et al.*, 1981				

This review of the Australian data relating to the concentrations of heavy metals in biota, waters and sediments leads to the following conclusions:

(a) The largest amount of data relates to heavy metal concentrations in marine biota used for human consumption, particularly oysters, lobster and fish. Heavy metal contamination in freshwater biota has been largely neglected.

(b) There are considerable data on concentrations of heavy metals in sediments and waters, with particular emphasis on a small number of well studied systems (e.g. Derwent estuary, Spencer's Gulf, Port Phillip Bay—Table 13.2 and 13.3).

(c) Considering those Australian ecosystems that have been the subject of any degree of examination for heavy metals, it is a sad fact that for no heavy metal do we possess a clear understanding of the pathways of metal movement. One of the major gaps in our knowledge in this area concerns the pathways connecting the abiotic environmental repositories of the heavy metals with the biotic components in any ecosystem. Thus, for no animal investigated in terms of bioaccumulation and biomagnification, do we have a full appreciation of the various pathways and mechanisms of metal uptake.

(d) In the terrestrial environment, the analysis of biota for heavy metal contamination has been limited in scope. The relatively small number of studies of concentrations of heavy metals in human beings in Australia have concentrated heavily on lead; these data are well summarized in the document 'Health and Environmental Lead in Australia' by the Australian Academy of Science (AAS 1981). Concentrations of other heavy metals in humans (e.g. cadmium; Spickett and Lazner, 1979), have received little attention. In terms of medical research effort, very little is carried out in Australia on the levels in humans and toxicity to humans of heavy metals.

BIOGEOCHEMICAL CYCLING OF HEAVY METALS

To assess properly the possible short-and long-term problems associated with the release of heavy metals to the aquatic environment, it is necessary to understand the biogeochemical cycling in such systems. The cycling of heavy metals can best be understood by using a conceptual model for an aquatic system which consists of a number of *compartments* or reservoirs coupled by *transfer pathways.* Figure 13.1 (Hart, 1982a) illustrates a possible model for heavy metal cycling in a lake, where the system is considered to consist of four main compartments:

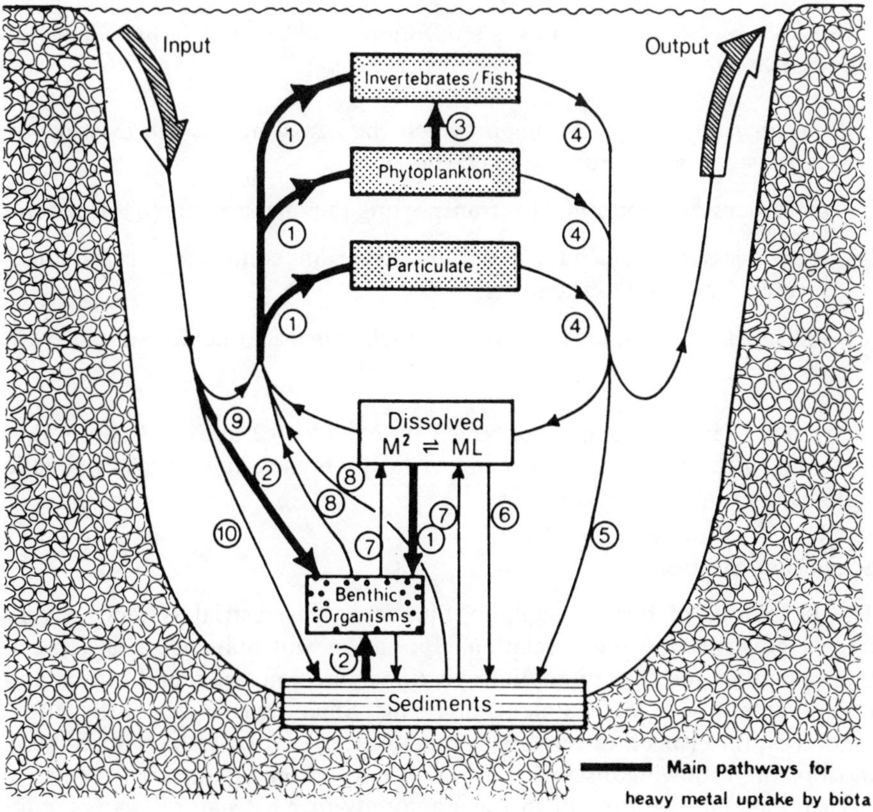

Figure 13.1 A simplified biogeochemical cycle for heavy metals in an aquatic system

(a) the 'dissolved' compartment—containing free metal ions and complexed and colloidally bound metal species;

(b) the (abiotic) particulate compartment—consisting of both inorganic and organic particulates;

(c) the (biotic) particulate compartment—consisting mainly of phytoplankton (and bacteria) in lakes and the deep ocean, littoral areas in estuaries and attached plants in streams;

(d) bottom sediments—the largest compartment of heavy metals in most aquatic systems.

The various transfer processes involved in the transformations and transporting of heavy metals from one compartment to another have been discussed in some detail by Hart (1982a) and Salomons and Forstner (1984).

In particular a detailed assessment of the potential problems associated with heavy metals released to the environment will include a knowledge of the following:

(a) the physico-chemical forms in which the heavy metals can exist in the various compartments (i.e. speciation);

(b) the processes responsible for transporting the metals through the system;

(c) the processes by which the metals in the various compartments are transformed from one to the other;

(d) the most important pathways by which the trace metals interact with the biota.

A quick scan of the data contained in Tables 13.2 and 13.3 shows that the study of biogeochemical cycling of heavy metals in Australian aquatic systems is noticeably lacking.

Speciation Studies

The behaviour of heavy metals in aquatic and terrestrial systems is very largely governed by the speciation. Speciation will influence heavy metal bioavailability and toxicity to biota, its transportation and mobilization, and its interaction with the sediments and soils. Since there is very little information on this aspect for terrestrial systems in either Australia or elsewhere, most of the following discussion will concentrate on aquatic ecosystems.

Heavy metals are present in trace amounts in most natural waters, concentrations generally being less than 1 μg/litre. Florence and Batley (1980) have noted that while the concentrations of heavy metals in unpolluted waters around the globe are relatively similar, the same conclusion does not necessarily hold for the physico-chemical forms in which the metals exist. The actual metal speciation is influenced by factors such as pH and the types and concentrations of inorganic ligands, organic ligands and colloidal species present.

There is increasing evidence to suggest that the formation of metal-organic complexes is very important in determining the speciation of many heavy metals. These complexes may form by the direct complexation of heavy metal ions by natural organic compounds, such as fulvic acid, or by association of metal ions with colloidal matter, either organic colloids or inorganic colloids with their surfaces coated with natural organics. There is now very good evidence that all colloidal and particulate matter in natural waters is so coated (Hart 1982b; Salomons and Forstner, 1984).

Chemical analysis at these low concentrations requires great care and skill (Florence and Batley, 1980; Florence, 1982a). Obviously these difficulties are magnified if, rather than just determining the total concentration, it is required to separate the total into a number of individual fractions. The experimental techniques presently available to speciate natural water samples are generally limited because they are either insufficiently sensitive or selective. Despite the many analytical limitations, considerable insight into the heavy metal speciation in natural fresh, estuarine and marine waters has been obtained by use of analytical techniques such as anodic stripping voltammetry, ultrafiltration, ion-exchange or chelating resins, dialysis and UV irradiation, either singly or in combination (for recent reviews on metal speciation studies see Florence and Batley, 1980; Florence, 1982a and Hart, 1982c).

Australian workers have been particularly prominent in the development of analytical methods for heavy metal speciation. Table 13.4 summarizes some of the speciation studies conducted on Australian waters. These generally represent analysis of a single sample only from the aquatic system. Additionally, the data available are almost specifically for the four metals cadmium, copper, lead and zinc that are best studied by anodic stripping voltammetry.

Transfer Processes

There have been very few systematic investigations of either the *biologically available forms* or the *mechanisms* or *processes* responsible for the transfer of heavy metals from the abiotic to biotic compartments in Australian ecosystems (Tables 13.2 and 13.3).

One of the best studied systems has been Magela Creek in northern Australia. The impetus for these studies was the establishment of a uranium mining and processing plant in the catchment (Fox *et al.*, 1977). Studies by Hart *et al.* (1982a, 1986) have shown that only relatively small amounts of heavy metals are presently transported by Magela Creek with a considerable proportion of this being associated with particulate matter (Fe 80%; Mn 55%; Cu 55%). There is also published work relating to the complexing capacity of these waters (Hart and Davies, 1981c; Hart and Jones, 1984); the transfer of copper from the water column to macrophytes (Hart *et al.*, 1983); and the processes important in cycling of heavy metals added to billabongs in the system (plastic enclosure experiments—Hart *et al.*, 1982b, 1985).

Work on transfer mechanisms between soils and plants and plants and animals has been reported by Davy (1975) for the Finniss River system in Northern Territory. Also there have been a number of reported studies of the transfer of heavy metals between sediments and water and seagrasses in Western Port Bay in Victoria (Shapiro, 1975; Harris *et al.*, 1979; Fabris *et al.*, 1982).

Table 13.4 Australian trace metal speciation studies

System	Metals studied	Reference
Streams		
Yarra river (Vic.)	Cd, Cu, Pb, Zn	Hart and Davies, 1981a
Magela creek (NT)	Cd, Cu, Zn	Hart *et al.*, 1982b
South creek (NSW)	Cd, Cu, Pb, Zn	Florence, 1982b
Lakes		
East Basin Lake (Vic.)	Cd, Cu, Pb, Zn	Hart and Davies, 1981b
Lake Tarli Karng (Vic.)	Cd, Cu, Pb, Zn	Hart and Davies, 1981b
Tarago reservoir (Vic.)	Cd, Cu, Pb, Zn	Hart and Davies, 1981b
Albert Park Lake (Vic.)	Cd	Blustein and Shaw, 1981
Woronora reservoir (NSW)	Cd, Cu, Pb, Zn	Florence, 1977
Magela Creek billabongs (NT)	Cu	Hart and Davies, 1981c
Estuaries		
Derwent estuary (Tas.)	Zn	Fisher, 1981
Yarra estuary (Vic.)	Cd, Cu, Pb, Zn	Hart and Davies, 1981a
	Cd, Cu, Pb, Zn	Blustein and Smith, 1978
Georges R. estuary (NSW)	Cd, Cu, Pb	Batley and Gardner, 1978
	Cr	Batley and Matousek, 1980
Corio bay (Vic.)	Cd, Cu, Zn	Fisher and Frood, 1980
Coastal waters		
Bass Strait (Vic.)	Cu, Zn	Fisher, 1981
	Cd, Cu, Zn	Fisher and Frood, 1980
South Eastern Australia	Cd, Cu, Pb, Zn	Batley and Florence, 1976
	Cu	Florence and Batley, 1977
	Cd, Cu, Pb	Batley and Gardner, 1978
	Cd, Cu, Pb, Zn	Florence, 1982b
Northwest shelf	Cd, Cu, Zn, Ni	
	Cr	Mackey, 1983
	Cd, Cu, Zn, Ni	Mackey, 1984

EFFECTS OF HEAVY METALS ON BIOTA

Two types of studies can produce information relating to the effects of heavy metals on biota: laboratory based toxicity studies (acute and chronic); and *in situ* studies of polluted systems where biota are studied at the level of population, communities or ecosystems. Australian studies in both these areas are summarized in Table 13.3.

Toxicity Studies

The greatest effort in this area has been on marine biota as a result of the Victorian government studies in Port Phillip and Westernport Bays. These studies are listed in Table 13.3; the results have been summarized by Lake and Hart (1986). Most of the tests were static and not continuous flow. Even though nine series of acute toxicity tests have been performed on a variety of marine animals, the taxonomic coverage has been very incomplete. Crustaceans have received considerable attention, but it is perturbing to note that only one toxicity study on marine fish has been published. All of the above tests have been performed on temperate marine animals, with the emphasis very strongly on the metals zinc, cadmium and copper (Table 13.3). A small number of acute toxicity and uptake studies using tropical marine species have been reported (Burdon-Jones *et al.*, 1975; Denton and Burdon-Jones, 1981).

There are no chronic toxicity studies of heavy metals with Australian marine animals.

Acute toxicity data on Australian freshwater invertebrates and fish have been summarized by Firth (1981), Hart (1982a) and Skidmore and Firth (1983). For fish (due largely to the work of Firth (1981) and Skidmore and Firth (1983)), copper is the metal that has been most tested, followed by zinc. Much of the toxicity testing with fish has been stimulated by the introduction of uranium mining and processing in the Northern Territory. Apart from the tests of Cassidy and Lake (1975) and Firth (1981), all the fish tests have been static.

For freshwater invertebrates, cadmium is the metal that has received most attention followed by copper, then zinc (Table 13.3). Most tests have been static.

The range of acute sensitivity to Australian freshwater animals is 0.04–21 mg/litre (525-fold) for copper, 0.14–38 mg/litre (271-fold) for zinc, 0.0148–2000 mg/litre (135 000-fold) for cadmium and 0.05–32 mg/litre (640-fold) for lead. Firth (1981) and Skidmore and Firth (1983) concluded that all these ranges, except for cadmium, are within the lower part of the corresponding ranges for northern hemisphere animals.

Ecosystem Studies

There have been very few attempts in Australia to study the ecological effects of heavy metal pollution. A limited number of freshwater and marine studies have attempted to relate physico-chemical data on metal levels and transport to the ecological impact on the system under investigation. Clearly such system studies need to be encouraged if we are going to be able to predict the impact of heavy metals on terrestrial or aquatic ecosystems. Ideally, such studies should have as their goal a full understanding of the pathways of heavy metal movement, the relative importance of such pathways and a comprehension of the relationships between metal movements and pattern of biota distribution and abundance.

The best marine case studied in terms of these criteria involves the inshore marine area in Spencer's Gulf close to the lead–zinc smelter at Port Pirie, South Australia. The sediments around Port Pirie are contaminated with cadmium, copper, manganese, lead and zinc (Ward and Young, 1981; Ward *et al.*, 1984). The epibenthic fauna of the seagrass beds and the intertidal mud flats around the smelter were studied by Ward and Young (1982, 1983). They found that both contaminant metals and sediment particle size had significant effects on the community structure. The contaminant metals had more effect on fish than crustaceans, while sediment particle size affected the distribution of 15 mostly less common crustaceans. Ward and Young (1983, 1984) also studied the effects of both sediment metal concentrations and particle size on the epifauna associated with the common bivalve mollusc *Pinna bicolor.* They found the different epifaunal taxa had quite different sensitivities to metal concentrations and particle size, and recommended four species that were sensitive to metal pollution but not sediment particle size as species suitable for monitoring the effects of metal contamination on the epifauna.

In a more recent study, Ward (1984) found that acute toxicity tests using dissolved metals were unable to predict ecological effects relating to the seagrass faunal communities in the contaminated area.

Freshwater systems investigated, albeit incompletely in terms of the above criteria, include the Molonglo River in New South Wales, Lake Burley Griffin in the A.C.T., the South Esk River and King River in Tasmania, and the Finniss River in the Northern Territory (Table 13.3).

Zinc pollution of the Molonglo River has greatly reduced the abundance and diversity of the invertebrate fauna (Weatherley and Dawson, 1973; Weatherley *et al.*, 1967, 1975, 1980; Sharley, 1982; Norris, 1983). This zinc pollution has also adversely affected primary production, zooplankton community structure and benthic community structure in Lake Burley Griffin (Hillman, 1974; Weatherley *et al.*, 1980). Fish are not found in most of the Molonglo River (Weatherley *et al.*, 1967). Recent surveys of the invertebrate fauna of the Molonglo River by Nicholas and Thomas (1978), Sharley (1982) and Norris (1983) have shown that there has been no substantial alleviation in the effects of pollution in this river since the first survey in 1963, even though extensive engineering works at the major sources (Captains Flat) have been carried out in an effort to reduce the pollution.

The South Esk River in north-east Tasmania is polluted by zinc and cadmium from tin and wolfram mining operations (Thorp and Lake, 1973; Tyler and Buckney, 1973; Norris *et al.* 1980, 1981, 1982). Thorp and Lake (1973) found that both the species richness and total abundance of macroinvertebrates were low near where the pollution entered. The fauna recovered somewhat with progressive distance downstream; however, even 50 km downstream there was still a marked lack of groups such as Crustacea and

Mollusca. These, together with the Annelida, were least tolerant of the pollution. Thorp and Lake (1973) found relatively tolerant groups to include Hemiptera, Arachnida and trichopteran larvae (Leptoceridae). In a more recent study, Norris *et al.* (1982), found that as far as 80 km downstream from the pollution source the numbers of both individuals and taxa were reduced relative to upstream control sites. They were able to classify the macroinvertebrate species as:

- those relatively abundant at both uncontaminated and contaminated sites (one species of leptocerid caddisfly and one species of baetid mayfly);
- those most abundant at sites above the pollution input (two molluscs, four species of leptophlebiid mayflies and five species of caddisflies);
- those whose numbers were highest at sites below the source of contamination (two species of dipteran larvae, four species of caddisflies, one mollusc species, one amphipod species and one species of water mite).

Norris *et al.* (1982) concluded from their study that there was little correlation between the concentrations found to be lethal in laboratory studies and those found to produce harmful effects in the South Esk River.

The King River in western Tasmania is severely polluted with copper and zinc, and also with lesser amounts of lead and cadmium, from acidic drainage, from disused mine shafts and from waste rock dumps that have resulted from the mining operations of the Mt. Lyell Mining and Railway Company (Lake *et al.*, 1977; HEC, 1979). Two biological surveys have shown that Linda and Comstock Creeks, and the King River for a considerable distance downstream of the junctions with these creeks, are totally devoid of trout and yabbies (*Astacopis* spp.), and macroinvertebrate fauna are significantly reduced in numbers (Lake *et al.*, 1977; Swain *et al.*, 1981). Even as far as 15 km downstream of the pollution inputs there was little recovery in the number of benthic macroinvertebrate taxa or numbers of individuals per square metre.

The Finniss River in the Northern Territory has also been studied ecologically (Davy, 1975; Jeffree and Williams, 1980). The pollution source is an abandoned uranium-copper mine at Rum Jungle. Although mining and processing of copper and uranium ceased in 1971, there is still considerable pollution from overburden heaps, open-cut mine pits and heaps of unprocessed ore as a result of bacterial action on pyritic minerals (Davy, 1975). The main pollution load, low pH water containing high concentrations of copper, zinc, manganese and sulphate, is discharged to the Finniss River during the wet season and at the beginning of the dry season. Jeffree and Williams (1975, 1980) found the biota were significantly affected by this polluted water. There were very few aquatic macrofaunal species that survived in the East Branch downstream of the mining area. Fish and decapod crustacea were also reduced in numbers of species and individuals in the Finniss River downstream of the confluence with the East Branch, although by 30 km downstream of the confluence the effects of the heavy metal pollution were not observable. Fish kills were observed in the Finniss River when moderate inflows from the east Branch coincided with low flow rates in the river itself, producing high pollutant concentrations in the main stream.

DISTINCTIVE FEATURES OF AUSTRALIAN AQUATIC SYSTEMS

The lack of any adequate understanding of the processes controlling the behaviour of heavy metals is of particular concern in Australia because, as is becoming increasingly apparent, many Australian aquatic ecosystems (certainly the freshwater ones) have a number of distinctive features, the most obvious of which are summarized below (Hart, 1982a; DeDeckker and Williams, 1986). These distinctive features preclude the simple use of overseas information relating to cycling and biological effects of heavy metals and make it imperative that we develop new methods for managing river ecosystems in Australia that take account of their individualistic and stochastic nature (Lake, 1985).

Rivers and Streams

A large proportion of Australia is without rivers or permanent lentic (standing) waters because of low and uncertain rainfall, flat topography, high evaporative losses, and a lack of permanent snow fields from which melting snow can replenish rivers and streams during summer. Most of the permanent streams are confined to a narrow strip along the northern and eastern coasts, along small stretches of the southern coast and in Tasmania (AWRC, 1976).

Flow Variability

A direct consequence of the highly variable rainfall is the variability in annual stream flow of many Australian rivers. Stream flows can vary from no flow up to hundreds of times the long-term mean annual discharge; ratios of highest to lowest discharge of up to 300 are common in Australian rivers. In terms of annual flow variability, Australian rivers form a distinctive and extreme group in comparison to world average streams (McMahon, 1982). Because of the great variability in stream flow, many of Australia's major rivers have been impounded, causing drastic changes to their ecology.

Many of Australia's rivers of major economic importance have their source in the Eastern Highlands, or Great Dividing Range, and from here flow either east or west (in some parts of Victoria also north and south). The easterly flowing streams have relatively short courses, descend rapidly from the mountains and frequently change their grade as they make the transition from steep mountain streams to slower coastal rivers. The streams flowing west or south west are very different. Typically they are wide, slow moving and muddy. Rivers such as the Darling and Lachlan may, after several dry seasons, dry up to a chain of waterholes extending over several hundred kilometres. After heavy rains these rivers tend to flood over extensive areas of land because their catchments are flat. This flooding pattern is extremely important to the distribution and reproduction of native fish (Weatherley, 1967).

Tropical rivers, such as those northerly flowing rivers in the Northern Territory and the northern part of Queensland, have received some attention in recent years. They are influenced by a monsoonal climate so that most flow only in the wet season between December and April, and dry up to a series of waterholes during the dry season.

Water Quality

With such a wide range of river systems, it is rather difficult to generalize about water quality (Hart and McKelvie, 1986). When flowing, most rivers have total dissolved solids (TDS) concentrations of less than 500 mg/litre; this equates with a conductivity of around 800 μS/cm. The highest TDS concentrations are found in the slower moving inland streams and the lowest in the fast moving coastal streams (Garman, 1983). There is considerable concern over the increasing salinity in many Australian streams, particularly the River Murray, and rivers in Western Australia and inland Victoria (Garman, 1983; Loh *et al.*, 1984).

Gibbs (1970) postulated that three main mechanisms—rock dominance, atmospheric precipitation and evaporation—control the surface water chemistry in river systems. Precipitation chemistry appears to dominate the water chemistry of the fast flowing, permanent coastal streams in eastern Australia, including Tasmania (Reinson, 1976; Bek and Bruton, 1979; Buckney, 1980). In these cases, sodium is always the dominant cation, with chloride or bicarbonate the dominant anion; the coastal flowing streams of eastern Victoria have the following ionic dominance $Na > Ca = Mg > K$, $Cl = HCO_3 > SO_4$. Reinson (1976) and Buckney (1977, 1980) have shown that, while rainfall is the dominant source of soluble ions (mainly sodium and chloride) in these streams, rock weathering can contribute increased amounts of calcium, magnesium and bicarbonate ions.

Inland flowing streams can, however, have a quite different ionic composition, particularly in their lower reaches. Many of these streams have their headwaters in the Great Dividing Range, and here show an ionic dominance similar to the easterly flowing streams. For example, the upper reaches of the River Murray have an ionic dominance Na > Ca = Mg and HCO_3 = Cl > SO_4 (Walker and Hillman, 1977; RMC, 1980), and the Cudgegong River Na > Mg > Ca and Cl > HCO_3 (Muir and Johnson, 1979). After the Cudgegong River has flowed some 240 km northeast, magnesium becomes the dominant cation and bicarbonate the dominant anion, mainly due to the weathering of magnesium minerals in the catchment.

Buckney (1980) has suggested that diluted seawater is a more appropriate reference water for Australia than the world standard freshwater given by Livingstone (1963). This may well be true for coastal streams but is clearly not appropriate for inland flowing streams.

Australian streams vary enormously in their turbidity or suspended solids concentrations. As a generality, inland streams have relatively high turbidity and coastal streams mostly low turbidity; the turbidity levels are very much influenced by flow and land-uses within the catchment.

Lakes

Except for the Antarctic, Australia is the continent most deficient in permanent lakes, lagoons and ponds. In fact, the only extensive area of permanent lakes is the Tasmanian Central Plateau. There are a very large number of farm dams in Australia and it seems that as time passes, the overwhelming bulk of standing water in Australia will be in artificial storages rather than in natural lakes.

Many of the lakes in Australia are saline, with TDS values greater than 3000 mg/litre (Bayly and Williams, 1973; DeDekker, 1983; Hart and McKelvie, 1986); in fact, the ratio of saline to freshwater standing water bodies is higher in Australia than for most other countries in the world. The high TDS concentrations reflect the general aridity of the continent. In Victoria, the best watered State on the mainland, more than half the natural lakes of significant size contain water with a TDS greater than 100 mg/litre, and fewer than 3% contain water with a TDS of less than 200 mg/litre. The range of TDS concentrations recorded for natural lentic waters in Australia is from 2.4 mg/litre in Lake Cootapatamba situated in the Kosciusko Plateau to 352 000 mg/litre in Lake McDonnell in South Australia (Williams, 1967; Williams *et al.*, 1970).

In the great majority of Australian lentic waters, sodium is the dominant cation and chloride the dominant anion. Sometimes bicarbonate becomes the dominant anion and, less often, calcium the dominant cation (Buckney, 1980). In arid areas, where evaporation is significant, the precipitation of calcium and magnesium carbonates probably accounts for the reduced proportions of calcium, magnesium and bicarbonate compared with sodium and chloride.

Many Australian lakes receive water inputs very sporadically, often after a considerable dry period which may be in excess of one year. Under such conditions the quality of the 'first flush' of water will be very much different from the rest of the input (Hart and McKelvie, 1986). Buckney (1980) has suggested these first flushes will be typically highly coloured and turbid, and have low pH, a predominance of sodium and chloride ions, and often high concentrations of bicarbonate and nutrients.

Commonly, lakes situated in the more temperate regions of Australia stratify only *once* per year with complete mixing taking place in the winter; such lakes are called monomictic. This is quite different from the seasonal pattern observed in northern hemisphere temperate lakes where two mixing periods occur, one after marked stratification during the summer and the other after ice cover in the winter; these lakes are called dimictic lakes. There are very few dimictic Australian lakes; one or two exist in Tasmania. Many of the water bodies in Northern Australia appear to be polymictic, in that they have no persistent periods of stratification (Hart and McGregor, 1980).

Biota

A large part of the aquatic fauna in Australian fresh waters has distinctive characteristics. Most of Australia's 130 native freshwater fish are endemic. Australia's long separation from other lands has ensured that none of the truly freshwater groups of fish from the northern hemisphere reached Australia by natural means. With very few exceptions, the freshwater fish fauna of Australia has a closer evolutionary relationship with marine fish than occurs in other continents. Until European settlement, Australia lacked salmonids (e.g. trout), cyprinids (e.g. carp) and true freshwater percids (Lake, 1971). There is a generally low diversity of freshwater fishes in Australia, which may be due to the scarcity of permanent water bodies. Many of the Australian native fish have evolved remarkable reproductive strategies suitable for the regimes of flood and drought. Adaptations include mouth brooding of eggs to survive low dissolved oxygen levels, and spawning on temporarily inundated land.

More than 2000 species of fish have been recorded from Australian marine waters. Less than 10% of these are exploited commercially.

Many of the Australian freshwater invertebrates are also endemic, probably with a large number still unknown to science. This endemicity is not restricted to the species level since the nature of the Australian freshwater fauna is also largely different at the generic levels. Most of the genera of Australian mayflies, stoneflies, caddisflies and odonatans are endemic, and the composition of fauna in Australian streams is very different from that in northern hemisphere streams. Thus, outside Australia, the Gripopterygidae, the most abundant stoneflies in Australian streams, are found only in New Zealand, Fiji and South America. Similarly, Australia's predaceous stone flies, the Eustheniidae, occur outside this country only in New Zealand and South America. Within the Ephemeroptera, the Leptophlebiidae dominate the Australian fauna in lotic habitats both in numbers of species and individuals. Of the 72 described species of Australian mayflies, 43 are leptophlebiids. This situation contrasts with that in North America north of Mexico, where the leptophlebiids are a relatively insignificant part of the fauna with some 64 leptophlebiids described out of a total mayfly fauna of 622 species. The above data suggest that there are considerable physiological differences between Australian fauna and the northern hemisphere fauna.

Lake (1982) has stated that the Australian stream invertebrate fauna is quite diverse and not depauperate as has been suggested by some. Additionally, he and others (e.g. Williams, 1980) have observed that the fauna are characterized by flexible life histories. Williams (1980) has related this to the unpredictable nature of the Australian climate and to the lack of a seasonal leaf fall in Australian forests. Lake (1982) has also suggested that the flexibility in life history of Australian stream invertebrates is fundamentally related to temperature and flow variations, but he casts doubt on the influence of the supposed 'aseasonal' litter input. In fact, Lake (1982) has collated data showing that Australian forests, from paperbark woodlands and wet sclerophyll forests to temperate and subtropical rain forests, do have a very distinctive seasonal leaf fall, with most (up to 70%—Attiwill *et al.*, 1978) occurring in summer. This is in contrast with the northern hemisphere where most occurs in the autumn. Perennial and temporary Australian streams are at least as rich in stream invertebrates as their northern hemisphere equivalents (Lake *et al.*, 1986).

Schemes of longitudinal stream zonation proposed from a temperate northern hemisphere background have been found wanting in their application to Australian streams (Lake *et al.*, 1986). This applies to both the longitudinal zonation scheme of Illies (1961) and to the longitudinal zonation scheme arising from the River Continuum Concept (Vannote *et al.*, 1980).

According to Williams (1980), the Australian freshwater flora is more like the freshwater flora in other continents than the fauna is like the fauna elsewhere. In the case of algae, the fact that they are cosmopolitan and easily transported and dispersed probably contributes to their similarity to algae in other continents. Macrophytes are considerably more distinctive, with Aston (1973) reporting that 40% of the 200 or more species found in Australia are endemic.

Implications for Aquatic Ecosystem Management

Lake (1985) has suggested that the currently accepted perception of the ecological nature of river communities in particular may have to be revised and that this revision could result in major changes in the management of rivers. In particular, he argues that stream communities are *individualistic*, that is the community is a collection of species that simply happen to exist together through converging accidents of space, time and similar environmental needs (Richardson, 1980).

Lake (1985) also argues that stream communities are *stochastic* rather than deterministic (see also Grossman, 1982, and Grossman *et al.*, 1982). Stochastic communities are conceived as collections of species, not in equilibrium, but changing with environmental changes in a probabilistic rather than predictive way. Deterministic communities are collections of species in equilibrium and with a predictable structure.

The generally accepted principles in water quality management involve the use of water quality criteria to protect various beneficial uses. These criteria are obviously based on a deterministic view of streams.

It seems clear that, particularly for Australian rivers, stream management principles must change to consider the stream ecosystem as individualistic and stochastic. This will also involve a change in the way streams are studied; a multivariate and probabilistic approach will be needed to predict the impact of pollution. If indeed stream communities are stochastic, considerably more long-term research will be needed to obtain the necessary understanding of them.

ACKNOWLEDGEMENT

The continued support of the Office of the Supervising Scientist and the Australian Research Grants Scheme is gratefully acknowledged by the authors.

REFERENCES

AAS (1981). *Health and Environmental Lead in Australia.* Aust. Academy Science, Canberra.

Ahsanullah, M. (1976). Acute toxicity of cadmium and zinc to seven invertebrate species from Westernport, Victoria. *Aust. J. Mar. Freshwater Res.*, **27**, 187–196.

Ahsanullah, M. (1982). Acute toxicity of chromium, mercury, molybdenum and nickel to the amphipod *Allorchestes compressa. Aust. J. Mar. Freshwater Res.*, **33**, 465–474.

Ahsanullah, M., and Arnott, G. H. (1978). Acute toxicity of copper, cadmium and zinc to larvae of *Paragrapsus quadridentatus* (H. Milne Edwards) and implications for water quality criteria. *Aust. J. Mar. Freshwater Res.*, **29**, 1–8.

Ahsanullah, M., Negilski, D. S., and Mobley, M. C. (1981a). Toxicity of zinc, cadmium and copper to the shrimp *Callianassa australiensis*, 1. Effects of individual metals. *Mar. Biol.*, **64**, 299–304.

Ahsanullah, M., Negilski, D. S., and Mobley, M. C. (1981b). Toxicity of zinc, cadmium and copper to the shrimp *Callianassa australiensis*, III. Accumulation of metals. *Mar. Biol.*, **64**, 311–316.

Anon. (1974). *Mine Waste Pollution of the Molonglo River.* Aust. Govt. Publishing Service, Canberra.

Arnott, G. H., and Ahsanullah, M. (1979). Acute toxicity of copper, cadmium and zinc to three species of marine copepods. *Aust. J. Mar. Freshwater Res.*, **30**, 63–71.

Aston, H. I. (1973). *Aquatic Plants of Australia.* Melbourne University Press, Melbourne.

Attiwill, P. M., Guthrie, H. B., and Leuning, R. (1978). Nutrient cycling from *Eucalyptus obliqua* (L'Herit.) forest, I. Litter production and nutrient return. *Aust. J. Botany*, **26**, 74–91.

AWRC (1976). *Review of Australia's Water Resources*, 1975. Aust. Govt. Publishing Service, Canberra.

AWRC (1981). *Characterisation of Pollution in Urban Storm Water Runoff.* Aust. Water Resources Council, Tech. Paper No. 60, Aust. Govt. Publishing Service, Canberra, 104 pages.

Ayling, G. M., and Bloom, H. (1976). Heavy metal analyses to characterise and estimate distribution of heavy metals in dust fallout. *Atmos. Environ.*, **10**, 61–64.

Batley, G. E., and Florence, T. M. (1976). Determination of chemical forms of dissolved cadmium, lead, zinc and copper in seawater. *Mar. Chem.*, **4**, 343–363.

Batley, G. E., and Gardner, D. (1978). A study of copper, lead and cadmium in some estuarine and coastal marine waters. *Estuar. Coastal Mar. Sci.*, **7**, 59–70.

Batley, G. E., and Matousek, J. P. (1980). Determination of chromium speciation in natural waters by electrodeposition on graphite tubes for electrothermal atomisation. *Anal. Chem.*, **52**, 1570–1574.

Bayly, I. A. E., and Williams, W. D. (1973). *Inland Waters and their Ecology*, Longman, Melbourne, 316 pages.

Beavington, F. (1973). Contamination of soil with zinc, lead and cadmium in the Wollongong city area. *Aust. J. Soil Res.*, **11**, 27–31.

Bebbington, G. N., Mackay, N. J., Chovjka, R., Williams, R. J., Dunn, A., and Auty, E. H. (1977). Heavy metals, selenium and arsenic in nine species of Australian commercial fish. *Aust. J. Mar. Freshwater Res.*, **28**, 277–286.

Bek, P., and Bruton, G. (1979). *Gippsland Lakes and Streams Hydrochemistry.* Environ. Studies Series No. 238, Ministry for Conservation, Melbourne.

Benson, A. A., and Summons, V. E. (1981). Arsenic accumulation in Great Barrier Reef invertebrates. *Science*, **211**, 482–487.

Beumer, J. P., and Bacher, G. J. (1982). Species of *Anguilla* as indicators of mercury in the coastal rivers and lakes of Victoria, Australia. *J. Fish. Biol.*, **21**, 87–94.

Bloom, H. (1975). *Heavy Metals in the Derwent Estuary.* Chem. Dept., Univ. Tasmania, Hobart.

Bloom, H., and Ayling, G. M. (1977). Heavy metals in the Derwent estuary. *Environ. Geol.*, **2**, 3–22.

Blutstein, H., and Shaw, R. F. (1981). Characterisation of copper binding in lake water. *Environ. Sci. Tech.*, **15**, 1100–1102.

Blutstein, H., and Smith, J. D. (1978). Distribution of species of Cu, Pb, Zn and Cd in a water profile of the Yarra River estuary. *Water Res.*, **12**, 119–125.

BMR (1979). Arsenic. In *Australian Mineral Industry Review*, Bureau of Mineral Resources, Aust. Govt. Publishing Service, Canberra.

Bos, C. (1984). Arsenic in sediments from the Lerderderg river, unpublished report, Water Studies Centre, Chisholm Inst. Tech., Melbourne.

Buckney, R. T. (1977). Chemical dynamics in a Tasmanian river. *Aust. J. Mar. Freshwater Res.*, **28**, 261–268.

Buckney, R. T. (1980). Chemistry of Australian waters: the basic pattern with comments on some ecological implications. In Williams, W. D. (Ed.), *An Ecological Basis for Water Resources Management*, Aust. Nat. Univ. Press, Canberra.

Burdon-Jones, C., Denton, G. W., Jones, G. B., and McPhie, K. A. (1975). *Long term sub-lethal Effects of Metals on Marine Organisms.* I. Baseline survey. Dept. Marine Biology, James Cook Univ., Townsville.

Bycroft, B. M., Coller, B. A. W., Deacon, G. B., Coleman, D. J., and Lake, P. S. (1982). Mercury contamination of the Lerderderg River, Victoria, from an abandoned gold field. *Environ. Pollut.*, **28**A, 135–148.

Cannon, J. R., Edmonds, J. S., Francesconi, K. A., and Langsford, J. B. (1979). Arsenic in marine fauna. In *Internat. Conf. Management and Control of Heavy Metals in the Environment*, pp. 283–286, CEP Consultants Ltd. Edinburgh.

Cartwright, B., Merry, R. H., and Tiller, R. G. (1977). Heavy metal contamination of soils around a lead smelter at Port Pirie, South Australia. *Aust. J. Soil Res.*, **15**, 69–81.

Cassidy, M., and Lake, P. S. (1975). The acute toxicity of cadmium for the fish *Galaxias maculatus* and changes in acute toxicity with season and temperature. *Bull. Aust. Soc. Limnol.*, **6**, 27–32.

Coller, B. A. W. (1983). Sources of mercury in the environment. In *Mercury in the Environment*, pp. 1–4, Royal Aust. Chem. Inst., Melbourne.

Davies, P. J. (1974). Arsenic in sediments on the continental shelf of southeast Australia. *Search*, **5**, 394–397.

Davy, D. R. (Ed.) (1975). *Rum Jungle Environmental Study*, Report AAEC/E365, Aust. Atomic Energy Commission, Sydney.

DeDeckker, P. (1983). Australian salt lakes: their history, chemistry and biota: a review. *Hydrobiol.*, **105**, 231–244.

DeDeckker, P., and Williams, W. D. (Eds.) (1986). *Limnology in Australia*. CSIRO, Australia, 671 pages.

DeFilippis, L. F., and Pallaghy, C. K. (1976a). The effect of sub-lethal concentrations of mercury and zinc on *Chlorella*, I. Growth characteristics and uptake of metals. *Z. Pflanzen Physiol.*, **78**, 197–207.

DeFilippis, L. F., and Pallaghy, C. K. (1976b). The effect of sub-lethal concentrations of mercury and zinc on *Chlorella*, II. Photosynthesis and pigment composition. *Z. Pflanzen Physiol.*, **78**, 314–322.

DeFilippis, L. F., and Pallaghy, C. K. (1976c). The effect of sub-lethal concentrations of mercury and zinc on *Chlorella*, III. Development and possible mechanism of resistance to plants. *Z. Pflanzen Physiol.*, **79**, 323–335.

Denton, G. R. W., and Breck, W. G. (1981). Mercury in tropical marine organisms in North Queensland. *Mar. Pollut. Bull.*, **12**, 116–121.

Denton, G. R. W., and Burdon-Jones, C. (1981). Influences of temperature and salinity on the uptake, distribution and depuration of mercury, cadmium and lead by the black-lip oyster *Saccostrea echinata. Mar. Biol.*, **64**, 317–326.

DOE (1973). Heavy metals in the marine environment of the northwest coast of Tasmania: Report on a survey November 1973, unpublished report, Dept. Environment, Hobart.

Doolan, K. J., and Smythe, L. E. (1973). Cadmium content of some New South Wales waters. *Search*, **4**, 162–163.

Dossis, P., and Warren, L. S. (1980). Zinc and lead in background sediments from Spencer Gulf, South Australia. *Environ. Sci. Tech.*, **15**, 1451–1456.

Ealey, E. H. M., Deacon, G. B., Coller, B. A. W., Bird, G. J., Bos-van der Zalm, C. H. and Raper, W. G. C. (1983). *Mercury in the Food Web of Raspberry Creek. I.*

An Integrated Report on Mercury Chemistry and Biology of Raspberry Creek. EPA Report, Environmental Protection Authority, Melbourne.

Edmonds, J. S., and Francesconi, K. A. (1977). Methylated arsenic from marine fauna. *Nature*, **265**, 436.

Edmonds, J. S., and Francesconi, K. A. (1981). The origin and chemical form of arsenic in the school whiting. *Mar. Pollut. Bull.*, **12**, 92–96.

Edmonds, J. S., Francesconi, K. A., Cannon, J. R., Raston, C. L., Skelton, B. W., and White, A. H. (1977). Isolation, crystal structure and synthesis of arsenobetaine, the arsenical constituent of the western rock lobster, *Panulirus longipes cygnus.* George. *Tetrahedron. Lett.*, **18**, 1543–1546.

Ehl, A. E. (1975). The effect of cadmium on the survival, growth and reproduction of the firetail gudgeon, *Hypseleotris galii.* PhD Thesis, University of Sydney.

Ellaway, M., Hart, B. T., and Beckett, R. (1982). Trace metals in sediments from the Yarra River. *Aust. J. Mar. Freshwater Res.*, **33**, 761–778.

Fabris, G. J., Harris, J. E., and Smith, J. D. (1982). Uptake of cadmium by the seagrass *Heterozostera tasmanica* from Corio bay and Westernport, Victoria. *Aust. J. Mar. Freshwater Res.*, **33**, 829–836.

Farrah, H., and Pickering, W. F. (1977). Influence of clay-solute interactions on aqueous heavy metal ion levels. *Water, Air, Soil Pollut.*, **8**, 189–197.

Ferguson, J. (1983). Concentration and speciation of lead, zinc and cadmium in seawater-like smelter effluent and adjacent marine environments, Port Pirie, South Australia. *Aust. J. Mar. Freshwater Res.*, **34**, 375–385.

Firth, I. (1981). The comparative sensitivity of selected Australian freshwater animals to acute exposure to copper and zinc. MSc Thesis, NSW Inst. Tech., Sydney.

Fisher, N. S. (1981). On the selection of heavy metal tolerance in diatoms from the Derwent estuary, Tasmania. *Aust. J. Mar. Freshwater Res.*, **32**, 556–561.

Fisher, N. S., and Frood, D. (1980). Heavy metals and marine diatoms: influence of dissolved organic compounds on toxicity and selection for metal tolerance among four species. *Mar. Biol.*, **59**, 85–93.

Florence, T. M. (1977). Trace metal species in natural waters. *Water Res.*, **11**, 681–687.

Florence, T. M. (1982a). The speciation of trace metals in natural water. *Talanta*, **29**, 345–364.

Florence, T. M. (1982b). Development of physico-chemical speciation procedures to investigate the toxicity of copper, lead, cadmium and zinc towards aquatic biota. *Anal. Chim. Acta*, **141**, 73–94.

Florence, T. M., and Batley, G. E. (1977). Determination of copper in seawater by anodic stripping voltammetry. *J. Electroanal. Chem.*, **75**, 791–798.

Florence, T. M., and Batley, G. E. (1980). Chemical speciation in natural waters. *CRC Critical Reviews Anal. Chem.*, **9**, 219–296.

Forstner, U., and Wittmann, G. T. W. (1981). *Metal Pollution in the Aquatic Environment*, Springer-Verlag, Heidelberg, 486 pages.

Fox, R. W., Kelleher, G. G., and Kerr, C. B. (1977). *Ranger Uranium Environmental Inquiry*, Second Report. Aust. Govt. Publishing Service, Canberra.

Garman, D. E. J. (1983). *Water Quality Issues in Australia. Water 2000*: Consultants Report No. 7, Aust. Govt. Publishing Service, Canberra.

Gibbs, R. J. (1970). Mechanisms controlling world water chemistry. *Science*, **17**, 1088–1090.

Gipps, J. F., and Coller, B. A. W. (1980). Effect of physical and culture conditions on uptake of cadmium by *Chlorella pyrenoidosa. Aust. J. Mar. Freshwater Res.*, **31**, 747–755.

Gipps, J. F., and Coller, B. A. W. (1982). Effect of some nutrient cations on the uptake of cadmium by *Chlorella pyrenoidosa. Aust. J. Mar. Freshwater Res.*, **33**, 979–988.

Grossman, G. D. (1982). Dynamics and organisation of a rocky intertidal fish assemblage: The persistence and resilience of taxocene structure. *Am. Nat.*, **119**, 611–637.

Grossman, G. D., Moyle, P. B., and Whitaker, J. O. (1982). Stochasticity in structural and functional characteristics of an Indiana stream fish assemblage: a test of community theory. *Am. Nat.*, **120**, 423–454.

Harris, J. E., Fabris, G. J., Statham, P. J., and Tawfik, F. (1979). Biogeochemistry of selected heavy metals in Westernport, Victoria, and the use of invertebrates as indicators with emphasis on *Mytilus edulis planulatus. Aust. J. Mar. Freshwater Res.*, **30**, 159–178.

Harrison, A. J., and Dix, T. G. (1973). Environmental surveys in connection with north west acid waste disposal; predumping (1970) and postdumping (1971) survey. *Tas. Fish. Res.*, **7**, 1–7.

Hart, B. T. (1982a). *Australian Water Quality Criteria for Heavy Metals.* Aust. Water Resources Council Tech. Paper No. 77, Aust. Govt. Publishing Service, Canberra.

Hart, B. T. (1982b). Uptake of heavy metals by sediments and suspended particulates: a review. *Hydrobiol.*, **9**, 299–313.

Hart, B. T. (1982c). Trace metals in natural waters. I. Speciation. *Chem. Aust.*, **49**, 260–265.

Hart, B. T., and Davies, S. H. R. (1979). *A Study of the Physico-chemical Forms of Trace Metals in Natural Waters and Wastewaters.* Aust. Water Resource Council, Tech. Paper No. 35, Aust. Govt. Publishing Service, Canberra.

Hart, B. T., and Davies, S. H. R. (1981a). Trace metal speciation in the freshwater and estuarine regions of the Yarra River, Victoria. *Estuar. Coastal Shelf Sci.*, **12**, 353–374.

Hart, B. T., and Davies, S. H. R. (1981b). Trace metal speciation in three Victorian lakes. *Aust. J. Mar. Freshwater Res.*, **32**, 175–189.

Hart, B. T., and Davies, S. H. R. (1981c). Copper complexing capacity of waters in the Magela Creek systems, northern Australia. *Environ. Tech. Letts.*, **2**, 205–214.

Hart, B. T., and Jones, M. J. (1984). Measurement of trace metal complexing capacity of Magela Creek waters. In Kramer, C. J. M., and Duinker, J. C. (Eds), *Complexation of Trace Metals in Natural Waters*, D. Reidel Publ. Co., Netherlands.

Hart, B. T., and McGregor, R. G. (1980). Limnological survey of eight billabongs in the Magela Creek catchment, Northern Territory. *Aust. J. Mar. Freshwater Res.*, **31**, 611–626.

Hart, B. T., and McKelvie, I. D. (1986). Chemical limnology in Australia. In DeDeckker, P., and Williams, W. D. (Eds), *Limnology in Australia*. CSIRO, Australia.

Hart, B. T., Davies, S. H. R., and Thomas, P. A. (1982a). Transport of iron, manganese, cadmium, copper and zinc by Magela Creek, Northern Territory, Australia. *Water Res.*, **16**, 605-612.

Hart, B. T., Jones, M. J., and Bek, P. (1982b). *Fate of Trace Metals in the Magela Creek Systems, Northern Australia: Experiments with Plastic Enclosures Placed in Island Billabong During the 1980 Dry Season—Trace Metals.* Water Studies Centre Tech. Rep. 23, Chisholm Inst. Tech., Caulfield East.

Hart, B. T., Jones, M. J., and Breen, P. (1983). Copper uptake by the aquatic macrophyte *Najas tenuifolia*. *Environ. Tech. Letts.*, **4**, 217–222.

Hart, B. T., Ottaway, E. M., and Noller, B. N. (1986). The Magela Creek system, northern Australia. I. 1982/83 Wet season water quality, II. Material budget for the floodplain. *Aust. J. Mar. Freshwater Res.* (in press).

Hart, B. T., Jones, M. J., and Currey, N. A. (1985). *Fate of Trace Metals in the Magela Creek Systems, Northern Australia: Experiments Involving the Addition of Copper, Manganese and Zinc to Plastic Enclosures Placed in Island Billabong.* Water Studies Centre Tech. Rep. 24, Chisholm Inst. Tech., Caulfield East.

HEC (1979). King-Franklin investigation: heavy metal pollution study. Geological Rep. No. 64-38-1, Hydro-Electric Commission, Tasmania.

Hillman, T. J. (1974). Planktonic primary production in a turbid zinc-polluted lake. PhD Thesis, Aust. National Univ., Canberra.

Illies, J. (1961). Versuch einer allgemeinen biotzonotischer Gliederung der Fliebgewässer. *Int. Revue. ges. Hydrobiol.*, **46**, 205–213.

Jeffree, R. A., and Williams, N. J. (1975). Biological indicators of pollution of the Finniss river system, especially fish density and abundance. In Davy, D. R. (Ed.), *Rum Jungle Environmental Studies* AAEC/E365, Aust. Atomic Energy Commission, Sydney.

Jeffree, R. A., and Williams, N. J. (1980). Mining pollution and the diet of the purple striped gudgeon *Morurnda morurnda* Richardson (Eleotridae) in the Finniss river, Northern Territory, *Aust. Ecol. Monogr.*, **50**, 457–485.

Jones, W. G., and Walker, K. F. (1979). Accumulation of iron, manganese, zinc and cadmium by the Australian freshwater mussel *Velesunio ambiguus* (Phillipi) and its potential as a biological monitor. *Aust. J. Mar. Freshwater Res.*, **30**, 740–751.

Lake, J. S. (1971). *Freshwater Fishes and Rivers of Australia*, T. Nelson (Aust) Ltd., Melbourne, 61 pages.

Lake, P. S. (1982). Ecology of macroinvertebrates of Australian upland streams: a review of current knowledge. *Bull. Aust. Soc. Limnol.*, **8**, 1–16.

Lake, P. S. (1985). Water quality management and the ecological nature of rivers. *Bull. Aust. Soc. Limnol.*, **10**, 41–50.

Lake, P. S., and Hart, B. T. (1986). Ecotoxicology of heavy metals in Australia—A survey of the published material. In Waid, J. (Ed.), *Ecotoxicology in Australia*, Aust. Academy Sci., Canberra (in press).

Lake, P. S., Coleman, D., Mills, B., and Norris, R. (1977). A reconnaissance of pollution of the King River in the Comstock and Crotty areas, Western Tasmania. In Banks, M. R., and Kirkpatrick, J. B. (Eds), *Landscape and Man: The Interactions between Man and Environment in Western Tasmania*, pp. 157–173, Royal Society of Tasmania, Hobart.

Lake, P. S., Swain, R., and Mills, B. (1979). *Lethal and Sub-lethal Effects of Cadmium on Freshwater Crustaceans.* Aust. Water Resources Council, Tech. Paper No. 37, Aust. Govt. Publishing Service, Canberra.

Lake, P. S., Barmuta, L., Boulton, A., Campbell, I. C., and St. Clair, R. (1986). Australian Streams and Northern Hemisphere Stream Ecology: Comparisons and Problems. In Dodson, J. R., and Westby, M. (Eds), *Are Australian Ecosystems Different?. Proc. Ecol. Soc. Aust.*, **14**, 61–62.

Livingstone, D. A. (1963). Chemical composition of rivers and lakes. In Fleischer, M. (Ed.), *Data on Geochemistry*, US Geological Survey, Prof. Paper 440-G, Washington, 64 pages.

Loh, I. C., Barrett, K. L., and Collins, P. D. K. (1984). Surface water resources assessment in Western Australia. In *Proc. Workshop on Surface Water Resources Data*, pp. 176–205, Aust. Govt. Publishing Service, Canberra.

Lyle, J. M. (1984). Mercury concentration in four carcharhinid and three hammerhead sharks from coastal waters of the Northern Territory. *Aust. J. Mar. Freshwater Res.*, **35**, 441–452.

McCredie, A. (1982). *Mercury and Mining Pollution in the Upper Goulburn River.* Environmental Report No. 9, Graduate School of Environmental Science, Monash University, Clayton.

McMahon, T. A. (1982). World hydrology–does Australia fit? In *Hydrology and Water Resources Symposium*, 1982 Institute of Engineers, Australia, Melbourne.

Mackay, N. J., Williams, R. J., Kacprzac, J. L., Kazacos, M. N., Collins, A. J., and Auty, E. (1975). Heavy metals in cultivated oysters (*Crassostrea commercialis* = *Saccostrea cucullata*) from the estuaries of New South Wales. *Aust. J. Mar. Freshwater Res.*, **26**, 31–46.

Mackey, D. (1983). Metal-organic complexes in seawater—an investigation of naturally occurring complexes of Cu, Zn, Fe, Mg, Ni, Cr, Mn and Cd using high performance liquid chromatography with atomic fluorescence detection. *Mar. Chem.*, **13**, 169–180.

Mackey, D. (1984). Trace metals and the productivity of shelf water off north west Australia. *Aust. J. Mar. Freshwater Res.*, **35**, 505–516.

Marshall, A. T., and Talbot, V. (1979). Accumulation of cadmium and lead in the gills of *Mytilus edulis*: X-ray microanalysis and chemical analysis. *Chem-Biol. Interactions*, **27**, 111–123.

Merry, R. H., and Tiller, K. G. (1978). The contamination of pasture by a lead smelter in a semi-arid environment. *Aust. J. Exp. Agric., and Animal Husb.*, **18**, 89–96.

Miller, G. J., Wylie, M. J., and McKeown, D. (1976). Cadmium exposure and renal accumulation in an Australian urban population. *Med. J. Aust.*, **1**, 20–22.

Muir, G. L., and Johnson, W. D. (1979). Chemistry of the Cudgegong river, New South Wales. *Aust. J. Mar. Freshwater Res.*, **30**, 325–341.

Negilski, D. S. (1976). Acute toxicity of zinc, cadmium and chromium to the marine fishes, yellow-eye mullet (*Aldrichetta forsteri C., and V.*) and small-mouthed hardyhead (*Atherinasoma microstoma Whitley*). *Aust. J. Mar. Freshwater Res.*, **27**, 137–149.

Nicholas, W. L., and Thomas, M. (1978). *Biological Release and Recycling of Toxic Metals from Lake and River Sediments.* Aust. Water Resources Council, Tech. Paper No. 33, Aust. Govt. Publishing Service, Canberra.

Noller, B. N., and Smythe, L. E. (1974). Distribution of lead in vegetation bordering roads in the Sydney metropolitan area. *Search*, **5**, 108–110.

Norris, R. H. (1983). *Water Quality Monitoring of the Molonglo River and Lake Burley Griffin.* National Capital Development Commission, Canberra.

Norris, R. H., and Lake, P. S. (1984). Trace metal concentrations in fish from the South Esk River, north-eastern Tasmania, Australia. *Bull. Environ. Contamin. Toxicol.*, **33**, 348–354.

Norris, R. H., Lake, P. S., and Swain, R. (1980). Ecological effects of mine effluents on the South Esk River, north-eastern Tasmania. I. Study area and basic water characteristics. *Aust. J. Mar. Freshwater Res.*, **31**, 817–827.

Norris, R. H., Swain, R., and Lake, P. S. (1981). Ecological effects of mine effluents on the South Esk River, north-eastern Tasmania. II. Heavy metals. *Aust. J. Mar. Freshwater Res.*, **32**, 165–173.

Norris, R. H., Lake, P. S., and Swain, R. (1982). Ecological effects of mine effluents on the South Esk River, north-eastern Tasmania. III. Benthic macroinvertebrates. *Aust. J. Mar. Freshwater Res.*, **33**, 789–809.

Reinson,, G. E. (1976). Hydrochemistry of the Genoa river basin, New South Wales-Victoria. *Aust. J. Mar. Freshwater Res.*, **27**, 165–186.

Richardson, J. L. (1980). *Dimensions of Ecology*, Oxford Univ. Press, Oxford.

RMC (1980). *Annual Report Year Ending June 1980.* River Murray Commission, Canberra.

Salomons, W., and Forstner, U. (1984). *Metals in the Hydrocycle.* Springer-Verlag, Heidelberg, 349 pages.

Shapiro, M. A. (1975). *Westernport Bay Environmental Study*, 1973–74. Ministry for Conservation, Melbourne.

Sharley, A. (1982). Changes in the composition of macroinvertebrate fauna drifting into a zinc-polluted river. Presented at Aust. Soc. Limnol. Annual Conf., Griffith.

Skidmore, J. F., and Firth, I. C. (1983). *Acute Sensitivity of Selected Australian Freshwater Animals to Copper and Zinc.* Aust. Water Resources Council, Tech. Paper No. 81, Aust. Govt. Publishing Service, Canberra.

Smith, J. D., Butler, E. C. V., Grant, B. R., Little, G. W., Mills, N., and Milne, P. J. (1981). Distribution and significance of copper, lead, zinc and cadmium in the Corio Bay ecosystem. *Aust. J. Mar. Freshwater Res.*, **32**, 151–164.

Smith, J. D. (1984). *The Distribution of Phosphates and Arsenic in Bass Strait Waters.* Report No VI 9, Vic. Inst. Marine Sciences, Melbourne (in press).

Spickett, J. T., and Lazner, J. (1979). Cadmium concentrations in human kidney and liver tissues from Western Australia. *Bull. Environ. Contam. Toxicol.*, **23**, 627–630.

Sullivan, J. K. (1977). Effects of salinity and temperature on the acute toxicity of cadmium to the estuarine crab *Paragrapsus gaimardii* (Milne Edwards). *Aust. J. Mar. Freshwater Res.*, **28**, 339–343.

Swain, R., White, R. W. G., Coleman, D. J., and Hortle, K. E. (1981). *A Biological Assessment of Pollution in the King River Catchment (1978–1979).* Report No. 64-38-2, Hydro-Electric Commission, Hobart.

Talbot, V. (1985). Heavy metal concentrations in the oyster *Saccostrea cucculLata* and *Saccostrea* sp. from the Dampier archipelago, Western Australia. *Aust. J. Mar. Freshwater Res.*, **36**, 169–175.

Talbot, V. W., and Chegwidden, A. (1982). Cadmium and other heavy metal concentrations in selected biota from Cockburn Sound, Western Australia. *Aust. J. Mar. Freshwater Res.*, **33**, 779–788.

Thomas, P. A., Davies, S. H. R., and Hart, B. T. (1980). *Transport of Trace Metals in the Magela Creek System, Northern Territory.* Water Studies Centre, Tech. Rep. No. 14, Chisholm Inst. Tech., Melbourne, 19 pages.

Thorp, V. J., and Lake, P. S. (1973). Pollution of a Tasmanian river by mine effluents. II. Distribution of macroinvertebrates. *Int. Revue ges. Hydrobiol.*, **58**, 885–892.

Thorp, V. J., and Lake, P. S. (1974). Toxicity bioassays of cadmium on selected freshwater invertebrates and the interaction of cadmium and zinc on freshwater shrimp, *Paratya tasmaniensis* Riek. *Aust. J. Mar. Freshwater Res.*, **25**, 92–104.

Tyler, P. A., and Buckney, R. T. (1973). Pollution of a Tasmanian river by mine effluent. I. Chemical evidence. *Int. Revue ges. Hydrobiol.*, **58**, 873–883.

Underwood, E. J. (1971). *Trace Elements in Human and Animal Nutrition* 3rd Edition, Academic Press, New York, 543 pages.

Vannote, R. L., Minshall, G. W., Cummins, K. W., Sedell, J. R., and Cushing, C. E. (1980). The river continuum concept. *Can. J. Fish. Aquatic Sci.*, **37**, 130–137.

Walker, K. F., and Hillman, T. J. (1977). *Limnological Survey of the River Murray in Relation to Albury-Wodonga, 1973–1976.* Albury-Wodonga Development Corporation, Albury.

Walker, T. I. (1976). Effects of species, sex, length and locality on the mercury content of school shark *Galeorhinus australis* (Macleay) and Gummy shark *Mustelus antarcticus* Guenther from south-eastern Australian waters. *Aust. J. Mar. Freshwater Res.*, **27**, 603–616.

Walker, T. I. (1981). Mercury content of edible flesh from Snapper *Chrysophrys auratus* (Bloch and Schneider) in the Victorian commercial catch. *Aust. J. Mar. Freshwater Res.*, **32**, 75–92.

Walker, T. I. (1982). Effects of length and locality on the mercury content of blacklip abalone, *Notohaliotis ruber* (Leach), blue mussel, *Mytilus edulis planulatus* (Lamarck), sand flathead, *Platycephalus bassensis* Cuvier and Valenciennes, and long-nosed flathead, *Platycephalus caeruleopunctatus* (McCulloch), from Port Phillip Bay, Victoria. *Aust. J. Mar. Freshwater Res.*, **33**, 553–560.

Walker, T. I., Glover, J. W., and Powell, D. G. M. (1982). Effects of length, locality and tissue type on mercury and cadmium content of the commercial scallop, *Pecten alba* Tate, from Port Phillip Bay, Victoria. *Aust. J. Mar. Freshwater Res.*, **33**, 547–552.

Ward, T. J. (1982a) Laboratory study of the accumulation and distribution of cadmium in the Sydney rock oyster *Saccostrea commercialis* (I and R). *Aust. J. Mar. Freshwater Res.*, **33**, 33–44.

Ward, T. J. (1982b). Effect of cadmium on particle clearance by the Sydney rock oyster *Saccostrea commercialis* (I and R). *Aust. J. Mar. Freshwater Res.*, **33**, 711–715.

Ward, T. J. (1984). Role of acute metal toxicity in structuring seagrass fauna near a lead smelter. *Mar. Ecol. Prog. Ser.*, **17**, 117–124.

Ward, T. J., and Young, P. C. (1981). Trace metal contaminants of shallow marine sediments near a lead smelter, Spencer Gulf, South Australia. *Aust. J. Mar. Freshwater Res.*, **32**, 45–56.

Ward, T. J., and Young, P. C. (1982). Effects of sediment trace metals and particle size on the community structure of epibenthic seagrass fauna near a lead smelter, South Australia. *Mar. Ecol. Prog. Ser.*, **9**, 137–146.

Ward, T. J., and Young, P. C. (1983). The depauperation of epifauna on *Pinna bicolor* near a lead smelter, Spencer Gulf, South Australia. *Environ. Pollut. Ser. A*, 30, 293–308.

Ward, T. J., and Young, P. C. (1984). Effects of metals and sediment particle size on the species composition of the epifauna of *Pinna bicolor* near a lead smelter, Spencer Gulf, South Australia. *Estuar. Coastal Shelf Sci.*, **18**, 79–95.

Ward, T. J., Warren, L. J., and Tiller, K. G. (1984). The distribution and effects of metals in the marine environment near a lead–zinc smelter, South Australia. In Nriagu, J. O. (Ed.), *Environmental Impacts of Smelters*, John Wiley, New York.

Weatherley, A. H. (1967). *Australian Inland Waters and their Fauna*, Aust. Nat. Univ. Press, Canberra, 287 pages.

Weatherley, A. H., and Dawson, P. (1973). Zinc pollution in a freshwater ecosystem: analysis and proposed solutions. *Search*, **4**, 471–476.

Weatherley, A. H., Beevers, J. R., and Lake, P. S. (1967). The ecology of a zinc-polluted river. In Weatherley, A. H. (Ed.), *Australian Inland Waters and their Fauna*, Aust. National University Press, Canberra.

Weatherley, A. H., Penridge, L., and Dawson, P. (1975). Assessment and eradication of heavy metal pollution in a planned urban environment. *Verh. Int. Verein. Limnol.*, **19**, 2112–2120.

Weatherley, A. H., Lake, P. S., and Rogers, S. C. (1980). Zinc pollution and the ecology of the freshwater environment. In Nriagu, J. O. (Ed.), *Zinc in the Environment, I. Ecological Cycling*, pp. 338–403, John Wiley, New York.

WGMF (1980). *Report on Mercury in Fish and Fish Products.* Working Group on Mercury in Fish, Aust. Govt. Publishing Service, Canberra, 371 pages.

Williams, W. D. (1967). The characteristics of surface lentic waters in Australia. In Weatherley, A. H. (Ed.), *Australian Inland Waters and their Fauna*, Aust. National Univ. Press, Canberra.

Williams, W. D. (1980). Distinctive features of Australian water resources. In Williams, W. D. (Ed.), *An Ecological Basis for Water Resources Management*, Aust. National Univ., Canberra.

Williams, W. D., Walker, K. F., and Brand, G. W. (1970). Chemical composition of some inland surface waters and lake deposits of New South Wales, Australia. *Aust. J. Mar. Freshwater Res.*, **21**, 103–116.

Wisely, B., and Blick, R. A. P. (1967). Mortality of marine invertebrate larvae in mercury, copper and zinc solution. *Aust. J. Mar. Freshwater Res.*, **18**, 63–72.

Lead, Mercury, Cadmium and Arsenic in the Environment
Edited by T. C. Hutchinson and K. M. Meema

CHAPTER 14

A Review on Lead Contamination of Hong Kong's Environment

M. H. WONG*

Department of Biology,
The Chinese University of Hong Kong,
Shatin, Hong Kong

ABSTRACT

With a population of 5.5 million and a limited amount of land (1046 km^2). Hong Kong is one of the most densely populated areas in the world. Due to the decentralization scheme to ease the densely populated centres, the rural area has been urbanized rapidly. The so-called 'satellite' cities in the New Territories have developed and expanded tremendously. As a result, housing estates are interspersed with agricultural and industrial areas.

The discharge of untreated industrial effluent and sewage is the cause of higher lead concentrations in the coastal waters and in marine sediment leading to the higher level of lead in aquatic organisms.

Combustion of leaded gasoline from automobiles is believed to be the major lead contamination in the environment. This is of concern for those who live next to traffic-congested streets. However, the airborne lead levels range from 0.6 to 3.7 $\mu g/m^3$ and the mean value of lead in human blood samples is 15μg/100 ml. These values are comparable to lead levels in other parts of the world.

The measurement of lead in dust has also been used to indicate the extent of lead contamination in several studies. In general, they all found that the level of lead in street dust (mean = 1627 μg/gm in one study) was of the same magnitude as values obtained in different cities throughout the world. As reported elsewhere, the lead content in soil samples fell off rapidly with increasing distance from highways.

Vegetables grown near roadways were found to contain higher lead concentrations than those growing at remote areas. Since many households and farms are close to traffic, and since vegetable growing for family consumption is traditional, a more detailed investigation of lead levels in food and drink consumed locally seems important.

Two reports showed that the aqueous extracts of roadside dust and soil inhibited root growth of crops. It has also been discovered that the roadside population of two grass species, *Eleusine indica* and *Cynodon dactylon* had a higher tolerance to lead than their normal counterparts. This suggests that the lead levels in roadside

* Present address: Department of Biology, Hong Kong Baptist College, Kowloon, Hong Kong.

soils are sufficiently high as to act as a relative factor in determining the survival of plants.

There is a common practice in Hong Kong, Taiwan, China and other Asian countries of recycling waste materials such as night-soil (human excreta), sewage sludge and animal manure for land application and as fishpond fertilizer or supplementary feeds. In general, various investigations demonstrated an increase in productivity with the appropriate application rates of these wastes. Nevertheless, the rather high concentrations of various heavy metals, including lead in the treated crops or fish, especially those treated with sewage sludge, might impose a health hazard.

GENERAL BACKGROUND

Hong Kong is located south of Kwantung Province, China, and consists of Hong Kong Island with its surrounding 200 smaller islands, Kowloon Peninsula, and part of mainland China named the New Territories (Figure 14.1). Out of the total land area of 1046 km^2, about 82% consists mainly of hill slopes.

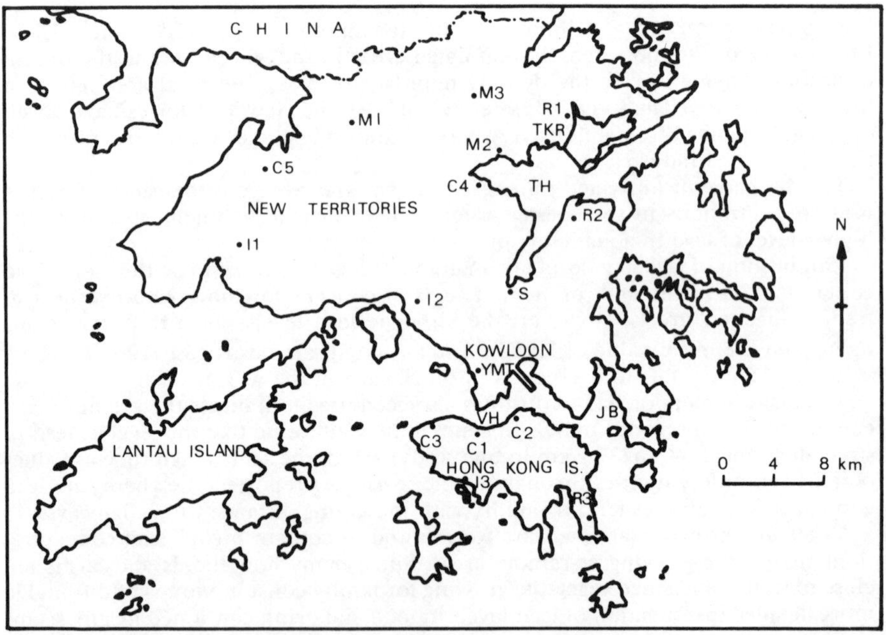

Figure 14.1 Map of Hong Kong showing various places mentioned in the text. JB: Junk Bay, VH: Victoria Harbour, TH: Tolo Harbour, TKR: Ting Kok Rd. S: Shatin, YMT: Yau Ma Tei. *Dust sampling sites*: Commercial and residential area C1: Central District, C2: North Point, C3: Kennedy Town, C4: Tai Po, C5: Yuen Long; *Minor agricultural area* M1: Mai Po, M2: Ha Hang, M3: Man Uk Pin; *Industrial area* I1: Tun Mun, I2: Tsuen Wan, I3: Aberdeen; *Recreational area* R1: Chung Pui, R2: Wu Kai Sha, R3: Shek O

A large portion of land in Hong Kong was reclaimed from the sea, for example, the commercial areas of the north coast of Hong Kong and the south of Kowloon Peninsula. Recently, a large scale land reclamation has been completed at Shatin New Town aiming to house over half a million people. Reclamations on a smaller scale have also been underway in other areas of the New Territories, e.g., Tai Po. This decentralization scheme of easing the densely populated areas led to rapid urbanization in the rural areas (the New Territories).

Although the number of registered vehicles had decreased to 327 803 by the end of 1983, a reduction of 3.5% compared with that at the end of 1982 after the introduction in May, 1982 of fiscal measures to restrain the growth in ownership of motor vehicles (Hong Kong Government, 1984), there is a high traffic density of over 200 vehicles per kilometre of road surface in Hong Kong with most of these vehicles concentrated in the small area of Hong Kong Island and also in 'metropolitan' Kowloon. Lower density traffic occurs in the New Territories.

LEAD FROM GASOLINE COMBUSTION

Hong Kong possesses some deposits of lead which have only been mined in small quantities. In addition, there is secondary lead production from recycled materials such as battery plates and lead pipes. The figures for recycled lead were 1887 tonnes for 1980, 2218 tonnes for 1981 and 2046 tonnes for 1982 (Hong Kong Environmental Protection Agency, 1984a). Secondary lead smelters have caused local neighbourhood health problems of lead in Europe and North America. Nevertheless, the emission of lead into the atmosphere from gasoline combustion is the greatest source of lead contamination although industrial emissions, for example, fossil fuel combustion from power plants, solid waste incineration, cement plants, etc., also emit a considerable amount of lead into the atmosphere. It has been estimated that the amount of lead emission from road transport was 123 tonnes/year (Kalma *et al.*, 1978).

Lead in Air

An earlier study using a computer model to predict atmospheric pollution in Hong Kong showed that the major source of lead emissions is the same as that for carbon dioxide and the estimated average lead concentrations at ground level ranged from 2 to 3 $\mu g/m^3$ under neutral atmospheric conditions. Maximum values at one of the most densely populated areas (Yau Ma Tei, Kowloon) were 8 $\mu g/m^3$, 3 $\mu g/m^3$ and 1 $\mu g/m^3$, respectively, for stable (where there is little turbulence and little wind), near-neutral (the most common condition of average wind speed and direction) and unstable conditions (where turbulence and wind speed are high) (Kalma *et al.*, 1978).

A subsequent survey indicated that the airborne lead levels ranged from 0.6 to 3.7 $\mu g/m^3$. It should be noted that the data were averages of seven consecutive twelve daylight-hour days while the International health standards are normally expressed as averages over three months to two years (Hong Kong Environmental Protection Agency 1982).

Lead in Dust and Soil

Lead in dust has also been used to indicate the levels of lead contamination derived from automobile exhaust. A significant correlation ($p < 0.005$) was found between lead in dust and average annual daily traffic. Table 14.1 lists the total concentration of lead in roadside dust and the values of the annual average daily traffic (AADT) in 14 sites throughout Hong Kong (Lau and Wong, 1982). The samples were tested by atomic absorption spectrophotometry after acid digestion (Allen *et al.*, 1974). The mean concentrations in these sites ranged from 132 to 3876 $\mu g/gm$ which are of the same order of magnitude as an earlier report (Ho, 1979a). Clearly a major factor determining the lead content of dust and soil at roadsides is traffic density.

Table 14.1 The total concentration ($\mu g/gm$) of lead in the roadside dust and the values of annual average daily traffic (AADT) in the 14 selected sites in Hong Kong (from Lau and Wong, 1982).

	C1	C2	C3	C4	C5	M1	M2	M3
Mean	3 876	2 572	2 461	2 002	2 145	1 324	2 001	1 681
SD	1 439	1 431	1 234	1 016	913	491	967	735
AADT	37 180	21 140	18 710	23 730	17 970	9 220	7 180	1 550

	I1	I2	I3	R1	R2	R3
Mean	748	617	1 415	386	132	423
SD	657	538	343	606	91	406
AADT	4 190	980	3 480	510	20	1 590

Commercial and residential area C1: Central District, C2: North Point, C3: Kennedy Town, C4: Tai Po, C5: Yuen Long. Minor agricultural area M1: Mai Po, M2: Ha Hang, M3: Man Uk Pin. Industrial area I1: Tun Mun, I2: Tsuen Wan, I3: Aberdeen. Recreational area R1: Chung Pui, R2: Wu Kai Sha, R3: Shek O.

Table 14.2 further compares the mean lead concentrations in the street dust and soil samples in different cities throughout the world. It indicates that the lead level of roadside dust (1627 $\mu g/gm$) in Hong Kong is not as high as Leeds (UK) (4400 $\mu g/gm$), or Urbana, Illinois (3600 $\mu g/gm$), but is comparable to that found in Birmingham, UK (1800 $\mu g/gm$), Lancaster, UK (1890 $\mu g/gm$), and some of the cities in the United States (1500 $\mu g/gm$).

However, comments on the causes of the differences between lead levels are unjustified due to the absence of knowledge regarding industrial activities in the various locations, as well as information related to the population density of each region.

Table 14.2 Comparison of mean Pb concentrations (μg/gm) in the street dust and soil samples in different cities (from Lau and Wong, 1982)

Location	Dust	Soil	Reference*
Hong Kong	1627	88	Lau and Wong (1982)
Hong Kong	2006	—	Ho (1979a)
Rio de Janeiro, Brazil	700	—	Branquinho (1973)
Birmingham, UK	1800	—	Environ. Prot. Unit. Bir. (1975)
Birmingham, UK	1700	—	Stephen, in Day *et al.* (1975)
Lancaster, UK	1890	—	Harrison (1979)
Leeds, UK	4400	—	Dunn and Bloxam (1933)
Liverpool UK	—	96	Roberts and Johnson (1978)
Manchester, UK	970	—	Day *et al.* (1975)
Detroit, USA	—	65	Tel Haar (1970)
Urbana, Ill., USA	3600	—	Solomon and Artford (1976)
USA (average of 77 cities)	240–1500	—	Hunt *et al.* (1971)

* Refer to Lau and Wong, 1982 for details of references.

A subsequent survey carried out by the Hong Kong Environmental Protection Agency obtained a lower level with an overall mean of 337 μg/gm (Hong Kong Environmental Protection Agency, 1984b). The reduction of lead in petrol from 0.84 to 0.6 gm/litre on 1 July 1981 might have contributed to this reduction. It is envisaged that there will be further reduction of lead in dust when the content of lead is limited to 0.25 gm/litre on 1 January 1985 and with a reduction in the number of cars. However, it seems likely that the lower values obtained by the latest survey might be misleading, or at least more impressive than warranted, due to the fact that samples were collected at an average distance of 10 metres from the curbside rather than along the curbside of the earlier studies. Furthermore, the study was also conducted largely in parks and gardens. In addition, a different sample preparation and analysis also resulted in a discrepancy, for example, the sieve pore size employed in this study was 500 μm compared with 750 μm (Ho, 1979a) and 106 μm (Lau and Wong, 1982) of the earlier studies. It was noted that dust particles with a size smaller than 125 μm belonged to a homogeneous group and did not have significant differences in metal content, including lead (Lau and Wong, 1983).

It is interesting to note that the soil next to a minor agricultural area contained an even higher concentration of trace elements, including lead, than a commercial–residential area (Lau and Wong, 1982). Figure 14.2 further illustrates the lead contents in different depths along the soil profile at three different sampling sites (Wong unpublished data). In general, there is a trend towards a decrease in lead content with increasing depth, with the highest concentrations being found at the soil surface. The lowest concentrations were found at the grassland site, followed by the commercial–residential area, while the highest lead levels occurred in the agricultural area. The latter is an abandoned piece of agricultural land next to an intensive vegetable-farming area. Besides the increased traffic volume, fertilizers, pesticides, animal manure and even night-soil have been suspected of causing heavy metal contamination of agricultural soils located near highways.

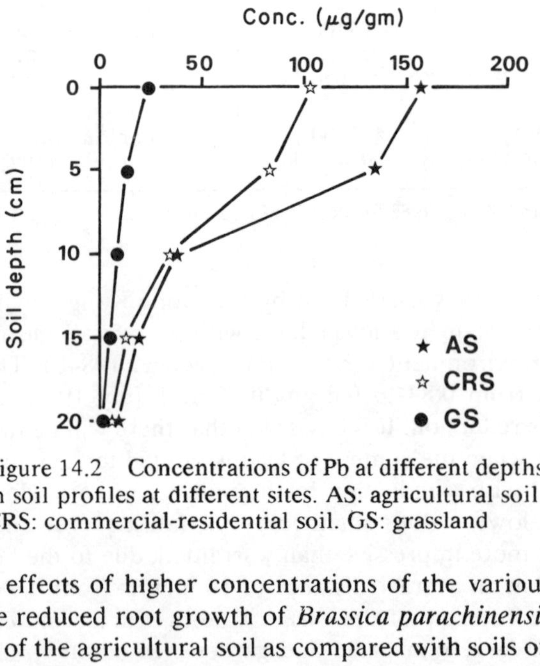

Figure 14.2 Concentrations of Pb at different depths in soil profiles at different sites. AS: agricultural soil. CRS: commercial-residential soil. GS: grassland

The adverse effects of higher concentrations of the various metals were reflected by the reduced root growth of *Brassica parachinensis* when grown on soil extracts of the agricultural soil as compared with soils of other origins (Wong and Lau, 1983).

When using *Brassica chinensis* and *B. parachinensis* to assess the phytotoxicity of the roadside dusts collected from different regions, it was also found that the higher concentrations of lead in water extracts of dust collected from Tai Po (AADT: 27 990) caused a greater reduction of root growth than at the other site (Ting Kok Road AADT: 16 040). Table 14.3 shows the higher concentrations of total and water-soluble lead, copper and zinc at the two sites (Wong *et al.*, 1984).

Table 14.3 The concentrations of Pb, Cu and Zn in dust samples from two different sites and the effects of dust extracts on root growth of *Brassica chinensis* and *B. parachinensis* (from Wong *et al.*, 1984)

Site		Pb	Cu (μg/gm)	Zn
Tai Po	Total content	1051*	100	124†
	Extractable content	777*	17	119‡
Ting Kok	Total content	258	75	83
	Extractable content	26	11	77

Comparisons are made between the two sites using Student's *t*-Test:
* $p < 0.001$,
† $p < 0.01$,
‡ $p < 0.05$

Lead in Vegetation

The lead content of soil samples fell off rapidly with the increase in distance from highways (Wong and Tam, 1978). Unfortunately for the Hong Kong residents, a high proportion live very close to the traffic-congested roads. Exposure may be especially high when there is extreme traffic congestion because of the high volume of traffic, magnified by the narrow canyon-type streets in Hong Kong Island and Kowloon. Furthermore, due to the establishment and expansion of new towns in the New Territories, agricultural areas are now interspersed with housing estates, recreational and industrial areas with highways linked to Kowloon. Vegetables (*Brassica alboglabra* and *B. parachinensis*) were noted to contain higher lead concentrations than those growing at remote areas although a portion of the lead remained as a topical coating on the foliage. Over 27% could be removed by repeated washing with detergent, while washing with distilled water alone reduced the content of lead to a lesser extent (2%) in leaves of *B. parachinensis* (Figure 14.3).

It has also been observed that lead deposition is facilitated on leaves with rough surfaces or hairs. Rain reduces lead tremendously and a drop of 70% in lead level was noted in *Alocasia odora* growing near highways after 13 mm of rain (Ho, 1979b).

Lead Tolerance in Plants

Populations of two grass species, *Eleusine indica* and *Cynodon dactylon*, collected from sites with high lead concentrations in soil and dust showed higher tolerance than those collected from sites with lower lead concentrations (Wong and Lau, 1985a). Table 14.4 shows the relative lead tolerance, as expressed by root elongation in the presence of lead, for three populations of the two grass species. The Tai Po population was the most

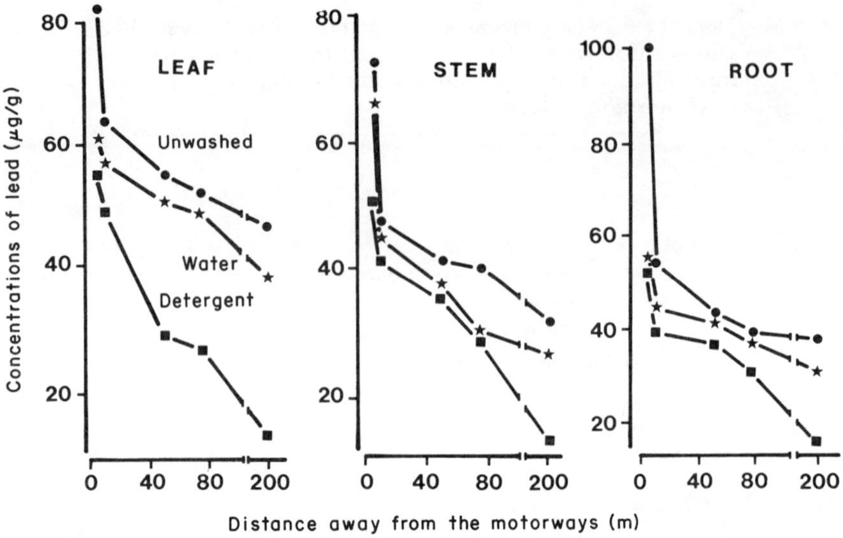

Figure 14.3 The concentrations of lead in different parts of *Brassica parachinensis* under different treatments (from Wong and Tam, 1978). 'Water' indicates that the plants were washed with distilled water and 'detergent' includes washing with detergent

tolerant. Lead tolerance in roadside populations of certain plant species has been reported by others (Wu and Antonovics, 1976; Artkins *et al.*, 1982). A genetic or physiological response to roadside lead appears to be evoked in some plant species suggesting that lead is acting as a selective factor, with advantages to those plant populations able to develop tolerance.

The special tolerance of vegetation growing on lead/zinc mines to these metals has been frequently reported (e.g. Jowett, 1964). At such sites lead levels are often very high. In the case of the Hong Kong grasses, it is interesting to note that these roadside grasses, growing at markedly lower concentrations of lead than vegetation of mine sites, can also evolve tolerance to lead.

While soils, dust and vegetation appear to have been influenced by the lead sources, a survey of lead levels in blood samples from the general human Hong Kong adult population indicates that the mean value of 15 μg/100 ml (Ng-Tan, 1980) is within the range found in other countries (WHO, 1977) and it is slightly lower than the figures reported for some cities in the US (United States Department of Health and Welfare, 1965; McLaughlin *et al.*, 1973). However, further studies are needed before a definite conclusion can be drawn. Human hair samples have also been used to indicate environmental exposure in Hong Kong (Chuang and Emery, 1978). However, lead was not included in this study.

Table 14.4 Index of tolerance of different populations of *Cynodon dactylon* and *Eleusine indica* at two different concentrations of Pb (Wong and Lau, in press a)

Conc. of Pb μg/gm	*Cynodon dactylon*		*Eleusine indica*	
	10	20	10	20
Tai Po	58.40c	35.80c	81.50c	13.10c
Shek O	34.60b	20.60b	73.30b	8.30b
Wu Kai Sha	10.50a	9.70a	25.60a	3.30a

$$\text{Index of tolerance} = \frac{\text{Mean length of longest root in solution with metal}}{\text{Mean length of longest root in solution without metal}} \times 100$$

Values followed by the same letter (a,b,c) under the same species and within the same Pb concentration are not significantly different at 5% level according to the results of Duncan's Multiple Range Test.

LEAD FROM OTHER SOURCES

Industrial Activities

The discharge of the relatively untreated effluent from the 20 000 factories in Hong Kong resulted in the lead concentration (0.66 μg/ml) of harbour waters being 160 times higher than those found in the open sea (Chan *et al.*, 1974). An analysis of 12 drain samples had lead concentrations on average 500 times higher and in one case 1000 times higher than those found in the open sea (Chan *et al.*, 1973).

Rather high concentrations of lead (20–77 μg/gm) were found in the sediments receiving runoff from several streams entering Tolo Harbour, an almost landlocked sea (Wong *et al.*, 1980).

Sewage Dumping

When analysing the marine sediments of Hong Kong for heavy metals it has been found that lead is one of the best indicators of human activities. The rather high values of lead in the sediments of Victoria Harbour are due to the fact that the harbour is one of the busiest in the world. The major source of pollution is due to the scraping off of the red lead-containing paints used in rust-proofing ships. This is emphasized as a source by the rather high concentrations of lead at Junk Bay where the shipbreaking industry is located (Yim and Fung, 1981). Furthermore, the discharge of 56 000 m^3 of sewage per day from the 18 sewage outfalls, which receive no pre-treatment apart from screening to remove large solid particles in the larger discharge, into Victoria Harbour also accounted for the elevated lead content in the sediment of the harbour (Watson and Watson, 1971).

Higher concentrations of various heavy metals including lead were also obtained in marine organisms such as the seaweed *Ulva lactuca* (green alga) sampled along the coastal areas within Tolo Harbour, than those collected from remote areas south of Hong Kong Island (Wong *et al.*, 1982b). In addition, the rather high concentrations of various heavy metals in the tissue of the Pacific oyster (*Crassostrea gigas*), cultured at Deep Bay, which receives water from rural areas where industries are interspersed with agricultural activities, have received much attention (Wong *et al.*, 1981). Another report indicated that although cadmium and arsenic were enriched in the Pacific oysters purchased from retail markets in Hong Kong, lead, zinc, and copper concentrations were within the range reported elsewhere (Phillips *et al.*, 1982a). Significantly elevated concentrations of cadmium and arsenic in Hong Kong seafood (fish, molluscs, lobsters, prawns, shrimps and crabs) were also noted (Phillips *et al.*, 1982b). However, the present limit for lead in Hong Kong seafood of 6 μg/gm wet weight was not exceeded by the various seafoods (including oysters) analysed in these reports. An attempt has been made to use the rock oysters (*Saccostrea glomerata*) as an indicator of trace metals in Hong Kong (Phillips, 1979). Unfortunately, the detection of lead was omitted.

Agricultural Wastes

Agricultural wastes are the most serious cause of pollution in Hong Kong. Pig effluents and poultry droppings contribute more than two-thirds of the total quantity of readily putrescible matter entering streams. Some of this is used by fish farmers and vegetable growers, some is used to feed back to the chickens after being passed through a drum-drier, some is dumped on land, and more than 50% is dumped into streams (Issac and Revell, 1977). It has been revealed that waste materials also possess a substantial amount of lead (total lead of chicken manure 20 μg/gm, pig manure 8.3 μg/gm), copper, zinc and manganese (Cheung and Wong, 1981) which might hinder their usage.

Iron Ore Tailings

The environmental impact of the iron ore tailings deposited along the coastal area of Tolo Harbour has been extensively studied. It is generally found that the comparatively higher level of various metals, including lead in seawater and sediment, results in higher concentrations of these metals in living organisms (Wong and Li, 1977, Wong *et al.*, 1979, etc.).

RECYCLING OF WASTE MATERIALS

Attempts have been made to recycle waste materials such as sewage sludge, animal manure and refuse compost to enhance biological production. It has been noted that the properties of sewage sludge from one sewage treatment plant fluctuated at different times of the year (Wong and Yip, 1980). The concentrations of lead in the activated and digested sludge from two separate treatment plants (the campus of the Chinese University of Hong Kong and the pilot sewage treatment plant at Shek Wu Hui, respectively) also exhibited a similar phenomenon (Wong and Yip, unpublished data).

Recycling of waste materials in a densely populated area such as Hong Kong is most important, especially as some of the streams in the New Territories of Hong Kong are grossly polluted by pig and chicken manure (Bine and Partners, 1974) and contamination of marine sediment is caused by the discharge of sewage.

Growing Vegetables

The addition of sewage sludge at an appropriate rate resulted in much higher productivities of vegetables when compared with the control without any addition of sludge (Wong and Yip, 1978; Wong and Lai, 1982). When activated sludge, digested sludge, chicken manure and pig manure were applied to soils used for growing *Brassica parachinensis*, it was revealed that there was no significant difference in the contents of total lead in the treated soil between the types of waste or as a result of the different concentrations of wastes applied. However, significant differences ($p > 0.05$) were found, among different wastes and different concentrations of wastes applied, in the contents of exchangeable lead, for example, ammonium acetate or dilute HCl extractable.

Sewage sludge, especially activated sludge-amended soil, had a lower pH and organic matter content than animal manure-amended soils. This led to the accumulation of lead and other heavy metals in the crops, although the total and exchangeable lead contents in the waste materials were in the same order of magnitude (Figure 14.4, Cheung and Wong, 1981; 1983).

Refuse compost from the refuse composting plant at Chai Wan, Hong Kong, has a lead level about four times as great as that in the activated sludge but resulted in a lower concentration of lead in the treated crops than those grown in sludge-amended soils. This may be due to the higher pH and organic matter content of the composted material.

It was noted that refuse compost and pig manure, with their higher organic matter, could reduce lead uptake in *Brassica chinensis* and *Raphanus sativus* by five-and seven-fold on average when compared with the control (without any application of waste materials). Na_4-pyrophosphate and $CaCO_3$ had similar effects in the reduction of lead uptake but the effectiveness was lower than with the two organic wastes (Wong and Lau, 1985b).

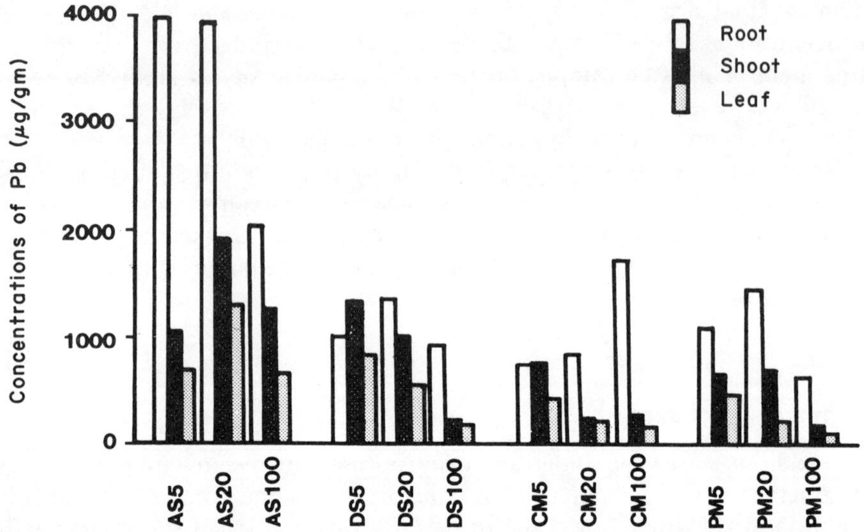

Figure 14.4 Concentrations of lead (μg/gm) in different parts of *Brassica parachinensis* harvested from soils treated with various concentrations of wastes (from Cheung and Wong, 1983). AS: activated sludge, DS: digested sludge, CM: chicken manure, PM: pig manure, rates of application: 5 = 0.1125 kg/m^2, 20 = 0.45 kg/m^2, 100 = 2.25 kg/m^2

Growing Algae

The aqueous extracts of sewage sludge and animal manure were found to be excellent culture media for growing *Chlorella pyrenoidosa, C. salina* and *Ulva lactuca* at suitable concentrations, even better than the artificially enriched media (Wong, 1977, 1981; Wong and Lau, 1979; etc.). However, the rather high levels of various heavy metals, including lead, in the sludge was believed to be the most important factor in inhibiting algal growth when the concentrations of sludge extracts exceeded 10% (v/v). Additions of 10^{-4} and 10^{-3} M EDTA greatly increased algal growth in activated sludge extract which possessed a rather high level of heavy metals (Figure 14.5, Wong *et al.*, 1984).

Since the public has reacted swiftly against food grown on wastes containing potentially objectionable inputs, such as those cultivated in sewage sludge, it is envisaged that algal products derived from food processing wastes would face less consumer resistance since these wastes, such as carrot, coconut, sugar-cane and soybean wastes, possess higher plant nutrients but a lower concentration of lead and other heavy metals (Wong *et al.*, 1985b).

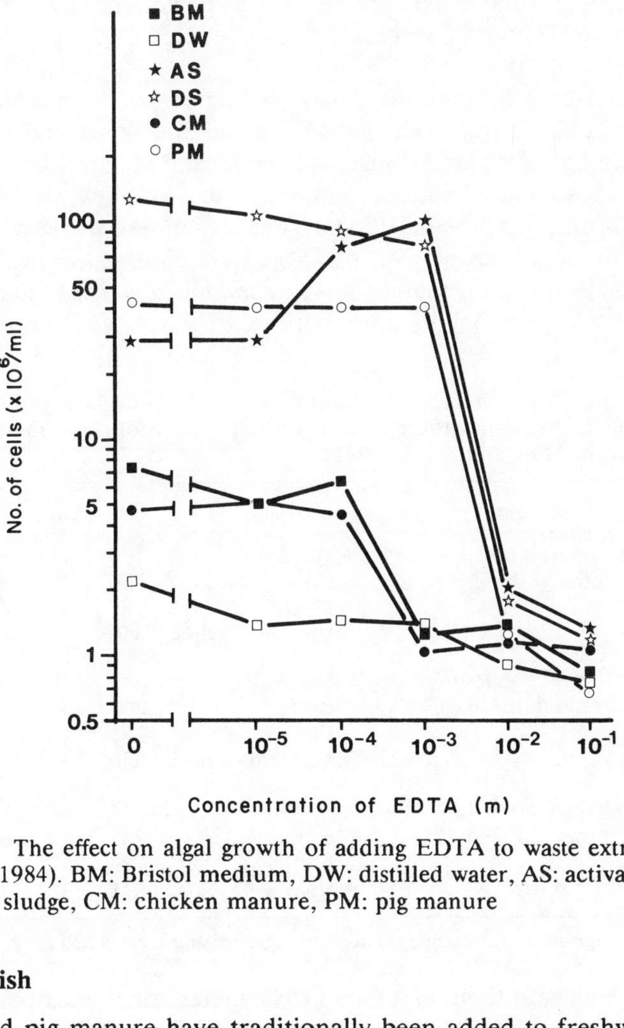

Figure 14.5 The effect on algal growth of adding EDTA to waste extracts (from Wong *et al.*, 1984). BM: Bristol medium, DW: distilled water, AS: activated sludge, DS: digested sludge, CM: chicken manure, PM: pig manure

Growing Fish

Chicken and pig manure have traditionally been added to freshwater fish-ponds to increase the productivity of fish mainly through the enrichment of the pond water, and used as supplementary feeds for fish to replace part of the expensive feedstuff in Hong Kong, Taiwan, China and other Asian countries.

Apart from the deoxygenation of the pond water due to organic matter utilization by microbes, accompanied by fish kills which are commonly observed in hot weather, the rather high concentrations of trace metals in the waste materials have also been of concern. A rather high concentration of lead was observed in the flesh of common carp (*Cyprinus carpio*) fed with the activated sludge supplementary diet, especially at the high ration (80%), as compared with those fed with carrot waste and chicken and pig manure which contained lower concentrations of lead (Wong and Cheung, 1980; Wong *et al.*, 1982a).

The bones of fish (common carp) feeding in ponds to which additions of 70% or 85% activated sludge were made and bones and heads of fish treated with 85% activated sludge had significantly higher concentrations of lead than those treated with a commercial diet (tubifex worm) alone (Table 14.5, Wong and Kwan, 1981). When examining the fish treated with sludge and manure, histological damages were observed in the gills and the hepatopancrease of fish (tilapia, *Sarotherdon mossambicus*) fed on diets supplemented with 60% pig manure, with 40 or 60% chicken manure (Wong *et al.*, 1985a).

Table 14.5 The concentrations of Pb, Cu and Zn (μg/gm dry weight basis) in different parts of carp (mean of three samples) (from Wong and Kwan, 1981)

Treatment	Part	Pb	Cu	Zn
Commercial feed alone (tubifex worms)	Head	106	18	932
	Bone	136	29	432
	Flesh	82	15	280
	Visceral organs	450	45	550
Activated sludge 70% plus commercial feed 30%	Head	98	16	859
	Bone	213*	31	432
	Flesh	81	16	400
	Visceral organs	446	133*	1166*
Activated sludge 85% plus commercial feed 15%	Head	188*	21	1008
	Bone	233*	43	601*
	Flesh	183	12	344
	Visceral organs	531	138*	1140*

* Significant difference between the control and treated fish ($p = 0.05$)

Higher concentrations of other heavy metals such as copper, zinc and manganese were also found in the waste materials and generally these elements were elevated in the treated fish.

GENERAL CONCLUSION

Although the data on the concentrations of lead in different ecological compartments of Hong Kong are fragmented, the preliminary studies indicated that lead contamination of air, soil, dust, marine sediment and living organisms are connected mainly to the combustion of leaded gasoline and the dumping of untreated sewage.

Knowledge of the sources of the environmental exposure and the pathways of this metal in the environment are urgently needed before any effective control strategy can be recommended. This is especially important in a densely populated area such as Hong Kong where a large portion of the population is exposed to the congested traffic as well as various industrial activities. Furthermore, surveys on lead emissions from industrial origins, for example, combustion of fossil fuel and incineration of municipal wastes, are required.

The common practice in the area of recycling waste materials such as night-soil (human excreta), animal manure, and sewage sludge which contain a substantial amount of heavy metals should not be ignored. Reports elsewhere have shown concern that continued use of sewage materials might lead to a hazardous accumulation of metals in soil and therefore restrictions on their uses are suggested. However, investigation on the usage of these waste materials as fishpond fertilizer and feed supplements should also be initiated in order to trace the fate of the various metals derived from the wastes.

REFERENCES

Allen, S. E., Grimshaw, H. W., Parkinson, J. A., and Quarmby, C. (1974). *Chemical Analysis of Ecological Materials*, Blackwell, Oxford, 565 pages.

Artkins, D. P., Trueman, I. C., and Clarke, C. B. (1982). The evolution of lead tolerance by *Festuca rubra* on a motorway verge. *Environ. Poll. Ser. A*, **27**, 233–241.

Bine and Partners (1974). *A Study of the Stream Pollution in New Territories*, Hong Kong. Hong Kong Government Press.

Chan, J. P., Cheung, M. T., and Li, F. P. (1973). Determination of lead and zinc content in drain water around the industrial areas of Hong Kong. Kowloon, and the New Territories. *J. Sci. Eng., Hong Kong Baptist College*, **1**, 9–14.

Chan, J. P., Cheung, M. T., and Li, F. P. (1974). Trace metals in Hong Kong waters. *Marine Pollut. Bull.*, **5**, 171–174.

Cheung, Y. H., and Wong, M. H. (1981). Properties of animal manures and sewage sludges and their utilization for algal growth. *Agric. Wastes*, **3**, 109–122.

Cheung, Y. H., and Wong, M. H. (1983). Utilization of animal manure and sewage sludge for growing vegetables. *Agric. Wastes*, **5**, 63–81.

Chu, L. M., and Wong, M. H. (1985). Utilization of refuse compost for food crop production. In Wong, M. H., Say, P. J., and Whitton, B. A. (Eds), *Ecological Aspects of Solid Waste Disposal*, Pergamon Press. Oxford.

Chuang, L. S., and Emery, J. F. (1978). Hair as an indicator of environmental exposure in Hong Kong. *J. Radioanal. Chem.*, **45**, 169–180.

Ho, Y. B. (1979a). Lead, copper and manganese in street dust in Hong Kong. *J. Asian Ecol.*, **1**, 95–101.

Ho, Y. B. (1979b). Effect of rain on lead levels in roadside vegetation in Hong Kong. *Bull. Environ. Contam. Toxicol.*, **23**, 658–660.

Hong Kong Environmental Protection Agency (1982). Air pollution from road traffic–EPA roadside on monitoring study. Unpublished report. HKEPA.

Hong Kong Environmental Protection Agency (1984a). *Environmental Protection in Hong Kong, 1983–84.* HKEPA.

Hong Kong Environmental Protection Agency (1984b). Lead levels in dust at roadside recreational areas of Hong Kong. Unpublished report. HKEPA.

Hong Kong Government (1984). *Hong Kong 1984, A Review of 1983.* Government Printer, Hong Kong.

Issac, P. V. H., and Revell, J. E. (1977). Pig and poultry wastes in Hong Kong. Seminar on Residue Utilization—Management of Agricultural and Agro-Industrial Wastes. Rome, Jan. 18–21, 1977.

Jowett, D. (1964). Population studies on lead tolerant *Agrostis tenuis. Evolution,* **18**, 70–80.

Kalma, J., Johnson, M., and Newcombe, K. (1978). Energy use and the atmospheric environment in Hong Kong: Part 1, Inventory of air pollutant emissions and prediction of ground-level concentrations of sulphur dioxide and carbon monoxide. *Urban Ecology,* **3**, 29–57.

Lau, W. M., and Wong, M. H. (1982). An ecological survey of lead contents in roadside dusts and soils in Hong Kong. *Environ. Res.,* **28**, 39–54.

Lau, W. M., and Wong, M. H. (1983). The effect of particle size and different extractants on the contents of heavy metals in roadside dusts. *Environ. Res.,* **31**, 229–242.

McLaughlin, M., Linch, A. L., and Snee, R. D. (1973). Longitudinal studies of lead levels in a US population. *Arch. Environ. Health,* **27**, 305.

Ng-Tan, L. H. (1980). A survey of blood lead levels in Hong Kong. *Modern Medicine of Asia,* **16**, 43–45.

Phillips, D. J. H. (1979). The rock oyster *Saccostrea glomerata* as an indicator of trace metals in Hong Kong. *Mar. Biol.,* **53**, 353–360.

Phillips, D. J. H., Ho, C. T., and Ng, L. H. (1982a). Trace elements in the Pacific Oyster in Hong Kong. *Arch. Environ. Contam. Toxicol.,* **11**, 533–537.

Phillips, D. J. H., Thompson, G. B., Gabuji, K. M., and Ho, C. T. (1982b). Trace metals of toxicological significance to man in Hong Kong seafood. *Envion. Poll. Ser. B,* **3**, 27–45.

United States Department of Health and Welfare (1965). *Survey of Lead in the Atmosphere of Three Urban Communities.* Report No. 999-AP-12 USDHEW.

Watson, J. P., and Watson, D. M. (1971). *Marine Investigations into Sewage Discharges, Report and Technical Appendices.* The Hong Kong Government.

Wong, M. H. (1977). The comparison of activated and digested sludge extracts in cultivating *Chlorella pyrenoidosa* and *C. salina. Environ. Pollut.,* **14**, 241–254.

Wong, M. H. (1981). Chicken manure and blood wastes for growing *Chlorella pyrenoidosa. Conserv. Recycl.,* **4**, 9–14.

Wong, M. H., and Cheung, S. P. (1980). Sewage sludge and carrot wastes as supplementary feed for the common carp, *Cyprinus carpio. Environ. Pollut. Ser. A,* **23**, 29–39.

Wong, M. H., and Kwan, S. H. (1981). The uptake of zinc, lead, copper and manganese by carp fed with activated sludge. *Toxicol. Letters.,* **7**, 367–372.

Wong, M. H., and Lai, K. K. (1982). Application of activated sludge for improving iron ore tailings. *Reclam. Vegetat.,* **1**, 83–97.

Wong, M. H., and Lau, K. K. (1979). Cultivation of *Ulva lactuca* in sewage. *Chemosphere.,* **4**, 217–224.

Wong, M. H., and Lau, W. M. (1983). The effects of roadside soil extracts on seed germination and root elongation of edible crops. *Environ. Poll. Ser. A.,* **31**, 203–215.

Wong, M. H., and Lau, W. M. (1985a). Root growth of *Cynodon dactylon* and *Eleusine indica* collected from motorways, at different concentrations of lead. *Environ. Res.*, **36**, 257–267.
Wong, M. H., and Lau, W. M. (1985b). The effects of applications of phosphate, lime, EDTA, refuse compost and pig manure on the lead contents of crops. *Agric. Wastes*, **12**, 61–75.
Wong, M. H., and Li, M. W. (1977). An ecological survey of the heavy metal contamination of the edible clam *Paphia spp.* on the iron ore tailings. *Hydrobiol.*, **56**, 265–273.
Wong, M. H., and Tam, F. Y. (1978). Lead contamination of soil and vegetables grown near motorways in Hong Kong. *Environ. Sci. Health*, **A13**, 13–22.
Wong, M. H., and Yip, S. W. (1978). The comparison of activated and digested sludge applied for Flowering Chinese Cabbage, *Brassica parachinensis*. *Environ. Sci. Health*, **A13**, 241–251.
Wong, M. H., and Yip, S. W. (1980). Seasonal fluctuation of the properties of sewage sludge. *Resource Recov. Conserv.*, **5**, 279–284.
Wong, M. H., Chan, K. Y., Kwan, S. H., and Mo, C. F. (1979). The heavy metal contents of two marine algae grown on the iron ore tailings. *Marine Pollut. Bull.*, **10**, 56–59.
Wong, M. H., Choy, C. K., Lau, W. M., and Cheung, Y. H. (1981). Heavy metal contamination of the Pacific oysters (*Crassostrea gigas*) cultured in Deep Bay, Hong Kong. *Environ. Res.* **25**, 302–309.
Wong, M. H., Ho, K. C., and Kwok, T. T. (1980). Degree of pollution of several major streams entering Tol Harbour, Hong Kong. *Marine Pollut. Bull.*, **11**, 46–50.
Wong, M. H., Cheung, Y. H., and Lau, W. M. (1982a). Toxic effects of animal manures and sewage sludge as supplementary feeds for the common carp, *Cyprinus carpio*. *Toxicol. Letters.*, **12**, 65–73.
Wong, M. H., Kwok, T. T., and Ho, K. C. (1982b). Heavy metals in *Ulva lactuca* collected within Tolo harbour, an almost land-locked sea. *Hydrobiol. Bull.*, **16**, 223–230.
Wong, M. H., Chu, L. M., and Chan, W. C. (1984). The effects of heavy metals and ammonia in sewage sludge and animal manure on the growth of *Chlorella pyrenoidosa*. *Environ. Poll. Ser. A*, **34**, 55–71.
Wong, M. H., Chan, K. M., and Liu, W. K. (1985a). Toxic effects of activated sludge and chicken manure as supplementary feeds for tilapia, *Sarotherdon mossambica*. In Wong, M. H., Say, P. J., and Whitton, B. A. (Eds), *Ecological Aspects of Solid Waste Disposal*, Pergamon Press, Oxford.
Wong, M. H., Chan, W. C., and Chu, L. M. (1985b). Food processing wastes as nutrient sources for algal growth. *Proc. Int. Sym. Industrial and Hazardous Solid Wastes.* Philadelphia, 1983.
World Health Organization (1977). *Environmental Health Criteria, 3. Lead.* United Nations Environ. Program., and WHO, 160 pages.
Wu, L., and Antonovics, J. (1976). Experimental ecological genetics in *Plantago* II. Lead tolerance in *Plantago lanceolata* and *Cynodon dactylon* from a roadside. *Ecology*, **57**, 205–208.
Yim, W. W. S., and Fung, F. W. (1981). Heavy metals in marine sediments of Hong Kong. *Hong Kong Engineer*, Oct. 1981, 33–39.

Lead, Mercury, Cadmium and Arsenic in the Environment
Edited by T. C. Hutchinson and K. M. Meema

CHAPTER 15

The Distribution of Trace Metals in Aswan High Dam Reservoir and River Nile Ecosystems

MOHAMED R. LASHEEN

Water Pollution Control Laboratory,
National Research Center,
Dokki, Cairo, Egypt

ABSTRACT

Since the main source of fresh water in Egypt is the River Nile, which originates from outside the country, a study was designed to assess the levels of trace metals and their effects on the river water quality and identify areas of diminished water quality. Trace elements, Cd, Co, Cu, Cr, Pb, Mn, and Zn, were determined in water, sediments and fish.

For the Aswan High Dam reservoir, the investigation revealed significant seasonal, spatial and vertical variations in trace element content. Sediments and suspended matter play a significant role in the overall distribution of trace elements in the reservoir.

For the River Nile, the results revealed that river water contains concentrations of these metals far below the levels suggested for potable water. However, the trace elements have accumulated in high concentrations in river sediments indicating the existence of localized pollution.

Fish samples showed that metal concentrations varied according to the species and the sampling area. Metal concentrations in fish samples followed the order: Zn > Mn > Pb > Cu > Cd.

The findings contribute to the understanding of the distribution and dynamics of metals in the reservoir and River Nile ecosystems.

INTRODUCTION

One of the most important projects in Egypt was the construction of the Aswan High Dam (1959–1969). The Aswan High Dam Reservoir (AHDR) extends over an area of 5000 km^2 with its northern two-thirds in Egypt (known as Lake Nasser) and southern third in Sudan (known as Lake Nubia).

When this reservoir is completely filled, it will extend from the High Dam at Aswan (Egypt) to the Dal Cataract (Sudan).

As shown in Figure 15.1, the reservoir has a long narrow shape with a number of side arms called khors. The number of khors is 100, 58 located on the eastern shore and 42 on the western shore. The reservoir is mostly surrounded by rocky terrain. To the west is the great Sahara Desert, and the Eastern Desert extending to the Red Sea is in the east.

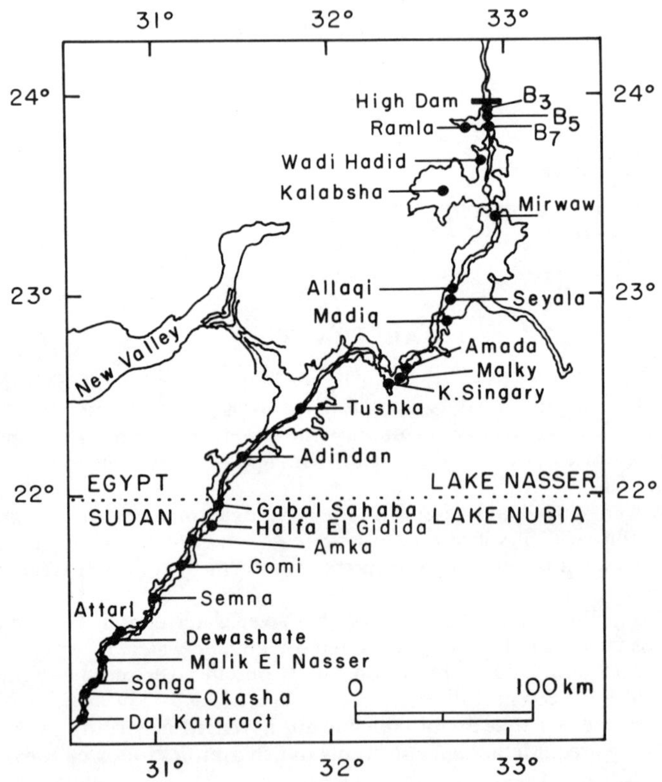

Figure 15.1 The Aswan High Dam Reservoir (Lake Nasser in Egypt and Lake Nubia in Sudan)

The Aswan High Dam Reservoir reached its operating level of 175 metres above sea level in 1975 with a total of 121.3 billion m^3 of stored water. Of this amount 31.6 billion m^3 are dead storage. The main characteristics of the reservoir morphology are summarized in Table 15.1 (Entz, 1974).

Table 15.1 Main morphological characteristics of the Aswan High Dam Reservoir

Characteristic	Lake Nasser water level 160 m	Lake Nasser water level 180 m	Total reservoir water level 160 m	Total reservoir water level 180 m
Length (km)	291.8	291.8	430	495.8
Surface area (km^2)	2585	5248	3057	6216
Volume (km^3)	55.6	132.5	65.9	156.9
Shoreline length (km)	5380	7844	6027	9250
Mean width (km)	8.9	18	7.1	12.5
Mean depth (m)	21.5	25.2	21.6	25.2
Maximal depth (m)	110	130	110	130

Siltation

The Nile receives its water from two main watershed areas, the Equatorial East African Plateau which feeds the White Nile, and the Ethiopian Highlands which feed the Sobat, Atbara and the Blue Nile (Figure 15.2). The contribution of the Blue Nile is most important since 86% of the main Nile's annual discharge is derived from the Ethiopian Highlands, while only 14% is contributed by the Equatorial Lakes (Mancy and Hafez, 1979a). Furthermore, the fluvium carried by the main Nile originates exclusively in the Ethiopian Highlands and is carried by the Blue Nile. The annual mud load of the main Nile is about 134 million m^3, consisting of a mixture of sand, silt, clay and organic matter.

Siltation in the reservoir depends on (a) the volume of the flood water, (b) water level in the reservoir at the onset of the flood period; and (c) the distribution of the flood water in the reservoir. When the reservoir was filling, the distribution of siltation covered wider areas depending on the size of the flood. More recently siltation has been confined to the most southern 200 km of the reservoir (between Akasha and Abu Simble) (Mancy and Hafez, 1979b). Particularly heavy siltation occurs in the area 360 km south of Aswan High Dam, that is between Amka and El-Deweshat. In this region siltation beds have already emerged in the life storage zone. Nevertheless, the sediment transport phenomenon tends to be in a northerly direction. Geomorphological changes due to siltation will undoubtedly affect the reservoir's water quality and ecosystem characteristics.

Occurrence of Metals in the Aquatic Environment

Metals enter rivers and lakes from a variety of sources, such as: (1) rocks and soils directly exposed to surface waters; this is the largest natural source, (2)

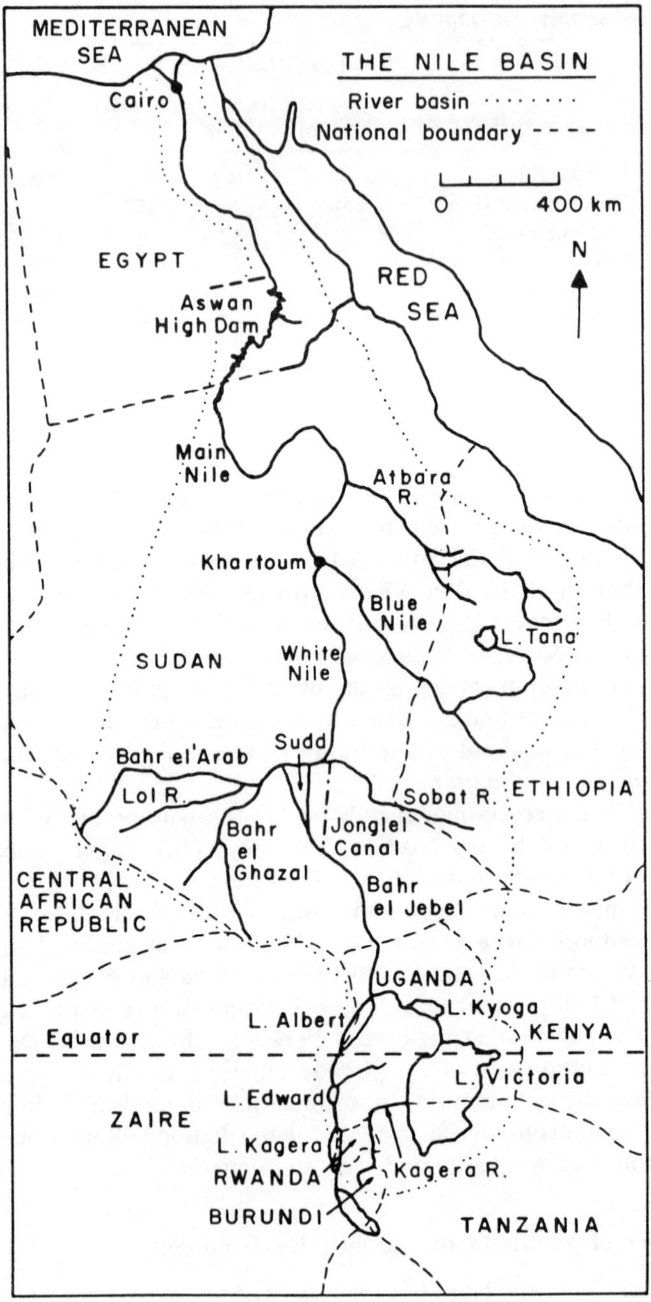

Figure 15.2 The Nile Basin

dead and decomposing vegetation and animal matter, (3) wet and dry fallout of atmospheric particulate matter, and (4) from man's activities, including the discharge of various treated and untreated liquid wastes into the water body.

Trace metals such as chromium, manganese, cobalt, copper and zinc play a biochemical role in the life processes of aquatic plants and animals, and their presence in trace amounts in the aquatic environment is essential. However, at high concentrations, these trace metals become toxic (Nurnberg, 1982).

A further special feature of toxic metals is that they are not biodegradable. Instead they undergo a biogeochemical cycle with substantially different residence times in the various spheres and compartments of the environment. Within this cycle they will be taken up also by man, predominantly from food and drinking water. In this respect toxic metals constitute a particular risk, because, although a certain fraction of the ingested amount is again excreted, they have a tendency to accumulate in vital organs. Thus, they will exert progressively growing toxic actions over long periods of the life-span depending also on the cumulative magnitude of the dose as a function of the long term exposure of the individual to its particular environment (Nurnberg, 1982).

Heavy metals in the aquatic environment are generally classified as existing in a number of chemical forms. The usual classification includes metals in the sediment, metals in suspended particulates (>0.45 μm), metals adsorbed onto hydrous oxides and humic colloids, and metals in true solution. The metals in true solution can be further divided into free ionic species, inorganic complexes, and complexes with naturally occurring organic ligands (amino acids, citric acid, etc.) (Guy and Kean, 1980).

Of particular importance for the regulation of trace constituents or pollutants in natural waters are the particles that act as 'scrubbing agents' by scavenging. All substances that are strongly adsorbed on suspended matter or become incorporated into settling biomass have short relative residence times. Obviously, the relative residence time of a compound decreases with increasing adsorbability (Stumm and Morgan, 1981). Particles are also of great importance in rivers and about 75% of the matter carried by the rivers to the ocean is in the form of particulate matter (Martin and Whitfield, 1983).

The specific objectives of this study are:

1. To assess the effect of the physicochemical characteristics of the aqueous phase on the distribution dynamics of the trace metals;

2. To assess the level of trace metals and effects on the water quality of the river and identify areas of diminished water quality;

3. To understand the role of sediments in the transport, transformation, and distribution of trace elements in the aquatic environment;

4. To elucidate the correlation between the transport behaviour of the suspended matter and trace element distribution patterns within the reservoir;

5. To assess the level of metals in fish.

METHODS AND MATERIALS

In the period between July 1976 and August 1977 water and sediment samples from the Aswan High Dam Reservoir were collected. The frequency of surveys was quarterly for Lake Nasser and once a year for Lake Nubia.

1	In front of the High Dam (sub-surface).
2	Before the High Dam.
3	Between the High Dam and Aswan Dam.
4	Before Aswan city.
5	At Abu Rouash village.
6	In front of the Assiut barrage.
7	After the Assiut barrage.
8	After the Assiut barrage by 2 km.
9	In front of Mankbad fertilizer factory.
10	El-Shobak village.
11	In front of the Helwan iron and steel factory.
12	Farouk Corner.
13	The Merridian hotel.
14	Road El-Farag.
15	Before the bifurcation of the Delta barrage.
16	After the Delta barrage in the Damietta branch.
17	Before the Zefta barrage.
18	After the Zefta barrage.
19	Beginning of El-Mansoura city.
20	Talkha fertilizer factory.
21	Faraskora Sandy Dam.
22	After the Delta Barrage in the Rosetta branch.
23	Kafr El-Zayat.
24	Kafr El-Shiakh Aly.
25	Edfina Barrage.

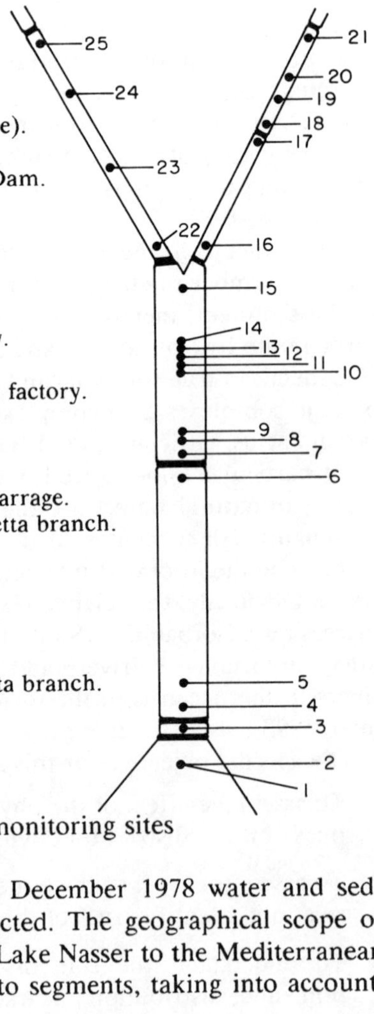

Figure 15.3 River Nile monitoring sites

In the period between April 1976 and December 1978 water and sediment samples from the river were collected. The geographical scope of this monitoring programme extends from Lake Nasser to the Mediterranean Sea (Figure 15.3). The river is divided into segments, taking into account:

(a) geographical features;
(b) administrative boundaries; and
(c) human activities.

Heavy metals in fish samples taken from the Nile River, its two branches and two main canals during the years 1981–1982 were assessed.

Trace metal analysis was done using a Perkin-Elmer Model 370 atomic absorption spectrophotometer equipped with a Heated Graphite Atomizer (HGA) Model 2100 and deuterium arc background corrector.

RESULTS AND DISCUSSION

There are two major processes affecting the distribution of trace metals in the AHDR.

1. The Transportation and Movement Patterns of Suspended Matter

Suspended matter plays a dominant role in the transport of trace metals in the reservoir. The results in Table 15.2 show that the suspended matter of Lake Nubia has high trace metal concentrations.

Table 15.2. Concentrations of selected trace metals in Lake Nubia suspended matter (July 1976).

Concentration ($\mu g/gm$)					
Co	Cr	Cu	Mn	Pb	Zn
116.3	90.0	28.7	416.7	95.0	91.0

Indeed, the partitioning of metals between particles and water is the key parameter in establishing the residence time and thus the residual concentrations of these elements in the lakes. Thus, the geochemical fate of metals is controlled by the chemical processes occurring between the solid surfaces and the water. The more reactive an element is in a lake the more it will be bound to particles, the more rapidly it will be removed, and the shorter will be its residence time (Whitfield, 1981).

A detailed insight into the movement of the suspended matter was obtained on the basis of the relationships between surface water manganese concentrations and transparency for selected stations along the reservoir. A standard Secchi disc (20 cm diameter) was used to measure water transparency. Readings were the average of the levels of disappearance and reappearance of the disc recorded in meters from the surface. As shown in Figure 15.4, loss of transparency was found to be inversely related with manganese concentrations, that is manganese concentrations increased when Secchi depth decreased. This relationship may be due to the fact that manganese has the ability to remain fixed to the suspended matter during transport in aerated water (De Groot and Allersma, 1975).

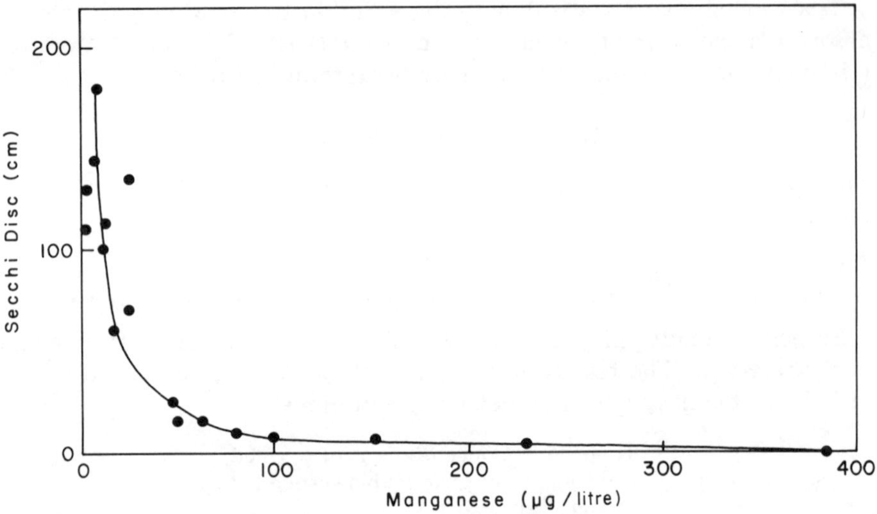

Figure 15.4 Relationship between manganese concentrations and Secchi disc depth in Aswan High Dam Reservoir during July–August 1976

2. Thermal Stratification

The thermal stratification and the hydrodynamic patterns in the reservoir have significant effects on trace metal distribution. During stratification, Lake Nasser develops a stagnant body of water at its bottom (Lasheen, 1982). The size of this stagnant layer and its existence depend on the morphology of the particular segment of the lake and the existing hydrodynamic regimes. The stagnant water layer at the bottom of the lake represents a condition in which the dissolved oxygen is completely removed. As a result of the disappearance of dissolved oxygen, free sulfide will be formed which influences the solubility of metals by forming insoluble metal sulfides. These sulfides will tend to strip the water of its metal content and enrich the bottom sediments with metals.

Stratification extends from May to September in Lake Nasser. In contrast, Lake Nubia does not exhibit any thermal stratification pattern.

In stratified water bodies the O_2 redoxcline is in the water column and the precipitation of iron and manganese takes place in the water column. This phenomenon has been extensively studied in the Black Sea and in a number of Swiss lakes (Sigg *et al.*, 1983). The precipitating iron hydroxides are able to scavenge and co-precipitate trace metals from the water column and transfer them to the sediment. In this way the O_2 redoxcline in stratified lakes not only acts as a barrier against upward diffusing metals but also actively removes some dissolved trace metals from the system (Salomons, 1984).

The quantitative effect of sediment on lake water composition depends to a large extent on the volume/surface area ratio. In shallow lakes the influence is expected to be much stronger compared with deep lakes.

Spatial and Seasonal Distribution of Trace Elements in the Reservoir Waters

1. Spatial Distribution

Analyses were performed for cadmium, cobalt, copper, manganese and zinc.

Tables 15.3 and 15.4 indicate that the mean concentrations of trace metals in Lake Nasser are lower than the mean values reported for Lake Nubia. The spatial distribution of soluble trace metal concentrations in August 1976 generally revealed high concentrations in Lake Nubia and in bottom waters of the reservoir.

Table 15.3 Mean concentrations of selected trace metals and pH values in Lake Nubia water (July–August 1976)

Depth	Concentration (μg/litre)					
	Cd	Co	Cu	Mn	Zn	pH
Surface	1.46	4.28	8.29	114.40	22.4	8.59
Bottom	0.81	4.90	8.34	130.81	20.7	8.27

Cadmium concentrations in Lake Nubia ranged from 0.74 to 4.89 μg/litre. In Lake Nasser cadmium concentrations varied from 0.4 to 3.14 μg/litre.

Cobalt concentrations in Lake Nubia ranged from 0.3 to 8.14 μg/litre. Lake Nasser water samples had between 0.6 to 5.64 μg/litre of cobalt.

Copper concentrations observed in Lake Nubia ranged from 3.38 to 18.92 μg/litre. The range of copper concentrations measured in Lake Nasser was between 0.87 and 13.8 μg/litre. The level of copper in the reservoir is considered to be low enough to have no effect on water quality.

Zinc concentrations in Lake Nubia ranged from 8.45 to 42.49 μg/litre. The observed zinc concentrations in Lake Nasser ranged from 0.67 to 24.96 μg/litre. The observed levels obviously do not reflect impaired suitability for wide usage of water.

Manganese concentrations observed in Lake Nubia ranged from 16.67 to 384.84 μg/litre and in Lake Nasser they ranged from 1.84 to 224.15 μg/litre. The greatest variation on vertical distributions of manganese occurred during the thermally stratified period.

Manganese concentrations in the bottom waters were much higher than in the surface waters. This difference is due to a combination of factors among which are surface assimilation by phytoplankton, the presence of manganese as soluble Mn^{2+} under the well-established anaerobic conditions in the bottom waters, and the release of manganese from suspended matter deposited on the sediment surface.

2. Seasonal Distribution

Table 15.4 represents the average concentrations in terms of season. There are variations between the data from the fall, winter and those from the summer. The results for manganese, zinc and copper concentrations produce the best evidence for the observed variation. Manganese concentrations are the highest of the metals analysed and cadmium are the lowest.

Table 15.4 Mean seasonal concentrations of selected trace metals and pH values in Lake Nasser water

		Concentration (μg/litre)					
Season	Depth	Cd	Co	Cu	Mn	Zn	pH
Summer	Surface	0.67	2.14	4.23	9.84	3.95	8.74
1976	Mid-depth	0.91	2.09	3.87	29.44	3.54	8.10
(August)	Bottom	1.03	2.37	5.40	74.49	6.33	8.01
Fall	Surface	1.18	3.30	2.81	38.61	8.98	8.30
1976	Mid-depth	1.22	3.54	2.68	40.78	4.75	8.11
(November)	Bottom	1.09	3.44	2.51	58.41	7.44	7.70
Winter	Surface	0.73	3.91	3.48	40.38	10.14	8.71
1977	Mid-depth	0.63	3.81	3.12	39.77	8.59	8.46
(February)	Bottom	0.62	3.76	3.43	40.52	6.98	8.31

The greatest variation in metal distribution patterns occurred during the thermally stratified period. With the exception of cadmium, metal concentrations were lower in the surface water than in deep waters indicating that biological activity in the photic zone removes any metals from the solution.

The peak values occurred at either the middle or bottom depth. In contrast, during the turnover period, concentrations of metals were mostly unchanged with depth, reflecting the mixing conditions.

The results show that trace metals are present in the parts-per-billion range, with the exception of manganese. Most of the metal concentrations are well below the recent proposed criteria for water quality (US Environmental Protection Agency, 1972).

Spatial and Seasonal Distribution of Trace Metals in the Reservoir Sediments

Table 15.5 indicates that the mean concentrations of Co, Cr, Mn and Zn in Lake Nubia sediments are higher than the mean values reported for Lake Nasser. Copper and lead were slightly higher in Lake Nasser sediments than in those of Lake Nubia.

Table 15.5 Means and ranges of metal concentrations of bottom sediments of Lake Nubia and Lake Nasser (July–August 1976)

	Concentration (μg/gm)			
	Lake Nubia		Lake Nasser	
Metal	Mean	Range	Mean	Range
Co	131.9	114–145	113.6	100–146
Cr	332.6	54–1250	158.3	88–251
Cu	54.7	26.7–68	59.9	50–80.5
Mn	1196	882–1395	784.0	623–1067
Pb	79.4	56–94	86.6	61–124
Zn	138.8	112–172	119.1	104–158

The concentrations of chromium and manganese were the highest of the metals in all sediments. Lead and copper, on the other hand, were in the lowest concentrations in most of the sediments.

Mean seasonal concentrations of the trace metals in Lake Nasser sediments are presented in Table 15.6. Manganese, lead and zinc concentrations were the highest in August 1976, but Co and Cr were the highest in February 1977. Copper content showed a slight increase in May 1977.

Where metal contamination is less severe, physical and chemical processes make spatial differences in the metal content of sediments difficult to interpret. However, the observed changes in the trace metal content of the sediments with distance from the High Dam suggest that changes in the environment of deposition do occur along the reservoir.

The Factors Influencing the Distribution Patterns of Trace Metals in the River Nile

The main factors which affect water quality characteristics of the River Nile include: (a) upstream changes south of Lake Nasser, (b) changes in Lake

Table 15.6 Mean seasonal concentrations of selected trace metals in Lake Nasser sediments

Season	Concentration (μm/gm)					
	Co	Cr	Cu	Mn	Pb	Zn
Summer (August 1976)	113.6	158.3	59.9	784	86.6	119.1
Fall (November 1976)	84.8	111.6	55.4	750.3	66.3	101.6
Winter (February 1977)	104.0	185.6	60.8	746.6	58.2	99
Spring (May 1977)	44.0	152.5	64.6	659	86.6	77.8
Summer (August 1977)	82.0	163.2	59.8	768.9	77.8	103.6

Nasser, and (c) localized changes in the river basin. Furthermore, the Nile receives increasing amounts of waste discharges, from point and non-point sources, as the river travels northward. The discharge of waste effluents is usually accompanied by localized effects of water quality deterioration immediately downstream from the waste outfall (Mancy and Hafez, 1979b). South of Cairo, at Helwan, there are iron, steel, coke and chemical fertilizer industries which discharge their waste into drains which finally reach the river.

The river exhibits its worst conditions in the Delta. This is due to a combination of factors: (a) reduction of the river velocity in certain locations to complete stagnation (b) industrial and domestic waste discharges and (c) return flow of agricultural drainage water.

At Kafr El Zayat there are three factories discharging their waste directly into the Rosetta branch. These factories are: Salt and Soda Company, Pesticides Company and Superphosphate fertilizer plant. As the river reaches the Edfina barrage, it slows down and bottom septic conditions occur during the summer months. Similarly, the Damietta branch receives industrial waste from the Talkha fertilizer plant, and stagnant waters at the Faraskour earth Dam exhibit septic conditions in the summer months.

Spatial Distribution of Trace Metals in the River Waters

Trace metal analyses were performed for cadmium, copper, lead and zinc. Figure 15.5 indicates that the mean concentrations of trace metals in the

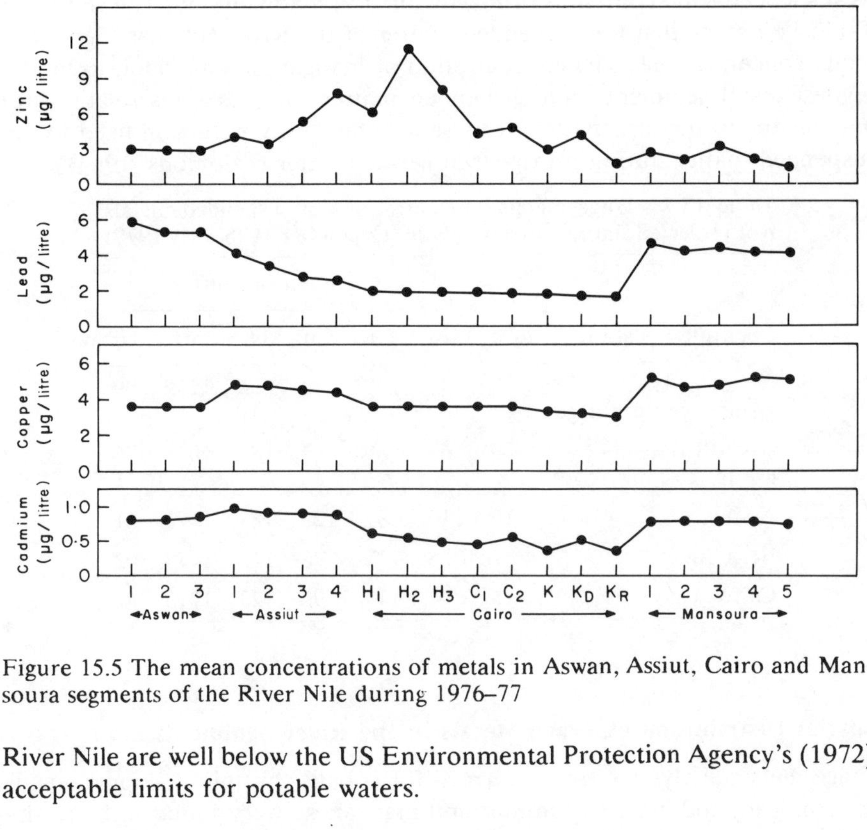

Figure 15.5 The mean concentrations of metals in Aswan, Assiut, Cairo and Mansoura segments of the River Nile during 1976–77

River Nile are well below the US Environmental Protection Agency's (1972) acceptable limits for potable waters.

The Importance of Suspended Solids in Controlling the Trace Metals Concentrations in Rivers

Knowledge of the origin and transport of the suspended solids is essential for an interpretation of regional variations in the trace metal contents. The suspended solids can act as scavengers for trace metals in the water. These suspended solids are introduced into the natural bodies of water by natural forces, for example, weathering, hydrothermal activities, and biosynthesis, or by man-made activities, for example, sewage discharge, dredging and agricultural excavation (Huange, 1978). The concentration of suspended solids in the aquatic environment varies from a few hundredths of a mg/litre in sewage effluent, to a few tenths of a mg/litre in rivers.

Since the construction of the Aswan High Dam, the Nile silt has been held in the reservoir. Recent studies indicated that siltation occurs in the most southern reaches of the reservoir (Mancy and Hafez, 1979b). As a result water released from the Aswan High Dam Reservoir is silt free, and its suspended solid content is primarily due to phytoplankton. The results in Table 15.7 show that the suspended matter of the River Nile has high trace metal concentrations. The concentration of manganese was found to be the highest in all sediments and suspended matter samples. This relationship may be due to the fact that manganese has the ability to remain fixed to the suspended matter during transport in aerated water (Salomons, 1984).

Table 15.7 Trace metals concentrations of suspended metals from selected stations in River Nile (December 1978–July 1979)

	Concentration (μg/gm)						
Sample Location	Cd	Co	Cr	Cu	Pb	Mn	Ni
Assiut El-Hagz group (west drainage)	9	47	79	72	204	1159	99
Assiut El-Maana group (east drainage)	9	61	89	62	87	1048	88
Assiut	10	110	87	125	185	1171	81
Beni-Suef	7	88	112	78	113	1474	92
Cairo	8	—	114	92	123	1715	82

Spatial Distribution of Trace Metals in the River Sediments

Trace metals analyses were performed for cobalt, chromium, copper, manganese, lead, and zinc. Chromium and manganese were found to be in the highest concentrations in all sediments (Table 15.8). Copper and lead, on the other hand, were found to be in the lowest concentrations in most of the sediments. The increase in mean metal concentrations in analysed sediment samples followed the order: Mn> Cr> Zn> Co> Pb> Cu (Table 15.9).

Although in the present study we found that the trace metals have accumulated in high concentrations in river sediments, and for the time being, remain relatively harmless, one cannot overlook the fact that these metals can be released by numerous processes (e.g. from anthropogenic sources and remobilization of trace metals from suspended material and sediments) and are potentially hazardous not only for the aquatic ecosystem, but also for the drinking water supply (Forstner and Wittmann, 1979).

Assessing Metal Pollution in Nile River Using Fish Studies

Muscle tissue from five species of fish taken from the Nile River were analysed for metals (Table 15.10). If whole fish had been digested, mean

Table 15.8 Mean concentrations of selected trace metals in River Nile sediments (1976–1978)

Sample location*	Concentration (μg/gm)					
	Co	Cr	Cu	Pb	Mn	Zn
Assiut (9)	95.4	622.6	27	61.4	961.2	71.2
Before Helwan iron and steel factory (9)	98.1	347.3	32.3	57.7	1049.3	86.3
In front of Helwan iron and steel factory (9)	92.3	187.7	34.5	49.7	1015.5	85.8
After Helwan iron and steel factory (6)	106.2	226.3	33.7	61.3	1116.8	90.5
Meridian (3)	103	193.7	42	50	1078	161.7
Road El-Farag (5)	88.4	246.8	78.4	94.6	1061.6	157.6
El-Manashi (6)	110.5	249	58.2	47	1082.5	99.5
El-Qanater (4)	96	276.3	57	69	1085.3	104
Kafr El-Zayat (7)	87	526.7	33.3	39.4	1334.7	90.9
Kafr El-Sheikh Ali (5)	96	254.2	47.2	82.8	806.8	179.4
Edfina (8)	104	354.5	62.9	75.1	1610.6	143.1
El-Mansoura (93)	81.3	144.7	47	74	1877.7	95
Talkha (9)	78.1	398.3	48.6	53.2	1329.1	93.4
Faraskur (9)	99.3	357.7	51.6	50.33	1105.9	89.4

* Numbers in parentheses indicates the number of samples analysed.

Table 15.9 Means and ranges of metal concentrations in sediments of the River Nile

Metal	Concentration (μg/gm)	
	Mean	Range
Co	95.1	11–186
Cr	343.7	44–1678
Cu	45	4–150
Pb	58.3	6–164
Mn	1163.6	291–3750
Zn	102.3	44–434

concentrations for each species would undoubtedly have been higher since metals are known to concentrate in certain target organs.

The concentration of zinc was higher than all the measured metals in the analysed fish. On the other hand, cadmium was lower than the other metals. Fish caught from El-Mahmodia canal showed the highest metal content followed by Abu-Elghaid canal.

By comparing metals concentrations in the Nile River water and bottom sediments with the fish data, it can be seen that metals concentrations in

Table 15.10 Median and range values for heavy metals in Nile fishes (μg/gm)

Sampling location	species	No. of samples*	Cd	Pb	Cu	Mn	Zn
Aswan	*Tilapia nilotica*	3	0.49 (0.12–0.57)	8.14 (5.6–13.72)	4.1 (3.55–4.64)	34.59 (23.99–38.64)	31.84 (22.84–48.97)
Assiut	*Tilapia nilotica*	5	0.49 (0.29–0.72)	4.89 (1.36–8.42)	3.16 (2.02–5.02)	10.29 (7.65–12.8)	30.69 (22.75–36.65)
	Chrysichthys auratus	1	1.08	9.09	5.94	15.68	34.72
Cairo	*Tilapia nilotica*	4	0.21 (0.14–0.29)	2.63 (1.14–4.45)	1.47 (1.13–1.94)	6.59 (4.89–8.28)	45.51 (33.43–75.6)
	Chrysichthys auratus	3	0.29 (0.14–0.43)	2.24 (1.14–3.12)	1.41 (0.92–1.7)	8.07 (5.14–11.79)	44.88 (42–50)
El-Mansoura	*Tilapia nilotica*	5	0.25 (0.13–0.38)	4.65 (2.5–7.0)	3.74 (3.21–4.14)	12.48 (5.34–19.94)	43.89 (35.22–51.88)
Faraskour	*Tilapia nilotica*	4	0.25 (ND–0.32)	5.61 (2.17–9.25)	4.09 (2.69–5.22)	13.21 (5.99–18.29)	49.7 (36.75–56.17)

Beginning of Rosetta branch	*Tilapia nilotica*	2	0.54	6.0 (5.52–6.47)	3.52 (2.97–3.66)	14.35 (11.59–17.1)	47.8 (46.6–49)
	Labco niloticus	4	0.18 (ND–0.54)	2.38 (1.94–2.87)	2.33 (2.24–2.4)	6.27 (5.65–6.58)	38.05 (32–45)
Edfina	*Tilapia nilotica*	4	0.52 (0.31–0.91)	3.58 (1.55–5.1)	1.99 (1.34–3.09)	12.45 (4.33–20.3)	53.3 (43.03–59.53)
El-Mahmodia Canal	*Tilapia zillii*	9	0.61 (0.25–0.89)	10.43 (6.88–15.25)	4.27 (3.37–8.4)	17.12 (13.04–20.7)	75.12 (44.03–108.27)
	Clarias lazera	3	0.18 (ND–0.18)	4.93 (1.8–9.93)	2.12 (1.05–3.5)	2.78 (2.38–3.43)	44.91 (23.63–56.71)
Abu-Elghaid	*Tilapia nilotica*	6	0.67 (0.43–0.9)	9.11 (4.98–12.8)	3.46 (3.2–4.05)	17.12 (11.65–21.08)	71.67 (53.25–86.75)
	Clarias lazera	3	0.40 (0.33–0.44)	3.15 (2.23–4.73)	2.31 (1.6–2.3)	1.81 (0.83–3.28)	50 (46.5–55)

* Each analysed sample represents at least one fish of each species.
ND: not detected

fish flesh were considerably lower than concentrations in bottom sediments. However, all metals were more highly concentrated in fish muscle tissue than in Nile water.

In recent years, intensive pollution resulting from the discharge of increasing amounts of agricultural, domestic and industrial waste-water to the Nile River is reported to have had a marked effect on the fish production from the river (Mancy and Hafez, 1979b).

We conclude that the threat of pollution to the fisheries of Egypt is sensed although not yet adequately defined. It is obvious that considerable work needs to be done in monitoring of heavy metal concentrations to improve our understanding of their cycle in the Nile environment.

CONCLUSIONS

The important findings of this study are summarized as follows:

1. For all of the metals studied the average level found in the reservoir water is below the value set by the United States EPA as harmless to aquatic life.

2. The high metal concentrations in sediments relative to that in water demonstrate the importance of the suspended matter as a transport pathway for metals.

3. The dissolved oxygen content and oxidation-reduction conditions to a large extent control the amount and species of metal ions.

4. Some elevated concentrations of trace metals were noted in the region of Lake Nubia. In general though, the trace metal content and suspended matter of the water samples revealed significant correlation.

5. The significant drop in metal concentrations in River Nile water might result from dilution or from precipitation and adsorption reactions which would tend to reduce the metals levels.

6. Trace metals have accumulated in high concentrations in river sediments indicating the existence of localized pollution. These metals can be released by various processes of remobilization and move up the biological chain, thereby reaching humans and possibly produce chronic and acute ailments.

Recommendations

It is now accepted by environmental scientists that to fully understand the biological availability, toxicity and transport of heavy metals in natural water systems, measurements of total metal concentrations are no longer adequate, the chemical forms of the metals in solution must be considered. Electroanalytical techniques should be applied to measure the concentrations of various metal species and to permit the determination of the equilibrium concentrations of metal complexes.

A more detailed insight into the relations between heavy metals characteristics and physical properties of the solid constituents of the aquatic environment is urgently needed, especially in view of processes of bioaccumulation.

Partial extraction techniques of metals from aquatic sediments should be applied to measure the sediment fraction that has a detrimental effect on water quality and the associated biota.

The presence of high concentrations of heavy metals in Nile sediments should prompt the development of dredged material disposal criteria in Egypt.

Since trace metals are concentrated in biological material a study should be made to ascertain whether metals reach excessive levels in the higher members of the biological food chains in AHDR and River Nile ecosystems.

As yet the total scope of latent toxic effects of trace metals is unknown. Thus until all doubt about the effect of potentially toxic metals on the biological food chain and drinking water supply is removed, trace metals should be kept to a minimum in all aquatic systems.

ACKNOWLEDGEMENT

This study was supported in part by the funds provided by the 'Water Quality Studies on the River Nile and Lake Nasser' research project administered by the Egyptian Academy of Scientific Research and Technology and the University of Michigan, USA.

REFERENCES

De Groot, A. J., and Allersma, E. (1975). Field observations on the transport of heavy metals in sediments. In Krenkel, P. A. (Ed.), *Heavy Metals in the Aquatic Environment*, Pergamon Press, New York, 352 pages.

Entz, B. (1974). *The Morphometry of Lake Nasser and Lake Nubia.* Lake Nasser Development Centre, Aswan.

Forstner, U., and Whittmann, G. T. W. (1979). *Metal Pollution in the Aquatic Environment*, Springer-Verlag, Berlin, 486 pages.

Guy, R. D., and Kean, A. R. (1980). Algae as a chemical speciation monitor. 1A. Comparison of algal growth and computer calculated speciation. *Water Res.*, **14**, 891–899.

Huange, C. P. (1978). Solid-solution interface: its role in regulating the chemical composition of natural waters. In Gibbs, R. J. (Ed.), *Transport Processes in Lakes and Oceans*, pp. 9–34, Plenum Press, New York.

Lasheen, M. R. (1982). Water Quality Study on Aswan High Dam Reservoir, *N. R. C. Bull.*, Vol. 7, Egypt.

Mancy, K. H., and Hafez, M. (1979a). An overview of the Nile physical and hydrological characteristics and its major control programs. *The Water Quality Bulletin.* The World Health Organization, April, 1979.
Mancy, K. H., and Hafez, M. (1979b). Water Quality and Ecosystem Consideration in Integrated Nile Resources Management. *Proc. of the 1st Water Conf. on Water Res. Management in Egypt.* Ministry of Irrigation and Cairo Univ., Cairo, Egypt.
Martin, J. M., and Whitfield, M. (1983). The significance of the river input of chemical elements to the ocean. In Wong, C. S., Boyle, E., Bruland, K. W., Button, J. D., and Goldberg, E. D. (Eds), *Trace Metals in Sea Water*, pp. 265–296, Plenum Press, New York.
Nurnberg, H. W. (1982). Voltametric trace analysis in ecological chemistry of toxic metals. *Pure and Appl. Chem.*, **54**(4), 853–878.
Salomons, W. (1984). Contamination in sediment, out of sight out of mind? *Int. Conf. Env. Contamination*, London, July 1984, pp. 766–774, CEP Cons. Pub., London.
Sigg, L., Sturm, M., Davis, U., and Stumm, W. (1983). *Metal Transfer Mechanisms in Lakes.* Thalassia, Yugoslavia.
Stumm, W., and Morgan, J. J. (1981). *Aquatic Chemistry*, Wiley Interscience, NY, 780 pages.
US Environmental Protection Agency (1972). *Water Quality Criteria*. National Academy of Sciences.
Whitfield, M. (1981). The world ocean mechanism or machination? *Interdiscipl. Sci. Rev.*, **6**, 20.

Lead, Mercury, Cadmium and Arsenic in the Environment
Edited by T. C. Hutchinson and K. M. Meema

CHAPTER 16

Bioaccumulation of Mercury by Aquatic Biota in Hydroelectric Reservoirs: A Review and Consideration of Mechanisms

P. M. STOKES AND C. D. WREN

Institute for Environmental Studies,
University of Toronto,
Toronto, Ontario, Canada

ABSTRACT

Methylation is accepted to be a major process controlling the biological availability of mercury, and a number of recent articles and reviews have addressed this process. Recently, the occurrence of elevated mercury in fish tissues from systems in regions considered to be remote from point or local sources of mercury have been documented. These appear to be related to acidification of surface waters and to recent impoundments, usually in connection with hydroelectric dam construction. The present chapter addresses this second phenomenon, which is being investigated for Canadian Reservoirs by Environmental and Social Systems Analysts (ESSA) Ltd, LGL Associated and the University of Toronto in an ongoing study funded by the Canadian Electrical Association.

Elevated mercury levels have been detected in fish from a number of Canadian reservoirs. A preliminary study in 1976 of approximately 6000 fish throughout Labrador revealed that the highest mercury levels were found in fish from the Smallwood Reservoir and waters of the Churchill River downstream of the Control Structure. The maximum mercury levels in burbot (1.93 μg/gm) and in lake trout (3.9 μg/gm) were in fish from the reservoir.

Southern Indian Lake has the largest commercial fishery in northern Manitoba, and is of particular interest since the water level was raised 3 metres in order to divert water from the Churchill River. Mercury levels in commercially caught fish appear to have increased along the Churchill–Rat–Burntwood River system from the early 1970s to 1978 and 1979. The average mercury content of commercially caught northern pike and pickerel (walleye) in 1971 was 0.26 μg/gm and 0.19 μg/gm respectively. In 1979, the mean mercury level was 0.75 μg/gm in pickerel and 0.88

μg/gm in pike. The maximum allowable mercury contents for commercial fish are 1.0 μg/gm for shipments exported to the United States, and 0.5 μg/gm for domestic markets.

In Saskatchewan, the government has recommended that fish from the Cookson Reservoir not be consumed by humans due to elevated mercury levels in the fish.

These three examples may suggest that elevated mercury levels in Canadian reservoirs are a well documented phenomenon. However, both the source of mercury in new impoundments, and the mechanisms resulting in high fish mercury level have yet to be clearly identified.

Hypotheses relating specifically to the factors controlling mercury levels in fish from recent impoundments (in contrast to factors more generally related to mercury concentration in biota) include:

1. The amount and quality of organic matter in flooded sediments will affect microbial activity.
2. The mercury concentration of flooded soils, rocks or vegetation is not limiting for fish uptake.
3. Rates of methylation in the flooded area determine the concentration of methyl mercury in water or in food items for fish.
4. Shoreline erosion influences the amount of organic content in flooded sediment.
5. The composition of upper soil horizons and vegetation in the flooded area influences the organic content of flooded sediment.
6. Physical characteristics of the reservoir influence the sedimentation of organic matter, which influences the organic content of flooded sediment.

These hypotheses are discussed with emphasis on mechanisms, practical consequences and duration in time of the phenomenon, and possible amelioration—*a priori* or *a posteriori*—of the fish mercury problem in reservoirs.

INTRODUCTION

The recognition of mercury as a risk to human health from environmental rather than from occupational exposure dates from the late 1950s and includes a number of examples relating to mercury from specific sources (D'Itri and D'Itri, 1977). Mercury readily accumulates in tissue and is also biomagnified through the aquatic food chain. Methylation of mercury is accepted to be a major process controlling its biological availability in aquatic ecosystems, and a number of recent articles and reviews have addressed this process (e.g. Nriagu, 1979).

More recently, occurrences of elevated mercury levels in tissues of fish from regions considered to be remote from point or local sources of mercury have been documented. These appear to be related to: (1) acidification of surface waters (Lindqvist *et al.*, 1984) and, (2) recent impoundments, usually in connection with hydroelectric dam construction (e.g. Bodaly *et al.*, 1984). The present chapter briefly reviews the second phenomenon and attempts to explain possible mechanisms leading to elevated mercury levels in fish from reservoirs.

MERCURY UPTAKE BY FISH

Fish can accumulate mercury either through food or directly from the water. In natural ecosystems, it is most likely that fish accumulate mercury via these two vectors concurrently. However, the relative importance of these two uptake processes has not been clearly established even under laboratory conditions and is probably related to species and site-specific conditions.

The efficiency of mercury assimilation from food appears to vary among species. Experimental studies have shown that 68–80% of mercury ingested is assimilated by rainbow trout (Phillips and Buhler, 1978; Rodgers and Beamish, 1982), about 20% in northern pike (Phillips and Gregory, 1979), and an efficiency of 80% has been used in uptake models for yellow perch (Norstrom *et al.*, 1976). However, it has also been shown that mercury assimilation across the gut will begin to decrease above certain critical concentrations (Rodgers and Beamish, 1982).

The amount of mercury accumulated directly from the water by fish is more difficult to calculate due to problems in accurately measuring mercury concentrations in surface waters. Mercury uptake from the water will be determined by the water concentration, fish metabolic rate and efficiency of uptake (bioavailability) as determined by the ambient water characteristics. The latter factor is the least understood and perhaps the most important condition governing mercury uptake in fish living under natural conditions.

Detailed models have been developed relating mercury uptake in fish to fish metabolic rate (Norstrom *et al.*, 1976; Rodgers and Beamish, 1981). Biotic and abiotic variables affecting fish metabolic rate and mercury uptake include water temperature (MacLeod and Pessah, 1973; Reinhart *et al.*, 1974) and fish body size (Defreitas and Hart, 1975; Sharpe *et al.*, 1977). Rodgers and Beamish (1981) showed that uptake of waterborne methylmercury by rainbow trout was linearly related to oxygen consumption and methylmercury concentration at values less than 8.0 μg/litre.

The uptake efficiency of methylmercury from water by salmonids has been estimated at 7.1% (Boddington *et al.*, 1979) and 8.3% (Rodgers and Beamish, 1981). The latter value was obtained while running the experiment in hard water (385 mg/litre $CaCO_3$). When the experiment was repeated in soft water (30 mg/litre $CaCO_3$), the uptake efficiency was increased to 25% (Rodgers, 1982). Increased uptake efficiency may therefore partially explain elevated mercury levels in fish from low pH lakes. The rate of accumulation and toxicity of several metals to fish are known to be reduced by increased water hardness or calcium levels (Pickering, 1974; Davies *et al.*, 1976; Howarth and Sprague, 1978; Carrol *et al.*, 1979). The mechanism to account for this effect is thought to be increased gill permeability at low calcium levels (Spry *et al.*, 1981), or competition between metals and calcium for cellular binding sites (Zitko and Carson, 1976). Observations from field studies also show correlations between fish mercury level and water calcium concentration (MacFarlane and Franzin 1980; Wren and MacCrimmon, 1983).

A variety of other physiochemical conditions within lakes have been empirically related to mercury levels in fish. Prominent among these relationships are elevated mercury levels at lower lake pH (Jernelov *et al.*, 1975; Brouzes *et al.*, 1977; Håkanson, 1980; Suns *et al.*, 1980; Wren and MacCrimmon 1983). Sloan and Schofield (1983) reported that mercury levels were higher in brook trout from drainage lakes (pH $\leq$ 5) than in brook trout from seepage and bog-type lakes (pH $>$ 5.0). However, they also state that trout mercury levels did not change after liming one lake, and this conflicts with the pattern of other data presented. A recent evaluation of data from over 200 lakes in Sweden suggests that mercury levels are elevated in fish within specific regions due to increased atmospheric deposition of mercury, and also lake acidification (Bjorklund *et al.*, 1984). Despite these studies, no plausible mechanism to account for the effect of water pH on fish mercury levels has yet been put forth. The original hypothesis concerning the production of monomethylmercury from sediments at pH $<$ 6.0 (e.g. Beijer and Jernelov, 1979) is almost certainly an oversimplification of the processes occurring within natural ecosystems. To elucidate the mechanisms governing mercury uptake in fish in freshwater systems there remain at least two important tasks:

1. development of a reliable technique to accurately measure mercury and methylmercury concentrations in water.
2. identification of the function and relative importance of chemical variables (e.g. calcium, humic acids) on the bioavailability of mercury to fish in water.

MERCURY IN FISH FROM RESERVOIRS

The earliest documentation in the literature of mercury in fish from impoundments appeared in 1974, when Smith *et al.* (1974) reported high mercury levels in fish from Willard Bay Reservoir, Utah. This is a shallow reservoir with a pH of 8.1–8.3. As shown in Table 16.1, predatory fish in particular contained high total mercury (from 0.27–7.3 mg/kg). However, only two of the fish species from Willard were common to other Utah water bodies reported by the same authors, and the authors at that time provided some evidence that Willard was in fact 'naturally' polluted with mercury. They stated that 'a major contribution of mercury to Willard Bay Reservoir may be made from insoluble mercury salts or sulfides in the muds deposited from the time of Lake Bonneville which receded to form the highly saline Great Salt Lake'. The only geochemical measurements referred to were those of the 1971 geological survey which reported 'The greatest concentration of mercury found in a sort sample was from Summit Country, Utah'. Without any more specific data it is not possible to speculate further on the significance of the observations on fish mercury; the study is included as an example of the high Hg levels attained in predatory fish in reservoirs.

Table 16.1 Mercury levels in various fish species for reservoir studies in the USA (All values in mg/kg wet weight)

	Carp	Walleye	Largemouth bass	Brown trout	Reference
Willard Bay Reservoir, Utah	0.520–1.210	0.160–3.890	0.270–7.300	1.330 (1 sample)	Smith *et al.*, 1974
Reference area, Utah	0.040–0.820	–	–	0.130–0.430	Smith *et al.*, 1974
Lake Powell, N. Mexico	0.181–0.339	0.209–0.763	0.192–0.688	0.073–0.101	Potter *et al.*, 1975

More comprehensive are the results of a study on mercury in a man-made desert reservoir (Lake Powell) in New Mexico reported by Potter *et al.* (1975). Flooding commenced in 1963 and by 1971 the volume of the reservoir was $17\,860 \times 10^6$ m^3. These workers measured mercury in seven different fish species, for eleven different tissues. Ranges in mercury concentration for four of the species for axial muscle (which was the tissue highest in mercury for most samples) are given in Table 16.1. Predatory fish commonly had mercury >0.5 mg/kg but some exceeded 1.0 mg/kg. Fish mercury levels were related to body weight and trophic status. There are no pre-impoundment data nor reference areas, and as shown in Table 16.1, the mercury levels in fish, where there are species in common, were generally much lower than for the Utah study of Smith *et al.* (1974). Nevertheless the Lake Powell study is valuable because the authors present a mercury budget for the reservoir, which indicates that apart from fish, the highest mercury levels were in 'water transported' plant debris, which contained up to 0.145 mg Hg/kg (dry weight), significantly higher than for live plants in the terrestrial and aquatic systems, which never exceeded 0.034 mg/kg in leaves.

The authors gave evidence for 'bioamplification' of mercury, and considered that in this man-made reservoir, remote from pollution sources and with 'only very slight and very local pollution', mercury was released by 'natural weathering' of the basin. They calculated that 800 kg/year might accumulate in the reservoir.

Abernathy and Cumbie (1977) reported on mercury accumulation in largemouth bass for three reservoirs in South Carolina. The waters contained very low mercury levels (ND—0.06 μg/litre). These authors showed relatively high fish mercury with differences among reservoirs of different ages (Table 16.2). Highest fish mercury levels were in a young, oligotrophic reservoir, and the lowest levels in the older, eutrophic reservoirs. However, the results do not conclusively relate fish mercury to impoundment or to reservoir age. There were other large differences among reservoirs, for example, in water quality parameters such as conductivity, alkalinity, pH and turbidity, all of which can affect mercury availability. The authors suggested that anaerobic conditions in the more nutrient-rich systems would result in binding of mercury to sulphur compounds and organics, which would remove the mercury from the food web. They refer briefly to methylation, which they consider would be favoured by aerobic conditions prevailing in the more oligotrophic systems, and conclude that the source of mercury to fish 'could easily have been provided by background levels of mercury which were present in topsoil which formed the sediments of the reservoirs after filling'.

In a follow-up study for Lake Jocassee, which had the highest fish mercury, Abernathy and Cumbie (1977) found a decrease in fish mercury levels between 1974 (1.9 ± 0.12 mg/kg) and 1975 (0.69 ± 0.06 mg/kg). Based on this and on the data for different aged systems, the authors concluded that 'elevated mercury in fish of new reservoirs is probably transitory, declining within 3–5 years after reservoir filling under conditions which exist in the systems we have studied'.

Meister *et al.* (1979) also identified soil from the flooded area which sedimented into the reservoir as the original source of mercury in crappie and largemouth bass in Cedar Lake Reservoir, Illinois. Some of the largest (3.5–5.5 kg) bass contained more than 0.5 ppm mercury (Table 16.2). Meister *et al.* (1979) considered that microbial methylation of the sediment mercury, combined with food chain biomagnification, resulted in elevated fish mercury particularly in predatory fish, and that 'anaerobic conditions that favour the uptake of mercury from soil by methylation can occur at the lake bottom'. They concluded that 'elevated mercury in predatory fish in new impoundments is to be expected and is independent of water quality'. Furthermore, they expected that eventually the concentration should drop because of constant removal of water, fish harvest and volatility of organic mercury. In a report on Cedar Lake, Cox *et al.* (1979) considered that the source of mercury was natural and cautioned that if reservoirs are to be used as fisheries, close monitoring of mercury in predatory fish should be carried out during the first few years.

A number of other studies on mercury levels in fish from reservoirs in the USA and Canada have been published during the late 1970s and early 1980s. In order to give the reader a sense of the extent of the studies and

Table 16.2 Mercury concentrations (μg/gm wet weight) in bass from a number of reservoir studies

Reservoir	Mercury concentration (μg/gm)	References
Lake Jocassee, S. Carolina (1 year)	1.87–4.49	Abernathy and Cumbie, 1977
Lake Hartwell, S. Carolina (11 years)	0.38–0.68	Abernathy and Cumbie, 1977
Lake Keowee, S. Carolina (3 years)	0.34–3.99	Abernathy and Cumbie, 1977
Cedar Lake Reservoir, Illinois	0.07–0.50	Meister *et al.*, 1979
Ress Bainett, Mississippi	>0.05–0.74	Knight and Herring, 1972
Occoquan Res., Virginia	0.00–0.8	Gawlick, 1979
Wallace Bay, Utah	0.27–7.3	Smith *et al.*, 1974
Lake Powell, New Mexico	0.192–0.688	Potter *et al.*, 1975

their respective limitations, these are summarized in Table 16.3. Even when one excludes those situations where a specific source of mercury is identified, for example, mining activity (Phillips *et al.*, 1980; Cooper, 1983) or contamination from a chloralkali plant (Hildebrand *et al.*, 1980), it is clear that fish mercury is frequently unacceptably high in impoundments remote from major point sources and that the phenomenon has been described for a variety of geographic regions, from east to west of the North American continent. Whether there are conspicuous exceptions remains to be seen; with one exception no studies report on data which show no change in fish mercury (Environment Canada, 1979). Furthermore, many of the earlier studies were commenced after flooding and time trends or pre-impoundment data are not available.

Within Canada, several recent studies which are fairly complete can be highlighted as examples of the geographical extent of the phenomena. In Labrador, preliminary studies in 1976 (Bruce and Spencer, 1979) on the Smallwood Reservoir which was filled in 1971 showed 'above background' levels of mercury in several fish species. A comparison of mercury in the different fish species from 88 locations in Labrador showed that the highest

Table 16.3 Summary of available reservoir studies

Water body	Location	References	Comments
S. Indian L (SIL)	Man.	Bodaly and Hecky, 1979 Bodaly *et al.*, 1984	Experimental + Se data
		McGregor, 1980 Env. Canada, 1979	Good informative statistics. 4 fish spp. Ref. to some sites where Hg(fish) *not* increased
Rat L. (RL)	Man.	McGregor, 1980 Bodaly *et al.*, 1984	Good statistics. 4 fish spp. (RL Hg fish) > SIL
Notigi L.	Man.	McGregor, 1980 Bodaly *et al.*, 1984	Good statistics. 4 fish spp.
Waskwatim	Man.	McGregor, 1980 Bodaly *et al.*, 1984	Good statistics. 4 fish spp. Waskwatim Hg(fish) >SIL
Proposal to B.C.	B.C.	Bodaly and Hecky, 1982	
Smallwood R.	Labrador	Bruce and Spencer, 1979	Highest in piscivores, 14 spp
Cookson	Sask.	Waite *et al.*, 1980	Walleye highest. Water Hg low
Strutt L.	N.W.T.	Weagle and Cameron, 1974 (ABST)	Abstract only Bodaly (1982) refers to Hg levels from this study
Various	N. Finland	Lodenius *et al.*, 1983	Increased Hg in reservoir and more in top predators. Highest hair Hg in people eating reservoir fish
Cedar L.	Illinois	Meister *et al.*, 1979	Soil probable origin of Hg
		Cox *et al.*, 1979	Samples dried at temperatures too high for Hg
Lake Powell	Arizona	Potter *et al.*, 1975	
Tongue R.	Montana	Phillips *et al.*, 1980	Sites contaminated from mining

Hartwell, Keowee and Jocassee	S. Carolina	Abernathy and Cumbie, 1977	High levels in fish from reservoir; differences not clearly related to impoundment. Jocassee and Keowee have highest fish Hg
Major reservoirs (9)	S. Carolina	Abernathy, 1979	
Willard Bay R.	Utah	Smith *et al.*, 1974	75% fish in reservoir > 0.55 ppm. Reference sites much lower
Pueblo R.	Colorado	Herrman and Mahan, 1977	No biological data
American Falls R.	Idaho	Kent and Johnson, 1979	Very high Hg in water > 1.78 ng/litre. Methods not given.
		Johnson *et al.*, 1978	Follow-up of 1974 paper 'Water Hg stillage'
Elephant Butte R.	N. Mexico	Garcia and Kidd, 1979	Great fluctuation in water level; difference in water Hg across reservoir
Lahonatan R.	Nevada	Cooper, 1983	Impounded 1915. weight/Hg correlation for some (11 spp. examined). Source of Hg is old mining
		Cooper *et al.*, 1983	Reservoir is eutrophic. Carp mortality in 1980 and 1981
B. Everett Joshia R.	N. Carolina	Shuman *et al.*, 1983	Municipal discharge contains Hg, and natural source
General	Canada	Baxter and Glaude, 1980	Has list of reservoirs across Canada
Various (80)	Illinois	Hullinger, 1975	Map of locations in Illinois. High Hg (water) seemes to be in certain geographical regions but only 1 sample 1 year

Table 16.3 Cont'd.

Boundary Dam	Saskatchewan	Allan and Richards, 1978	Original water quality poor, surface sediment high in Hg (as for other metals) TDS downstream improved after dam built
Liard River Branager (Prior to reservoir formalities)	B.C. Canada	Sekerak and Mace, 1982	Metals low in water, sediment and fish
La Grande 2 and Opinacca	Quebec James Bay	Boucher and Schetagne, 1983	Lots of raw data and good time trends. Fish Hg increased years 1–2 and more in year 4. Erosion negligible, pH low
Cherokee Reservoir	Virginia and Tennessee	Hildebrand *et al.*, 1980	Chloralkali plant was source of Hg.
B. Everett Jordan Falls L.	N. Carolina	Weiss and Bond, 1978	Over 140–150 days, Mercury increased. Decreased after suggest [Hg] associated with richness of organic content
B. Everett Jordan Falls L.	N. Carolina	Shuman *et al.*, 1983	Goldmining area 1032 sports fish sampled. One arm of system fish collected 28% had Hg > 0.5 ppm
Occaquan Reserve	Virginia	Gawlick, 1979	Agricultural and urban run off
La Grande 2 and Opinacca	Quebec	Boucher and Schetagne, 1983	

mercury levels were in fish from the Smallwood Reservoir and the Churchill Reservoir downstream of the Lobstick Control structure. The highest levels were for lake trout (3.9 μg/gm) and burbot (1.93 μg/gm) from the reservoir. Thus, although no pre-impoundment data were available, spatial comparisons indicated abnormal levels in the reservoir fish.

In Quebec, reservoirs constructed on La Grande river were studied over the period 1978–82 (Boucher and Schetagne, 1983). These authors compared mercury in fish for pre-impoundment with post-impoundment conditions. At all sites, mercury was consistently higher in the piscivorous pike and walleye. For all fish species there was a correlation between age and mercury, or between length and mercury, but a great deal of variability existed in the data. After impoundment, mercury in fish increased: for example, walleye year 2 (2×) and year 4 (3.5×), and for whitefish in year 2 (3×) and in year 4 (5.5×). Table 16.4 summarizes some of their results. Most fish in the reservoirs exceeded the federal 0.5 mg/kg limit, and it was recommended that fish consumption should be limited.

Table 16.4 Pre- and post-impoundment mercury levels (in μg/gm wet weight) in fish from Quebec reservoirs. (From Boucher and Schetagne, 1983)

Reservoir	Fish species	Pre-impoundment levels	Year 2	Year 4
La Grande 2	Lake whitefish	0.09–0.17	0.26	0.51
Opinacca	Longnose sucker	0.13–0.19		0.40
Caniapiscau	Pike	0.52–0.60		1.24
La Grande 2 Opinacca	Walleye	0.6	1.28	2.05

In a discussion of the possible mechanisms, Boucher and Schetagne considered the role of sediments resulting from erosion and of organic matter as factors mediating the influx of mercury and the methylation of mercury. For La Grande Reservoir, however, erosion had been negligible and no increase of organic matter had been observed. These authors considered that a major controlling factor for methylation in the specific systems under consideration was a decrease in pH of surface waters from about 6.5 to 6.0, with slightly lower (5.4–5.9) pH in subsurface water. Inundation of vegetation was also considered an important factor, since terrestrial plants, especially mosses, have a high metal binding capacity, and may be loaded with mercury (originally from the atmosphere). A third factor which these authors discussed was the concentration of nutrients, which could enhance methylation of mercury. After flooding, La Grande and Opinacca reservoirs did, in fact, have higher concentrations of nutrients than hitherto.

The Cookson Reservoir in Saskatchewan was created in 1976 in an area remote from anthropogenic sources of mercury. Data for mercury were collected in 1979 (Waite *et al.*, 1980). Walleye in the reservoir exceeded the 0.5 mg/kg limit, but there were no pre-impoundment or reference site data. For this study, there were insufficient data on other factors to permit speculation concerning the mechanism, but 'preliminary information on mercury concentrations in terrestrial soils and reservoir sediments do not show unusually high levels'. The authors speculate that 'if the mercury source is newly inundated soil containing average southern Saskatchewan mercury concentrations, and if there are no other sources, it is likely that mercury concentration in the fish of Cookson Reservoir will peak and then begin to decline'.

The most comprehensive studies to date in the mercury-reservoir problem come from a hydroelectric development in the Churchill and Nelson river valleys in Northern Manitoba. This is an immense development in which the combined flow of two of Canada's major rivers are utilized to provide hydroelectric power to Manitoba. Approximately 820 m^3/s of the 960 m^3/s of the Churchill River is now diverted into the Nelson River basin (Bodaly *et al.*, 1984). The ecosystems prior to development were riverine lakes (10–2000 km^2) on the precambrian shield, with bedrock overlain by glacier-lacustrine and glacial deposits of clay, silt and sand. The climate is subarctic and much of the soil is permanently frozen. In 1976, the diversion began, and details of the works and hydrology are described in Newbury *et al.* (1984). In 1976, the level of South Indian Lake was raised by 3 metres which flooded 414 km^2. Since then work proceeded on impoundments and diversion until 1980, when construction was suspended indefinitely.

In the context of the present review it is pertinent to note that part of the higher areas surrounding the lake are permanently frozen and the deposits in the flooded area are being severely eroded. Pre-and post-impoundment studies on a number of ecological and physical aspects are being conducted and have been described in a recent series of papers (Newbury and McCullough, 1984; Hecky and McCullough, 1984; Hecky, 1984; Hecky and Guildford, 1984; Bodaly *et al.*, 1984). Based on the reservoir paradigm (Rzoska, 1966) a number of predictions of physical changes were correct, but for the most part biological changes were not correctly predicted and in particular the measurement increases in the fish mercury concentrations had not been predicted. Table 16.5 gives some pre-and post-impoundment mercury levels for whitefish, pike and walleye. Other lakes in the same area show corresponding and comparable post-impoundment increases (i.e. in the order of 20-40%) (Bodaly *et al.*, 1984) also shown in Table 16.5. In the latter document, reference is made to spatial comparisons. Data for fish mercury levels were available for 30 other lakes in Northern Manitoba and during the 1970s these did not show any trends toward increasing concentrations.

Bodaly *et al.* reject the hypothesis that differences in deposition rates could account for the sudden surge in fish mercury beginning in 1976, although according to lake sediments there has been a very slow increase in mercury deposition, with an approximately 2× increase in flux sediments in pristine areas since 1900. It is also convincingly argued that anthropogenic point sources could not explain the surge, and mercury-rich geologic formations are also unlikely to be the cause.

Table 16.5 Pre- and post-impoundment mean Hg levels (in μg/gm wet weight) in three fish species in lakes flooded by the Churchill River Diversion, Northern Manitoba

Fish species	Lake	Pre-impoundment	Post-impoundment
Lake whitefish	SIL area 4	1975 0.05 (25)*	1978 0.22 (16)*
(*Coregonus*	SIL areas 2 and 6	1975 0.06 (50)*	1978 0.30 (17)*
clupeaformis)	IS	1975 0.15 (24)*	1978 0.32 (5)*
	WK	1970 0.08 (1)†	1970 0.33 (28)†
Northern pike	SIL	1971–73 0.29(12)*	1976–79 0.47 (33)*
(*Esox lucius*)	IS	1978 0.61 (5)†	1982 0.90 (26)†
	WK	1979 0.91 (75)†	1982 1.13 (1)†
Walleye	SIL	1971–73 0.25 (12)*	1975–79 0.40 (21)*
(*Stizostedion*	IS	1978 1.52 (5)†	1982 0.79 (19)†
vitreum)	WK	1970 0.34 (1)†	1981 0.89 (34)†

(N)
SIL = South Indian L, Churchill River, at point of diversion.
IS = Issett L., upper end of Notigi Reservoir.
WK = Waskwatin L., on Burntwood R. basin, Notigi Reservoir.
* Bodaly and Hecky, 1979
† Bodaly *et al.*, 1984

Mercury analysis of vegetation litter, other soil horizons, lake and suspended sediment and water are given, and none of these are high enough to be considered contaminated; rather they tend to be below average compared with similar materials elsewhere. The authors consider that the activity of flooding soils resulted in increased bacterial methylation of naturally-occurring mercury, with the primary source of mercury in the organic rich upper soil horizon. The study of Furutani and Rudd (1980) supports the idea that increased bacterial production would occur when terrestrial vegetation, moss, peat and humus is inundated.

While there are problems related to small sample size for some parts of the study (Table 16.5) referred to by Environment Canada (1979) there seems no reasonable doubt that the phenomenon of elevated fish mercury is real, and in a number of instances the differences between pre-and post-impoundment fish mercury are significant (Bodaly and Hecky, 1979). The post-impoundment levels of mercury in fish frequently exceeded 0.5 mg/kg for pike and walleye but not for whitefish. It is nevertheless of interest to note that in a summary of all available data for the Southern Indian Lake–Burntwood River system, comparing 1978 and 1979 data with previous years, Environment Canada (1979) identifies from commercial samples six sites where fish mercury levels have changed significantly and nine where they have not changed significantly. In no cases did they find significant decreases.

POSSIBLE MECHANISMS

There is little disagreement that regardless of the situation, methylmercury results in higher levels of uptake into fish than does inorganic mercury (e.g. Ribeyre and Boudou, 1983). This assumption can be applied to highly polluted situations such as the English Wabigoon River System (Jackson and Woychuk, 1980) as well as to lakes and rivers in remote areas which have not been impacted by man (Lindqvist *et al.*, 1984) and reservoirs which although impacted by man have no known local anthropogenic source of mercury. There is also general agreement that 70-80% of the mercury in fish muscle is methyl mercury (Westoo, 1966; Lockhart *et al.*, 1972). The explanation for the now widely recognized phenomenon of elevated fish mercury in recent impoundments can therefore be narrowed down to finding an explanation for mechanisms of increases in methyl mercury in the systems of interest.

Where mercury in the environmental compartments of the reservoirs and their adjacent environment has been measured, there is rarely any evidence of contamination or above background concentrations of mercury. (Exceptions to this have been noted in Table 16.3.) Abernathy (1979) in a rough calculation provided evidence that there was sufficient mercury in the flooded soils which formed the sediments of the reservoir to account for the fish mercury and others have referred to 'natural levels' of mercury as the source. This is probably true for any situation except the remote pre-industrial conditions described by Lindqvist *et al.* (1984). It is therefore reasonable to assume that the source of mercury is *either* the materials already in the lake or river prior to flooding *or* the flooded soils and vegetation, and that this mercury is either natural, or at least background even if not pre-industrial levels (Lindqvist *et al.*, 1984). There is no immediate atmospheric source in these situations. The question of mechanism then becomes one of mobilization and methylation of existing mercury resulting from the factors which change between pre-and post-impoundment.

From the studies which have been reviewed in the present paper, it is possible to reach a consensus that soils and vegetation in the flooded areas are the source of mercury. The Churchill River studies show a relationship between the area flooded and the mercury levels in predatory fish (Bodaly *et al.*, 1984). There is also strong evidence that methylation is promoted by vegetation and/or already decaying vegetation or other organic matter in the flooded area. Methylation may occur in the previously exposed soil and vegetation, or in the material which is washed into the reservoir and becomes bed sediment or suspended particulate matter.

Experimental studies are now underway at the Churchill River diversion site, using large enclosures (limnocorrals) in which additions of the various types of vegetation in the flooded area are made, and actual methylation rates or fish mercury as a surrogate for methylation rates are measured. Preliminary results (Hecky *et al.*, 1986) indicate that the best source of mercury and promoter of methylation is *Sphagnum* moss. In contrast, the addition of inorganic subsoil to the experimental enclosures did not have the same effect (Newbury *et al.*, 1984). While these studies can be considered site-specific, they are nevertheless consistent with the results of other experimental studies which have shown that additions of organic material—microbial substrate—increased rates of microbial mercury methylation (Furutani and Rudd, 1980). A recent review of the Swedish mercury problem (Lindqvist *et al.*, 1984) which did not in fact consider the reservoir phenomenon, but rather considered remote lakes and the related fish mercury problem, concluded that a major route of mercury for fish is transport from surrounding soils and vegetation (which in Southern Sweden are already contaminated by increased atmospheric fallout of Hg from long range transport). If this is so for unflooded systems, then flooding would probably further enhance this tendency. There is also some evidence from Japanese studies (Nagase *et al.*, 1984a,b), and from Swedish studies (Hultberg, personal communication) that non-microbial methylation (probably via a free radical reaction) is important in these systems. Non-microbial mercury methylation in experiments was also stimulated by organic matter and certain metals such as Fe, Mn, and Cu (Hultberg, personal communication).

REHABILITATION AND CONTROL MEASURES

With the recognition of the risks to humans associated with mercury in the aquatic environment, a number of measures have been proposed, or have been the subject of experimental manipulation, to ameliorate or control the mercury problem in contaminated lakes and river systems.

Several studies are summarized in Table 16.6. It is immediately apparent, based on the proposed explanation of the mechanism of mercury accumula-

Table 16.6 Summary of proposed amelioration methods to restore mercury-contaminated waterways

Major Concern/ impact	Source of Hg	Location	Type of study	Treatment	Rationale	References
Fish Hg (low O_2) organic pollution benthos effects	Bed sediment	Tokyo Bay, Japan		Dredging (7–40 mg/kg Hg)	Sediment the major source.	Sameshima, 1977
Fish Hg	Chisso chem Hg	Minaomata Bay, Japan		Dredging and settling	Dredge soil particles not likely to cause Hg release	Yoshida and Ikegaki, 1977
Fish Hg	Release from sediments	Wabigoon R. Ontario	Field studies to test sources and transport	Dredging most contaminated sediments	Removal of source	Jackson and Woychuk, 1980 Wilkins, 1978
Fish Hg	Contaminated sediments	Miscellaneous St.Clair R. Detroit R, Islair Cr.	Pilot experiments	Cover with crushed automobiles the sand or clay	Fe removes soluble Hg	Feick *et al.*, 1972
Fish Hg	Contaminated sediments	Wabigoon R.		Plough sediments	Dilute contaminated with clean sediment.	Wilkins, 1978
Fish Hg	Bottom sediments	Accumulate Hg in settling ponds, immobilize chemically and dredge	Control of source and remainder			
Fish Hg	Contaminated sediments	Wabigoon R, Clay L.	Limmocorrals	Resuspension of sediments and downstream deposit	Control Hg in water and at sediment–water interface	Rudd *et al.*, 1983
Fish Hg	S. Indian Lake		Limnocorrals	Add clay	Reduction of toxicity in fish	Hecky *et al.*, unp. data

Fish Hg	Contaminated sediments	Swedish Lakes	Proposed field studies	Cover sediments with organic sediment	Maintain reducing conditions and high rates of sulphide production	Jernelov and Lann, 1973
Fish Hg	Contaminated sediments	Swedish Lakes		Add lime	Prevent form of CH_3Hg^+ favour $(CH_3)_2Hg$	Jernelov and Lann, 1973
Fish Hg	Contaminated sediments	Swedish Lakes	Proposed field studies	Maintain or increase eutrophic conditions, add fertilizer	Dilute CH_3Hg^+ with large and rapidly growing biological community	Jernelov and Lann, 1973
Fish Hg	Contaminated sediments	Wabigoon R.	Field	accumulate Hg in settling ponds, immobilize chemically and dredge	Control of source	Jackson and Woychuk, 1980
Fish Hg	No point source of Hg	S. Sweden	Field experiments	Intensive fishing	Change in 'biological flow' of Hg	Gothberg, 1983
Fish Hg	Contaminated sediments	Wabigoon R, Clay L.	Limnocorrals	Addition of selenium	Decreased rate of Hg uptake by biota	Rudd *et al.*, 1983
Contaminated dredge spoil	Impounded spoil area	Dokai Bay Kitakyushu city	Line sea walls with Plasma and cover with rafts of bamboo, plastic sheets and sand			Fujino, 1977
Dredge spoils	Contaminated dredgate	Detroit R. St. Clair R.	Proposed experiments	Dredged sediments leached with hypochlorite	Dredgate cleaned and can be added to water or land fill	Feick *at al.*, 1972

tion in fish from reservoirs, that most of these approaches are not relevant to the problem of concern. Any measure which aims at sealing or removing sediments, appropriate enough for a system with contaminated sediment, will not be appropriate for a reservoir if the problem originates in the vegetation or soils of the watershed. Similarly, neither the diversion of a water course, nor a control of the source of mercury, are feasible in the reservoir situation.

Until it is determined whether the phenomenon is relatively transient and short-lived, in which case it would be proposed to limit or control fishing for the appropriate number of years, it is not economically realistic to propose remedial measures. However, assuming that the phenomenon is sufficiently serious and long-lived that manipulations towards decreasing fish mercury are realistic, only four approaches appear to be even remotely appropriate. The first involves removal of the vegetation and, possibly, the organic soil horizon from the area to be inundated, prior to flooding. The second involves addition of selenium to water, as proposed by Rudd *et al.* (1983). Thirdly, with the observation, also by Rudd *et al.* (1983) that resuspension of sediments decreased mercury uptake by biota in experimental enclosures in the English–Wabigoon River system, deliberate addition of suspended sediment, assuming that the sediment itself is not contaminated, could be investigated. However, since the assumption of this method in which mercury is scavenged from the water is that mercury *per se* becomes limiting, it is probably not applicable to the reservoir situations, where it appears that methylation rates are limiting, rather than mercury concentrations. The final possibility relates to Gothberg's (1983) observation that intensive fishing can result in decreased levels of mercury in fish. The most tenable explanation for this effect appears to be an alteration of the 'biological flow of mercury' since the fishing caused very little reduction in the total mercury content of the lake. Control of erosion would also be indicated as an adjunct to any of the above.

Of these four possibilities, only the first appears to be even remotely practical. For reservoirs much could be discussed concerning site selection from the point of view of watershed and soil characteristics, but again from a practical standpoint, hydrological, physical and engineering considerations would probably overwhelm those pertaining to mercury.

It has to be emphasized that pre-impoundment studies of the hydrology, soils, vegetation and water quality and fish Hg should be conducted, as well as ongoing monitoring of the fish mercury as an indicator and integration of methylation. Only by such careful and systematic studies can the extent and duration of the problem of mercury in fish in impoundments be evaluated. In many instances it is clearly too late to conduct such longitudinal studies, and spatial comparisons have to be used.

REFERENCES

Abernathy, A. R. (1979). *Heavy Metal Contamination of Surface Waters and Fish Flesh in South Carolina.* Office of Water Research and Technology, Washington, DC, PB-296, 215.

Abernathy, A. R., and Cumbie, P. M. (1977). Mercury accumulation by largemouth bass (*Micropterus salmoides*) in recently impounded reservoirs. *Bull. Environ. Contam. Toxicol.*, **17**, 595–602.

Allan, R. J., and Richards, D. J. (1978). Effect of thermal generating station on dissolved solids and heavy metals in a prairie reservoir. *Can. Wat. Res. Conf. on Environ. Aspects of Industrial Cooling in Northern Climates.* IWD Sci. Series 93 CA156.

Baxter, R. M., and Glaude, P. (1980). Environmental effects of dams and impoundments in Canada. *Can. Bull. Fish. Aquat. Sci.*, **205**, 34 pages.

Beijer, K., and Jernelov, A. (1979). Methylation of mercury in aquatic environments. In Nriagu, J. O. (Ed.), *The Biochemistry of Mercury in the Environment*, pp. 203–208, Elsevier North-Holland Biomedical Press, New York.

Bjorklund, I., Borg, H., and Johansson. K. (1984). Mercury in Swedish lakes—its regional distribution and causes. *Ambio*, **13**, 118–121.

Bodaly, R. A., and Hecky, R. E. (1979). *Post Impoundment Increases in Fish Mercury Levels in the Southern Indian Lake Reservoir, Manitoba.* Fisheries and Marine Service Manuscript Report 1531. Department of Fisheries and the Environment, 15 pages.

Bodaly, R. A., Hecky R. E., and Fudge, R. J. P. (1984) Increases in fish mercury levels in lakes flooded by the Churchill River diversion, Northern Manitoba. *Can. J. Fish. Aquat. Sci.*, **41**(4), 682–691.

Boddington, M., Defreitas, A. W. S., and Miller, D. R. (1979). The effect of benthic invertebrates on the clearance of mercury from sediments. *Ecotox. Environ. Safety*, **3**, 236–244.

Boucher, R., and Schetagne, R. (1983). *Repercussions de la Mise en eau des Reservoirs de la Grande 2 et Opinaca sur la Concentration du Mercure dans les Poissons.* Direction de L'Environment, Quebec.

Brouzes R. J., McLean, R. A., and Tomlinson, G. (1977). *The Link Between the pH of Natural Waters and the Mercury Content of Fish.* Domtar Res. Centre Report, May 3, 33 pages.

Bruce, W. J., and Spencer, K. D. (1979). *Mercury Levels in Labrador Fish.* Canadian Industry Report of Fisheries and Aquatic Sciences. No. 111, Dept. of Fisheries and Oceans.

Carroll, J. J., Ellis, S. J., and Oliver, W. S. (1979). Influences of hardness constituents on the acute toxicity of cadmium to brook trout (*Salvelinus fontinalis*). *Bull. Environ. Contam. Toxicol.*, **22**, 575–581.

Cooper, J. J. (1983). Total mercury in fishes and selected biota in Lahontan Reservoir. Nevada. *Bull. Environ. Contam. Toxicol.*, **31**, 9–17.

Cooper, J. J., Vigg, S. C., Bryce, R. W., and Jacobson, R. L. (1983). *Limnology of Lahontan Reservoir*, 1980–1982. Nevada Univ. System Reno Desert Res. Inst. Report, 173 pages.

Cox, J. A., Carnahan, J., Di'Nunzio, J., McCoy, J., and Meister, J. F. (1979). Source of mercury in fish in new impoundments. *Bull Environ. Contam. Toxicol.*, 23, 779–783.

Davies, P. H., Goettl, J. P., Sinley, J. R., and Smith, N. (1976). Acute and chronic toxicity of lead to rainbow trout (*Salmo gairdneri*) in hard and soft water. *Water Research,* **10**, 199–206.

Defreitas, A. S. W., and Hart, J. S. (1975). Effect of body weight on uptake of methylmercury by fish. *ASTM Symp.*, **573**, 356–363.

D'Itri, P. A., and D'Itri, F. M. (1977). *Mercury Contamination: A Human Tragedy.* John Wiley & Sons, New York. 311 pages,

Environment Canada (1979). *A Summary of all Available Mercury Data for the Southern Indian Lake–Rat–Buntwood River System.* Inlands Waters Directorate, Western and Northern Region, Regina, 115 pages.

Feick, G., Johansson E. E., and Yeaple, D. S. (1972). *Control of Mercury Contamination in Freshwater Sediments.* Report No. EPA-RS-72-077. US EPA, 160 pages.

Fujino, S. (1977). Using Sandfill to Cover Dredge Spoils Containing Mercury. In *Management of Bottom Sediments Containing Toxic Substances.* Proc. 2nd U. S.–Japan Experts Mtg. 1971.

Furutani, A., and Rudd, J. W. M. (1980). Measurement of mercury methylation in lake water and sediment samples. *Appl. and Environ. Microbio.* **40**(4), 770–776.

Garcia, J. D., and Kidd, D. E. (1979). Mercury levels in water and sediment of Elephant Butte Reservoir, New Mexico. *Bull. Environ. Contam. Toxicol.*, **21**, 624–630.

Gawlick, S. P. (1979). Heavy metals in fish from a reservoir receiving urban and agricultural runoff. MSc Thesis, Dept. Env. Sci., and Engineering, Virginia Polytechnic.

Gothberg, A. (1983). Intensive fishing—A way to reduce mercury levels in fish. *Ambio,* **13**, 256–261.

Håkansson, L. (1980). The quantitative impact of pH, bioproduction and Hg contamination on the Hg content of fish (pike). *Environ. Pollut. (B),* **1**, 285–304.

Hecky, R. E. (1984). Thermal and optimal characteristics of southern Indian Lake before, during and after impoundment and Churchill River diversion. *Can. J. Fish. Aquat. Sci.*, **41**(4), 579–590.

Hecky, R. E., and Guilford, S. J. (1984). Primary productivity of southern Indian Lake before, during and after impoundment and Churchill River diversion. *Can. J. Fish. Aquat. Sci.*, **41**, 591–604.

Hecky, R. E., and McCullough, G. K. (1984). Primary productivity of southern Indian Lake before, during and after impoundment and Churchill River diversion. *Can. J. Fish. Aquat. Sci.*, **41**(4), 591–604.

Hecky, R. E., Bodaly, R. A., Strange, N. E., Ramsey, D. J., Anema, C., and Fudge, R. J. P. (1986). Mercury bioaccumulation in yellow perch in limno corrals simulating the effects on reservoir function. *Can. Data Rep., Fish and Aquatic Sci.* (in press).

Herrmann, S. J., and Mahan, K. I. (1977). *Effects of Impoundments on Water and Sediments in the Arkansas River Pueblo Reservoir.* Bureau of Reclamation, Denver, Colorado Engineering and Research Centre, Report Rec-ERC-76-19.

Hildebrand, S. G., Lindberg, S. E., Turner, R. R., Huckabee, J. W., and Strand, R. H. (1980). *Biochemistry of Mercury in a River-Reservoir System: Impact of an Inactive Chloralkali Plant.* ORLN/TM 6141, Contact No. W-740, 5-ENG-26, Oak Ridge Nat. Lab., Environ. Sci. Div. Pub. 1519.

Howarth, R. S., and Sprague, J. P. (1978). Copper lethality to rainbow trout in waters of various hardness and pH. *Water Res.*, **12**, 455–462.

Hullinger, D. L. (1975). A study of heavy metals in Illinois impoundments. *J. Awwa Oct.*, **67**, 575–576.

Jackson, T. A., and Woychuk, R. N. (1980). The geochemistry and distribution of mercury in the Wabigoon River system. In *Mercury Pollution in the Wabigoon–English River System of N. W. Ontario, and Possible Remedial Measures.*

Jernelov, A., and Lann, H. (1973). Studies in Sweden on feasibility of some methods for restoration of mercury-contaminated bodies of water. *Environ. Sci. Tech.*, **7**(8), 712–718.

Jernelov, A., Landner, L., and Larsson, T. (1975). Swedish perspectives on mercury pollution. *J. Water Pollut. Control. Fed.*, **47**, 810–822.

Johnson, D. W., Kent, J. C., and Campbell, D. K. (1978). *Availability and Concentration of Pollutants from American Falls Reservoir Sediments to Forage and Predaceans Fish.* Off. of Water Res., and Tech., Washington, D. C., Report PB–275-564.

Kent, J. C., and Johnson, D. W. (1979). Mercury, arsenic and cadmium in fresh water and sediment of American Falls Reservoir Idaho, 1974. *Pest. Monit. J.*, **13**, 35–40.

Knight, L. A., Jr., and Herring,J. (1972). Total mercury in largemouth bass (*Micropterus salmoides*) in Ross Barnett Reservoir, Mississippi 1970 and 1971. *Pest. Monit. J.*, 6(2), 103–106.

Lindqvist, O., Jernelov, A., Johansson, K., and Rodhe, H. (1984). *Mercury in the Swedish Environment.* Nat. Swedish Env. Protect. Bd. Report SNV PM 1846, 105 pages.

Lockhart, W. L., Uthe, J. F. Kenny, A. R., and Mehrle, P. M. (1972). Methylmercury in northern pike (*Esox lucius*): distribution, elimination and some biogeochemical characteristics of contaminated fish. *J. Fish. Res. Bc. Can.*, 27, 677–684.

Lodenius, M., Seppanen, A., and Heeranen, M. (1983). Accumulation of mercury in fish and man from reservoirs in northern Finland. *Water, Soil and Air Poll.*, **19**, 237–246.

McFarlane, G. A., and Franzin, W. G. (1980). An examination of Cd, Cu and Hg concentrations in livers of northern pike and white suckers from five lakes near a base smelter at Flin Flon, Manitoba. *Can. J. Fish. Aquatic Sci.* **37**, 1573–1578.

McGregor, G. W. G. (1980). Summary of mercury levels in lakes on the Churchill–Rat–Burntwood and Nelson river system from 1970–1979. *Canada Data Repor. Fish. Aquatic Sci.*, **195**, 16 pages.

MacLeod, J. C., and Pessah, E. (1973). Temperature effects on mercury accumulation toxicity, and metabolic rate in rainbow trout (*Salmo gairdneri*). *J. Fish. Res. Bd. Can.*, **30**, 485–492.

Meister, J. F., Di'Nunzio, J., and Cox, J. A. (1979). Source and level of mercury in a new impoundment. *J. Am. Water Works Assoc.*, **71**, 574–576.

Nagase, H., Ose, Y., Sato, T., and Ishikawa, T. (1984a). Mercury methylation by compounds in human material. *Sci. Total. Envir.*, **32**, 147–156.

Nagase, H., Ose, Y., Sato, T., and Ishikawa, T. (1984b). Methylation of mercury by human substances in an aquatic environment. *Sci. Total Environ.*, **25**, 133–142.

Newbury, R. W., and McCullough, G. K. (1984). Shoreline erosion and restabilization in the southern Indian Lake Reservoir. *Can. J. Fish. Aquat. Sci.*, **41**(4), 558–565.

Newbury, R. W., McCullough, G. K., and Hecky, R. E. (1984). The Southern Indian Lake impoundment and Churchill River diversion. *Can. J. of Fish., and Aquat. Sci.*, **41**(4), 548–557.

Norstrom, R. J., McKinnon, A. F., and Defreitas, A. S. W. (1976). A bioenergenics based model for pollutant accumulation by fish. Simulation of PCB and methylmercury residues in Ottawa River perch. *J. Fish. Res. Bd. Can.*, **33**, 248–267.

Nriagu, J. O. (1979). *The Biochemistry of Mercury in the Environment.* Elsevier North–Holland Biomedical Press, New York, 696 pages.

Phillips, G. R., and Buhler, D. R. (1978). The relative contributions of methylmercury from food or water to rainbow trout in a controlled laboratory environment. *Trans. Am. Fish. Soc.*, **107**, 853–861.

Phillips, G. R., and Gregory, R. W. (1979). Assimilation efficiency of dietary methylmercury by northern pike. *J. Fish. Res. Bd. Can.*, **36**, 1516–1519.

Phillips, G. R., Lenhart, T. E., and Gregory, R. W. (1980). Relation between trophic position and mercury accumulation from the Tonge River Reservoir, Montana. *Environ. Res.*, **22**, 73–80.

Pickering, Q. H. (1974). Chronic toxicity of nickel to the fathead minnow. *J. Water Pollut. Control Fed.*, **46**, 760–765.

Potter, L., Kidd, D., and Standiford, D. (1975). Mercury in Lake Powell. Bioamplification of mercury in man made reservoir. *Environ. Sci. Technol.*, **9**, 41–46.

Reinhart, R. E., Stone. L. J., and Willford, W. A. (1974). Effect of temperature on accumulation of methylmercuric chloride and DDT by rainbow trout (*Salmo gairdneri*). *J. Fish. Res. Board Can.*, **31**, 1649–1652.

Ribeyre, F., and Boudou, A. (1983). Concepts and methodology for a systemic experimental approach towards the study of the bioaccumulation of mercury compounds in aquatic ecosystems. In *Proc. Internat. Conf. on Heavy metals in the Environment*, CEP Consultants, Heidelberg, Sept. 1983.

Rodgers, D. W. (1982). Dynamics of methylmercury accumulation in rainbow trout (*Salmo gairdneri*). PhD Thesis, Univ. of Guelph, Guelph. 129 pages.

Rodgers, D. W., and Beamish, F. W. (1981). Uptake of waterborne methylmercury by rainbow trout (*Salmo gairdneri*) in relation to oxygen consumption and methylmerury concentration. *Can. J. Fish. Aquat. Sci.*, **38**(11), 1309–1315.

Rodgers, D. W., and Beamish, F. W. (1982). Dynamics of dietary methylmercury in Rainbow trout, *Salmo gairdneri. Aquatic Toxicology*, **2**, 271–290.

Rudd, J. W., Turner, M. A., Furutani, A., Swick, A. L., and Townsend, B. E. (1983). The English–Wabigoon River system. A synthesis of recent research on the English–Wabigoon River system with a view towards merçury amelioration. *Can. J. Fish. Aquat. Sci.*, **40**, 2206–2217.

Rzoska, J. (1966). The Biology of Reservoirs in the USSR. In Lowe–McConnel (Ed.), *Manmade Lakes, Symposium of the Institute of Biology, 15.* pp. 149–154. Academic Press.

Sameshima, T. (1977). Dredgery of contaminated bed sediment in Japan. In *Management of Bottom Sediments Containing Toxic Substances.* pp. 272–684. Environ. Res. Lab. Oreg.

Sekerak, A. D., and Mace. T. F. (1982). *Baseline Survey of Heavy Metal Concentration in Water, Sediment, Bank Material and Fish in the Liard River Drainage, British Columbia.* LGL Associates Report to B. C. Hydro.

Sharpe, M. A., Defreitas, S. W., and McKinnon, A. E. (1977). The effects of body size on methylmercury clearance by goldfish (*Carassius auratus*). *Environ. Biol. Fish.*, **2**, 177–183.

Shuman, M. S., Robinson, K. G., and Stokes, C. E. (1983). *Metal Loading of the B. Everett Reservoir.* PB84-111913, Contact D1-14-34-0001-1135, NC Water Resources Res. Inst., Raleigh, NC.

Sloan, R., and Schofield, D. L. (1983). Mercury levels in brook trout from selected acid and limed Adirondack Lakes. *Northeast Env. Sci.*, **2**(3/4), 165-170.
Smith, F. A., Sharma, R. P., Lynn, R. I., and Low, J. B. (1974). Mercury and selected pesticide levels in fish and wildlife of Utah: Levels of mercury in fish. *Bull. Env. Toxicol.*, **12**, 153–157.
Spry, D. J., Wood, C. M., and Hodson, P. V. (1981). *Effect of environmental acid on freshwater fish with particular reference to the softwater lakes in Ont., and the modifying effects of heavy metals.* Tech. Rep. of Can. Fish. Aquat. Sci., **999**, 144 pages.
Suns, K., Curry, C., and Russell, D. (1980). *The Effects of Water Quality and Morphometric Parameters on Mercury Uptake by Yearling Yellow Perch.* Ont. Min. Environ. Tech. Rep. Lts BO-P, 16 pages.
Waite, D. T., Dunn, G. W., and Stedwill, R. J. (1980). *Mercury in Cookson Reservoir (East Poplar River).* Sask. Env. Report. 19 pages.
Weagle, K. V., and Cameron, R. A. (1974). *The Impact of the Strutt Lake Hydro Project on the Snare River, NWT (Abstr).* Dept. of the Environment, Fisheries and Marine Service, Tech. Report Series No. CEN/T-74-4.
Weiss, C. M., and Bond, A. E. (1978). *Mercury and New Impoundments. B. Everett Jordan and Falls Lake Sub-Impoundments.* US Army Corps of Eng., May 1978, 43 pages.
Westoo, G. (1966). Determination of methylmercury compounds in foodstuffs. *Acta. Chem. Scand.*, **20**, 211–213.
Wilkins, W. D. (1978). Engineering and economic evaluation of selected remedial measures. In *Mercury Pollution in the Wabigoon–English River System of N. W. Ontario and Possible Remedial Measures.*
Wren, C. D., and MacCrimmon, H. R. (1983). Mercury levels in sunfish (*Lepomis gibbosus*) relative to pH and other environmental variables of precambrian shield lakes. *Can. J. Fish. Aquat. Sci.*, **10**, 1737–1744.
Yoshida, T., and Ikegaki, Y. (1977). A study of the behaviour of mercury-contaminated sediments in Minimata Bay. In *Management of Bottom Sediments Containing Toxic Substances.* pp. 127–143. Corv. Environ. Res. Lab. Oreg. PB272 684.
Zitko, V., and Carson, W. G. (1976). A mechanism of the effects of water hardness on the lethality of heavy metals to fish. *Chemosphere*, **5**, 299–303.

Lead, Mercury, Cadmium and Arsenic in the Environment
Edited by T. C. Hutchinson and K. M. Meema

CHAPTER 17

Global Cycling of Arsenic

D. C. CHILVERS AND P. J. PETERSON

Monitoring and Assessment Research Centre,
The Octagon Building, Chelsea College, University of London,
459A Fulham Road, London SW10 OQX, England

ABSTRACT

In recent years, tentative estimates of arsenic fluxes among land, oceans, rivers, sediments and the atmosphere have highlighted the problems of obtaining a mass balance for arsenic on a global basis. Arsenic is mobilized by both natural and anthropogenic processes and these should be identified and located before quantification can be carried out. Much of the early data on arsenic concentrations and emission factors is outdated with regard to existing environmental control strategies and analytical techniques. Emission data for volcanoes can vary by more than two orders of magnitude, and the quantification of the natural low temperature volatilization of organo-arsenic compounds from soils is especially difficult.

In this study, current information on anthropogenic emission factors, including those for coal combustion, copper smelting and pesticide use, has been applied to present-day production data. Volcanic emissions have been updated and experimental data on the biological conversion of arsenate in soils have been used to estimate the extent to which methylated arsenic compounds contribute to the ambient atmospheric arsenic concentration.

The estimated ratio of natural to anthropogenic atmospheric inputs is 60 : 40. Copper smelting accounts for 40% of the anthropogenic input, while other non-ferrous metal production, coal combustion, agricultural chemicals and agricultural burning also emit significant amounts of arsenic. Low temperature volatilization and volcanism accounted for 97% of natural emissions. Atmospheric inputs were balanced with the calculated rain-out flux, while the net oceanic flux showed a small positive balance.

In this study, the uncertainties associated with published data have been noted and an attempt made at providing a more realistic basis for estimating present and future arsenic fluxes in the environment.

INTRODUCTION

In recent years, considerable attention has been directed towards the compilation of global inventories for natural and anthropogenic emissions of arsenic to the atmosphere (Lantzy and Mackenzie, 1979; Mackenzie *et al.*, 1979; NAS, 1977; Walsh *et al.*, 1979a). In some studies, important emission sources have not been adequately assessed and, in general, emission factors for industrial sources have not taken into account many of the variations in the world situation, including differences in source concentrations, industrial processes and environmental control strategies.

A previous study on global atmospheric fluxes of arsenic showed a gross imbalance which was attributed to biological volatilization of arsenic from the ocean surface to the atmosphere (Table 17.1) (Lantzy and Mackenzie, 1979). However, an examination of arsenic concentrations and speciation in sea water and atmospheric precipitation showed that this assumption was incorrect and that the tropospheric burden had been over-estimated (Andreae, 1980).

Table 17.1 A comparison of global estimates of natural arsenic emissions to the atmosphere (tonnes/year)

Source	Reference 1*	Reference 2†	Present study
Volcanoes	7 000	280	17 150
Low temperature volatilization	160	16 000	26 200
Wind erosion	240	2 500	1 980
Forest fires	160	—	125
Sea spray	28	194 790§	27
Total	7 800‡	213 570	45 480
Overall total (incl. anthropogenic emissions)	31 400	296 470	73 540

* Walsh *et al.*, 1979a.
† Mackenzie *et al.*, 1979.
‡ Also incorporates oceanic gas exchange, direct volatilization from earth's crust and particle emissions from plants.
§ Includes biological volatilization.
— Not determined.

The present study provides emission factors and a source inventory for anthropogenic inputs and indicates the magnitude of natural inputs of arsenic to the atmosphere. In addition, recent measurements of arsenic concentrations in the atmosphere have been applied to provide a rain-out flux. Information regarding anthropogenic inputs to rivers and coastal waters is limited but possible sources of arsenic are discussed as a guideline for future studies. A tentative estimate of the oceanic flux is provided.

The present chapter was prepared as part of an ongoing study to estimate anthropogenic and natural fluxes of arsenic to land, air and sea and to estimate future variations in the arsenic burden of the different environmental compartments. An outline of further research which is required to obtain a complete and accurate global cycle for arsenic is included.

CALCULATIONS OF ARSENIC EMISSIONS FROM ANTHROPOGENIC SOURCES

Copper Smelting

World smelter production for 1982 was 8.35×10^6 tonnes (World Bureau of Metal Statistics, 1984). However, this figure incorporated an estimated 300×10^3 tonnes of secondary blister produced from scrap. As such, smelter production from primary materials was 8.05×10^6 tonnes.

Air An emission factor of 3.9 kg As/tonne of Cu produced was calculated for the Ajo smelter in the USA from production data at 70 per cent of theoretical capacity (MINIMET, 1982) and an estimated emission of 163 tonnes of arsenic per year (Morse *et al.*, 1979). The copper concentrate at the Ajo smelter contains between 700 to 7000 ppm (average ~ 1300) of arsenic whereas the global average was assumed to be 500 ppm (Germani *et al.*, 1981); the emission factor was proportionately reduced to 1.5 kg As/tonne Cu produced. The estimated global emission from primary copper smelting is 12 080 tonnes/year (Table 17.2).

Water An emission factor of 3.5 kg As/tonne of Cu produced was calculated from production and emission data for a copper smelter in Sweden (Lindau, 1977). The estimated global emission in the various waste waters which result from copper smelting is 28 175 tonnes/year (Table 17.3).

Land The amount of arsenic which is disposed of or included in landfill was calculated proportionately from the amount of arsenic dispersed to land in the USA in 1979 (Lowenbach and Schlesinger, 1979), USA production figures for 1979 and world production figures for 1982 (World Bureau of Metal Statistics, 1984). The estimated global arsenic content of slag and sludges dumped on land is 47 290 tonnes/year (Table 17.3).

Copper Refining (Non-thermal)

In 1982 world refined production was 9.55×10^6 tonnes (World Bureau of Metal Statistics, 1984). Approximately 90% was produced by non-thermal methods (MINIMET, 1982), while 1.122×10^6 tonnes were from secondary production (World Bureau of Metal Statistics, 1984).

Table 17.2 Production and emission factors utilized in the estimation of arsenic emissions to the atmosphere from anthropogenic sources

Source	Global production (tonnes/year)	Emission factor	Total arsenic emission (tonnes/year)
Copper smelting	8.05×10^6	1.5 kg As/tonne Cu produced	12 080
Lead smelting	3.58×10^6	0.4 kg As/tonne Pb produced	1 430
Zinc smelting	1.20×10^6*	0.65 kg/As tonne Zn produced	780
Steel production	645×10^6	0.091 gm As/tonne steel produced	60
Coal combustion	3696×10^6	0.5, 1.0 and 2.0 gm As/tonne coal (dependent on use and control device)	6 240
Wood fuel	850×10^6	0.5 μg As/gm wood	425
Clearance of tropical forests	3200×10^6	0.5 μg/gm vegetation	1 600
Clearance of woodland and savannah	640×10^6	0.5 μg/gm vegetation	320
Burning of pasture land	2000×10^6	0.5 μg As/gm vegetation	1 000
Herbicides	8 000†	430 kg As/tonne As applied	3 440
Wood preservatives	150† (1976) 16 000† (1982)	100% released during burning of wood	150
Glass manufacture	4 000†	117 kg As/tonne As used	467
Waste incineration	150×10^6	0.52 gm As/tonne As waste	78

* Pyrometallurgical processes.
† Arsenic used.

Table 17.3 Estimated inputs of arsenic to land and water from anthropogenic sources

Source	Inputs to land (tonnes/year)	Inputs to water (tonnes/year)
Copper smelting	47 290	28 175
Copper refining	10 580	230
Lead smelting	3 580	NS
Zinc production	6 050	NS
Steel production	5 850	NS
Coal combustion	35 100	–
Herbicides	4 560	–
Desiccants	12 000	–
Waste incineration	–	–
Fertilizer application	–	–
Sludge application	–	–
Municipal waste water	–	–
Dumping of industrial wastes to sea	–	–

– Not determined.
NS = not significant.

Air Emissions were regarded as relatively insignificant, though local problems have been reported as a result of such As emissions.

Water An emission factor of 0.03 kg As/tonne of Cu produced (USEPA, 1976a) gives a global emission of 230 tonnes/year (Table 17.3).

Land The amount of arsenic which is landfilled was calculated proportionately from the amount of arsenic dispersed to land in the USA in 1979 (Lowenbach and Schlesinger, 1979), US non-thermal production of copper from primary materials in 1979 ($\sim 10^6$ tonnes) and world production figures for 1982 (MINIMET, 1982; World Bureau of Metal Statistics, 1984). The estimated global arsenic content of residues dumped on land is 10 580 tonnes/year (Table 17.3).

Lead Smelting

World production for 1982 was 3.578×10^6 tonnes (World Bureau of Metal Statistics 1984).

Air An emission factor of 0.4 kg/tonne of Pb produced (USEPA, 1976b) gives a global emission of 1430 tonnes/year (Table 17.2).

Land The amount of arsenic which is landfilled was calculated proportionately from the amount of arsenic dispersed to land in the USA in 1974, excluding residues entering the lead industry from electrolytic zinc production (Lowenbach and Schlesinger, 1979) and world production data for 1982 (World Bureau of Metal Statistics, 1984). The estimated global arsenic content of residues dumped on land is 3580 tonnes/year (Table 17.3).

Zinc Production

World production for 1982 was 6.51×10^6 tonnes (World Bureau of Metal Statistics, 1984) of which approximately 18% was produced by pyrometallurgical processes (International Lead and Zinc Study Group, 1983).

Air An emission factor of 0.65 kg/tonne of Zn produced gives a global emission of 780 tonnes/year (Table 17.2).

Land The total arsenic content of zinc ore smelted, less the amount of arsenic emitted to the atmosphere, was used to calculate the amount of arsenic dispersed to land as slag from smelters. Using an arsenic concentration of 1.05 gm/kg (USEPA, 1976a), approximately 586 tonnes/year are landfilled. Almost all of the arsenic present in residues from electrolytic refining is transferred to lead smelters where the ultimate fate is landfill. An emission factor of 1.05 kg/tonne (USEPA, 1976a) gives a total arsenic content for slag dispersed to land of 5466 tonnes/year. This gives a total to land of 6050 tonnes/year (Table 17.3).

Steel Production

In 1982 world steel production was 645×10^6 tonnes (International Iron and Steel Institute, 1984).

Air An emission factor of 0.091 gm/tonne of steel produced (USEPA, 1976a) gives a global arsenic emission of 60 tonnes/year (Table 17.2).

Land An emission factor of 9.1 gm/tonne of steel produced (USEPA, 1976a) gives the global amount of arsenic disposed to land as 5850 tonnes/year (Table 17.3).

Coal Combustion

Air Coal consumption was considered under five separate sections: electricity generation, metallurgical coal, other industrial consumption, retail use and coal conversion (Predicasts, 1982). The first three groups were categorized by country into those users which have good, moderate or poor emission control strategies. Assuming an average arsenic concentration of

15 gm/tonne of coal (Goetz *et al.*, 1981; Chadwick and Lindman, 1982) and removal efficiencies of 97% for electostatic precipitators, 95% for Venturi scrubbers and 80% for mechanical cyclone collectors (Table 17.4) (Klein *et al.*, 1975; Schwitzgebel *et al.*, 1975; Lowenbach and Schlesinger, 1979; Ondov *et al.*, 1979a,b; DOE, 1980; Sabbioni and Goetz, 1983; Elseewi *et al.* 1984), emission factors of 0.5, 1.0 and 2.0 gm/tonne were applied for the three groups of countries. Emission factors from electricity generation, metallurgical coal (coke production and use) and other industrial usage were considered to be the same. An emission factor of 2.0 gm/tonne was applied to residential/commercial consumption. Atmospheric emissions from coal conversion are not significant (Filby *et al.*, 1981). Total global estimates were 2410, 750, 1980 and 1100 tonnes/year for electricity generation, metallurgical coal, industrial use and retail use, respectively, giving a total atmospheric emission of 6240 tonnes (Table 17.2).

Table 17.4 A comparison of atmospheric emission factors for arsenic from coal combustion

Emission factor (gm As/gm As in coal)	Control device*	Reference
0.18	GA	DOE, 1980
0.01†	ESP	Elseewi *et al.*, 1984
0.01†	VS	Elseewi *et al.*, 1984
0.20†	CC	Elseewi *et al.*, 1984
0.03	ESP	Klein *et al.*, 1975
0.03	—	Lindau, 1977
0.25	—	Lowenbach and Schlesinger, 1979
0.01	ESP (99.8)‡	Ondov *et al.*, 1979a,b
0.12	ESP (97.0)‡	Ondov *et al.*, 1979a,b
0.02	—	Pacyna, 1984
0.03	ESP (99.5)‡	Sabbioni and Goetz, 1983
0.01	ESP	Schwitzgebel *et al.*, 1975
0.08	VS	Schwitzgebel *et al.*, 1975
0.21	CC	Schwitzgebel *et al.*, 1975

* GA—grit arrester; ESP—electrostatic precipitator; VS—Venturi scrubber; CC—cyclone collector.
† Based on fly ash removal efficiencies.
‡ Fly ash removal efficiencies (%).

Land Information is sparse regarding water-borne wastes from power plants and metallurgical industries, particularly those from control devices (e.g. wet scrubbers). It was assumed that contained arsenic was disposed to land, thereby giving an emission of (2.34×10^9 tonnes/year $\times$ 15 μg/gm) − 6240 tonnes = 35 100 tonnes/year to land (Table 17.3).

Wood Fuel

The consumption of wood fuel is estimated to be 850 $\times$ 10^6 tonnes/year (Logan et al., 1981). The emission factor is assumed to be the mean concentration of arsenic in vegetation, 0.5 μg/gm (NRCC, 1978). The same emission factor is used for tropical forests, grassland, woodland and savannah. The estimated global emission from fuel wood is 425 tonnes/year (Table 17.2).

Clearance of Land for Crops

Tropical forests An estimated 3200 $\times$ 10^6 tonnes of vegetation are burnt for agricultural purposes each year (Logan *et al.*, 1981). The estimated global emission is 1600 tonnes/year (Table 17.2).

Woodland and savannah An estimated 640 $\times$ 10^6 tonnes of vegetation are burnt for agricultural purposes each year (Logan *et al.*, 1981). The estimated global emission is 320 tonnes/year (Table 17.2).

Burning of pasture land In some parts of the world, the burning of savannah to maintain grassland for grazing involves large areas of land. An estimated 2000 $\times$ 10^6 tonnes/year are burnt (Logan *et al.*, 1981). The estimated global emission is 1000 tonnes/year (Table 17.2).

Arsenic Chemicals

World production and sales of refined arsenic trioxide are estimated to be approximately 50 000 tonnes/year (Metal Bulletin Ltd., 1977). Since 1976, wood treatment has increased from 7% to 33% of the market, while cotton desiccants (25%), herbicides (17%) and food additives (10%) utilize large amounts (Woolson, 1983). Alloys and glass utilize about 5% each and 5% is used in miscellaneous chemicals (Metal Bulletin, Ltd., 1977).

Herbicides An estimated 8000 tonnes of arsenic are used each year. An emission factor of 170 kg/tonne As applied has been reported (Lowenbach and Schlesinger, 1979), while an average of 260 kg/tonne As applied is volatilized per year. The total non-accumulative emission factor is 430 kg/tonne. An estimated 3440 tonnes of arsenic are emitted from herbicide application per year (Table 17.2).

Desiccants An estimated 12 000 tonnes of arsenic are used each year but reliable estimates of atmospheric emissions during ginning operations and the proportion of ginning wastes which are incinerated are not available. It is assumed that the majority of waste is dispersed to land (Table 17.3).

Table 17.5 A comparison of global estimates of anthropogenic arsenic emissions to the atmosphere (tonnes/year)

Source	Reference 1*	Reference 2†	Present study
Non-ferrous metal production	15 200	60 000	14 290
Copper smelting	13 000	–	12 080
Lead smelting	2 200 (Lead and Zinc smelting)	–	1 430
Zinc smelting		–	780
Iron and steel production	4 200	–	60‡
Coal	550	15 800	6 240
Electricity generation	–	–	2 410
Metallurgical use	–	–	750
Other industrial consumption	–	–	1 980
Retail use	–	–	1 100
Coal conversion	–	–	NS
Wood used as fuel	60	–	425
Clearance for crops	–	–	1 920
Tropical forests	–	–	1 600
Woodland and savannah	–	–	320
Burning of pasture land	–	–	1 000
Agricultural burning	60	–	–
Agricultural chemicals	1 920	5 000 (Agricultural chemicals and Wood preservatives)	3 440
Wood preservatives	–		140
Glass manufacture	–	–	468
Cement production	–	2 100	–
Waste incineration	430	–	78
Total	23 600§	82 900	28 060¶

* Walsh *et al.*, 1979a.
† Mackenzie *et al.*, 1979.
‡ Steel production only.
§ Includes emissions from the combustion of light and residual fuels, mining mineral ore and arsenic chemicals.
¶ An additional 2,000–3,000 tonnes could come from cement manufacture and agricultural burning.
– Not determined.
NS = Not significant.

Wood preservatives Although 16 000 tonnes/year of arsenic are used, the emissions from this source are low at present but could increase in the future as old treated timber is disposed of, usually by burning. Treatment of timbers from power plant cooling towers represented only 0.3% of the arsenic fungicide used in 1976 but this would represent an emission, if burnt, of 150 tonnes (DOE, 1980) (Table 17.2).

Glass manufacture An estimated 4000 tonnes As/year are used. An emission factor of 117 kg As/tonne of As used gives a global emission of 468 tonnes/year (USEPA, 1976a) (Table 17.2).

Waste Incineration

The global mass of waste material which is incinerated is approximately 150×10^6 tonnes (Nriagu and Davidson, 1980). An emission factor of 0.52 gm/tonne (Pacyna, 1984) gives a total emission of 78 tonnes/year (Table 17.1).

Summary

Table 17.5 compares the results of the present study with previous source inventories for anthropogenic sources.

CALCULATIONS OF ARSENIC EMISSIONS FROM NATURAL SOURCES

Sea-salt Spray

The total production of sea-salt particles was estimated to be 500×10^6 tonnes/year (Peterson and Junge, 1971). The arsenic concentration in sea water at the surface is approximately 1.6 μg/litre (Burton and Statham, 1982), while the salt content is approximately 3%. An emission factor of 0.053 μg As/gm sea-salt was derived to give a global emission of 26.5 tonnes/year (Table 17.6).

Table 17.6 Production and emission factors utilized in the estimation of arsenic emissions to the atmosphere from natural sources

Source	Global production (tonnes/year)	Emission factor	Total arsenic emission (tonnes/year)
Sea-salt spray	500×10^6	0.053 μg As/gm sea-salt	26.5
Wind erosion	330×10^6	6 mg As/kg soil	1 980
Forest fires	250×10^6	5 μg As/gm vegetation	125
Volcanoes	25×10^6	686 gm As/tonne ash	17 150
Low temperature volatilization	Land area = 59×10^{12} m^2	0.044 μg/cm^2/year	26 200

Wind Erosion

Wind erosion contributes 330×10^6 tonnes of particulates to the atmosphere per year (Junge, 1979). Because soils cover most of the land surface, an emission factor equivalent to an average soil concentration of 6 mg/kg was used. This gives a global emission of 1980 tonnes/year (Table 17.6).

Forest Fires

An estimated 250 × 10^6 tonnes of vegetation are burnt in forest fires each year (Logan *et al.*, 1981). The estimated global arsenic emission is 125 tonnes (Table 17.6).

Volcanoes

The total production of ≤5 μm diameter particles from volcanoes is estimated to be 25 × 10^6 tonnes/year (Peterson and Junge, 1971). The method of Walsh *et al.* (1979a) was applied to data from the Mount Augustine volcano, 1976 eruption (Lepel *et al.*, 1978). The product of the enrichment factor of 618 and an ash concentration of 1.11 μg/gm gives an emission factor of 686 μg/gm. The global arsenic emission from volcanoes was estimated to be 17 150 tonnes/year (Table 17.6).

Low Temperature Volatilization

The methylation and subsequent volatilization of arsenic to the atmosphere from soils is dependent on the presence of organic matter and microbes and the availability of labile arsenic (Woolson, 1983). Certain environmental factors, including water content, temperature and aeration, will also affect arsenic emissions from soils. As such, in estimating the land area to be assessed, soils with low productivities (<500 gm/m^2/year) and high water content were excluded. The area assumed to produce significant quantities of volatile arsenic compounds was 59 × 10^{12} m^2 (Ajtay *et al.*, 1979). Woolson (1977) showed that soils treated with 10 mg/kg ^{74}As–sodium arsenate released 1.0% of the activity after 160 days. It was also demonstrated that approximately 18% of the arsenate applied was extractable after this time. Under natural conditions, only 1% of the arsenic in most soils is labile (NAS, 1977). Assuming an average soil concentration of 5 mg/kg and a diffusion layer of 5 cm depth, an emission factor 0.044 × 10^{-6} gm/cm^2/year was derived from the above information, giving a total global arsenic emission of 26.2 × 10^3 tonnes/year (Table 17.6).

Summary

Table 17.1 compares the results of the present study with previous source inventories for natural sources.

CALCULATIONS OF ARSENIC RESERVOIRS AND FLUXES

Global Atmospheric Burden

Arsenic concentrations in the atmosphere differ for marine and continental areas and differences have also been noted between the northern and southern hemispheres (Walsh *et al.*, 1979b). It is necessary, therefore, to calculate the arsenic burdens separately for each area of the Earth. To calculate the global atmospheric arsenic burden, the method of Walsh *et al.* (1979b) was slightly modified. The height of the lower troposphere (5 km) was used in the calculation rather than 3.9 km (small particle scale height) and an arsenic concentration of 2.8 ng/m^3 for northern continental air was applied to the calculation of the arsenic burden for continental air in the northern hemisphere (Austin and Millward, 1984) (Table 17.7).

Table 17.7 Estimated global tropospheric burden of arsenic*

	Northern hemisphere		
	Land	Atlantic	Pacific/Indian
Area (m^2)	1.0×10^{14}	4.5×10^{13}	1.1×10^{14}
Regional surface arsenic concentration (ng/m^3)	2.8†	0.16	0.072
Burden (tonnes As)	1400	36	40

	Southern hemisphere	
	Land	Ocean
Area (m^2)	4.8×10^{13}	2.0×10^{14}
Regional surface arsenic concentration (ng/m^3)	1.0	0.018
Burden (tonnes As)	240	18
Northern hemisphere total (tonnes As)	1480	
Southern hemisphere total (tonnes As)	260	
Global tropospheric total (tonnes As)	1740	

* Based on Walsh *et al.*, 1979a.
† Austin and Millward, 1984.

The global burden was estimated to be 1740 tonnes with a partition ratio of 0.85 : 0.15 between the northern and southern hemispheres, thereby demonstrating the high anthropogenic input in the former. In addition, 94 per cent of the atmospheric burden was over land areas, suggesting that major sources are predominantly land-based.

Global Atmospheric Fluxes to Land and Oceans

Table 17.8 shows the atmosphere to land/ocean flux of arsenic calculated by three methods:

Model 1 This assumes that there are 40 rainfalls/year which remove atmospheric particulates (Bruland *et al.*, 1974), which gives a total flux of 69 600 tonnes/year.

Model 2 This utilizes a deposition velocity of 0.5 cm/sec which is equivalent to 0.16×10^6 m/year (Bennett, 1981) and the atmospheric levels shown in Table 17.7. Deposition rates of 0.44 mg/m^2/year (land), 0.025 mg/m^2/year (Atlantic), 0.011 mg/m^2/year (Pacific/Indian) were obtained for the northern hemisphere and 0.16 mg/m^2/year (land) and 0.003 mg/m^2/year (ocean) for the southern hemisphere, which gave a total global flux of 54 600 tonnes/year.

Model 3 This utilizes a washout factor which is the ratio of the rain concentration in μg/litre to the air concentration in μg/m^3. This factor has been estimated to be 355 (Cawse, 1974). When applied to the observed atmospheric concentrations of arsenic (Table 17.7) and rainfall data (Bowen, 1979), the continental flux was estimated to be 73.6×10^3 tonnes/year while that for the oceans was only 6.2×10^3 tonnes/year, giving a total flux of 79 800 tonnes/year.

Net Atmospheric Flux

The estimates shown in Table 17.8 are in close agreement with the estimated influx of arsenic to the atmosphere from the anthropogenic and natural sources (73 540 tonnes/year) illustrated in Tables 17.1 and 17.5.

Global Oceanic Burden and Flux

To estimate the total arsenic burden of the oceans, separate estimates of arsenic in dissolved form, particulates, skeletal particulates, phytomass and animal biomass must be made.

Dissolved The total volume of the oceans is 1.37×10^{21} litres (Mackenzie *et al.*, 1979) and the average arsenic concentration is 1.7 μg/litre (Burton and Statham, 1982), which gives a dissolved arsenic burden of 2.33×10^9 tonnes.

Table 17.8 Information used in the estimation of arsenic fluxes from the atmosphere to land and oceans

	Model 1		Model 2		
	Atmospheric burden (tonnes)	Flux (tonnes/year)	Land area (m^2)	Deposition rate ($mg/m^2/year$)	Flux (tonnes/year)
Northern Hemisphere	1 480	59 200	2.55×10^{14}		46 400
Land	1 400	56 000	1.00×10^{14}	0.440	44 000
Atlantic Ocean	36	1 400	0.45×10^{14}	0.025	1 100
Pacific/Indian Oceans	40	1 600	1.10×10^{14}	0.011	1 300
Southern Hemisphere	260	10 400	2.48×10^{14}		8 200
Land	240	9 600	0.48×10^{14}	0.160	7 600
Ocean	18	720	2.00×10^{14}	0.003	600
Total flux		69 600			54 600

Model 3

	Rainfall (litres/year)	Atmospheric concentration (ng/m^3)	Flux (tonnes/year)
Land	9.9×10^{16}	2.220*	73 600
Ocean	3.47×10^{17}	0.053*	6 200
Total flux			79 800

* Weighted average of concentrations from Table 17.7.

Particulate The mean suspended particulate concentration of the world's oceans is approximately 1 mg/litre and the total volume of the oceans is 1.37 $\times$ 10^{21} litres (Mackenzie *et al.*, 1979). A mean concentration of arsenic in shales and clays of 14.5 μg/gm (Woolson, 1983) was applied to these data to give a particulate arsenic burden of 19.87 $\times$ 10^6 tonnes.

Skeletal particulate An estimate of 16.2 $\times$ 10^3 tonnes of arsenic (Mackenzie *et al.*, 1979) was used.

Phytomass The world's oceans contain about 40 $\times$ 10^9 tonnes of plant material (Duce *et al.*, 1983) with an average arsenic concentration of 4 μg/gm (Boyle and Jonasson, 1973), which gives a phytomass burden of 0.16 $\times$ 10^6 tonnes.

Animal biomass The world's oceans contain 1 $\times$ 10^9 tonnes of animal material (Whittaker and Likens, 1973) with an average arsenic concentration of 3 μg/gm (Fowler, 1977), which gives a total animal arsenic burden of 3000 tonnes.

The total oceanic burden of arsenic is thus 2.35 $\times$ 10^9 tonnes.

Oceanic Influx

The oceanic influx was calculated from the summation of river (dissolved and suspended) inputs and the atmospheric flux.

Dissolved 3.7 $\times$ 10^{16} litres of water enter the ocean from rivers each year (Bowen, 1979). The average arsenic concentration in the world's rivers is 1.7 μg/litre (Boyle and Jonasson, 1973), which gives a total arsenic load of 62.9 $\times$ 10^3 tonnes/year.

Suspended Approximately 8 $\times$ 10^9 tonnes of sediment enter the world's oceans each year (Milliman, 1981). Using the mean arsenic concentration of shales and clays of 14.5 μg/gm (Woolson, 1983), a total arsenic load of 0.116 $\times$ 10^6 tonnes/year was obtained, thereby giving a total river flux of 178.9 $\times$ 10^3 tonnes/year.

Atmosphere The average flux of arsenic from the atmosphere to the oceans is 4310 tonnes/year (average of the fluxes calculated by the three models used previously).

The total influx of arsenic to the oceans is thus 183.2 $\times$ 10^3 tonnes/year.

Oceanic Outflux

The oceanic outflux was calculated from the summation of the amounts of arsenic which are lost as sea-salt spray and the amounts of arsenic precipitated in sediments each year.

Sedimentation Approximately 3.2×10^9 tonnes of sediments are deposited each year (Garrels and Mackenzie, 1971). By applying an average arsenic concentration for shales and clays of 14.5 μg/gm (Woolson, 1983), a total deposition rate of 46.4×10^3 tonnes/year is obtained.

Sea-salt spray An estimated 26.5 tonnes of arsenic per year is lost from the ocean surface by this process (cf. p. 288).

This gives a total oceanic outflux of 46.4×10^3 tonnes/year.

Net Oceanic Flux

The net ocean flux was calculated to be $+136.8 \times 10^3$ tonnes/year. If it is assumed that most of this arsenic has a long retention time in the oceanic surface layer (depth = 300 m) and the area of the oceanic surface layer is 360×10^{12} m^2 (volume of surface layer = 1.08×10^{17} m^3), then the rate of increase in arsenic concentration in the surface layer would be 1.27 ng/litre/year, which would represent a concentration increase of 0.1% per year (Andreae, 1979).

DISCUSSION

Anthropogenic Emissions

Over 40% of anthropogenic emissions to the atmosphere were accounted for by copper smelting. Other non-ferrous metal industries contributed less than 10%, while coal combustion contributed over 20% with approximately equal amounts coming from electricity generation, industrial consumption and retail use and less than 2% from metallurgical coal. The burning of vegetation for agricultural purposes contributed approximately 10% of the atmospheric burden, while the use of chemicals in agriculture added a similar amount.

Large inputs were expected from non-ferrous metal smelting due to the high arsenic content of concentrates and the high temperatures of stack gases. The removal efficiencies are therefore low, due to the fact that a large proportion of the arsenic is in the vapour phase or is associated with < 1 μm diameter particles (Piver, 1983). It is extremely difficult to calculate arsenic emissions from smelting operations; the extensive range of concentrations found in concentrates results in a range of emission factors (Table 17.9). In the USA, arsenic concentrations in copper concentrates range from 0.003 to 11% (Parker, 1978) but most contain between 0.02 and 0.15%. The average arsenic concentration used in the calculations shown above was based on an estimate of the relative importance of each ore, especially those ores from Canada, Peru and the Philippines which contain large amounts of arsenic

(Parker, 1978; World Bureau of Metal Statistics, 1984). Different emission factors were not applied to smelters on a regional basis because present control technology is not efficient in removing arsenic from copper smelter stack gases (Piver 1983).

Table 17.9 A comparison of atmospheric emission factors for arsenic from copper smelters*

Source (study dates)	Emission factor (kg As/tonne Cu produced)	Reference
Tacoma, Washington (1970–72)	3.2	Crecelius *et al.*, 1975
Tacoma, Washington (1971)	16.8	PSAPCA, 1980
Tacoma, Washington (1971–73)	1.8	PSAPCA, 1974
Tacoma, Washington (1974–76)	8.0	Lowenbach and Schlesinger, 1979
Tacoma, Washington (1980)	5.7	PSAPCA, 1980
Ajo, Arizona (1977)	3.9	Schwitzgebel *et al.*, 1978
USA smelters	1.5 (range 0.01–8.0)	Lowenbach and Schlesinger, 1979
Sweden (1976)	0.6	Lindau, 1977
European smelters	3.0	Pacyna, 1984
World average (1982)	1.5	Present study

* Production data assumed to be 70% of production capacity from MINIMET publications.

The processing of iron ore was not considered because almost all of the arsenic is transferred to the product during primary production (DOE, 1980). In steel production, 90% of the arsenic appears in the steel product, 7% occurs in iron and steel-making slags, 2% in slurries and solid wastes and less than 0.5% is discharged to the atmosphere (DOE, 1980).

In the calculations of arsenic emissions from coal combustion it is necessary to consider emissions on a regional basis. The implementation and use of control devices will vary by region and emission factors will also be dependent on the type of combustion process. For example, in China, 20%, 8%, 47% and 25% of coal consumed is used for electricity production, metallurgical coal, industrial and retail use, respectively, which compares with

81%, 9%, 9% and 1% in the USA (Predicasts 1982). As such, it is relatively more difficult to control arsenic emissions from coal combustion in China.

In this study, an emission factor of 3% was applied to those power plants which had electrostatic precipitators or other high efficiency particulate removal devices. However, it has been shown that as control equipment increases in age, a small reduction in the efficiency of particulate removal may result in a large increase in the amount of arsenic released in stack gases (Ondov *et al.*, 1979a,b). As such, we may have underestimated the amount of arsenic emitted from coal combustion. Information on the implementation of control devices and their operating efficiencies is therefore required.

In the future, arsenic emissions from energy production may increase because of the additional use of synfuel technology (Filby *et al.*, 1981; Ghassemi *et al.*, 1981) and geothermal power (Sabadell and Axtmann, 1975).

The large arsenic emission from the use of arsenic containing chemicals in agriculture is mainly due to poor application procedures but is also due to the initially high rate of volatilization. In addition, there may be considerable emissions from the disposal of cotton ginning wastes.

Future Trends in Anthropogenic Emissions

Copper smelting Long-term forecasts for the production and consumption of copper indicate that for the period 1986–91 an annual rate of growth of 1.1% per annum is expected (International Wrought Copper Council, 1984). With regard to atmospheric arsenic emissions, this is likely to be easily offset by improvements in control technology and the use of copper ores which contain less arsenic. However, with the implementation of control devices, particularly wet scrubbers, large volumes of contaminated water and sludges are produced, thereby shifting the arsenic burden to the aquatic compartment. Severe local aquatic pollution problems could be averted by the installation of adequate removal devices (Bhattacharyya *et al.*, 1979) but, as demonstrated in Sweden during the 1970s, it is probable that the initial result will be to discharge more arsenic to waters (Lindau, 1977). If world demand for arsenic compounds increases at its present rate, then more companies will install equipment to collect and refine the arsenic. However, the present aquatic input of 28 175 tonnes/year from copper smelting (Table 17.3) is likely to increase.

Other non-ferrous metal production The major factor which will affect arsenic emissions from the lead and zinc industries will be the reduction in the use of pyrometallurgical processes and a concomitant increase in the use of electrolytic techniques. As with copper production, this will result in an increase in arsenic emissions to the aquatic environment. The largest source of arsenic-bearing waste in the United Kingdom is the Capper Pass tin smelter (DOE, 1980) which discharges approximately 500 tonnes/year of arsenic to the North Sea (Mance and O'Donnell, 1984) while producing only 15 000 tonnes of tin per annum (Williamson and Fromson, 1982). As more primary tin smelters utilize low grade ore, it is likely that large quantities of arsenic will be emitted to the atmosphere, rivers and coastal waters.

Coal combustion Coal consumption is expected to increase by 3-4% per year until the year 2000 (Chase Manhattan Bank, 1981; IEA, 1982; Predicasts, 1982; Frisch, 1983). However, a considerable proportion of the increased coal utilization will be in new power plants and coal-fired heating systems which will presumably employ control devices. Older power plants may also have control devices installed. The increased utilization of coal is not expected to cause a dramatic increase in the atmospheric burden of arsenic in IEA countries. However, the higher growth rates for coal use in some Eastern European countries that lack adequate control strategies may result in large inputs, particularly from residential use. The development and diffusion of large-scale systems using fluidized bed technology, beginning in the early 1990s, will have some effect, due to a larger particulate load. The environmental impact of the future magnetohydrodynamic (MHD) process should also be investigated.

Arsenic chemicals World demand for arsenic is likely to increase due to the rapid growth of its use in wood preservatives (Woolson, 1983). This will not have an immediate effect on the environment but disposal of treated timber may pose a problem for the future. A major source of arsenic contamination may be the use of desiccants but reliable data do not exist on the fate of these chemicals.

Natural Emissions

This study illustrated the extent to which natural emissions can contribute to the global arsenic cycle. From our preliminary calculations, it appears that at least 60% of atmospheric inputs are derived from natural sources, with nearly 60% of the total natural flux coming from low temperature volatilization from soils and most of the remainder from volcanoes. These emissions can be regarded as minimum amounts due to the fact that it was assumed that (i) only 0.126% of soil arsenic is volatilized per year, and (ii) the volcanic emission was calculated for a year with no large eruptions.

CONCLUSIONS

The present chapter updates and improves our knowledge of global arsenic cycling and indicates areas where additional information is required, particularly for the aquatic compartment. A rapid growth in energy production may well increase arsenic emissions, particularly from geothermal power, synfuel processes and changes in coal combustion technology. At present, the activities of man result in large inputs of arsenic to the environment but anthropogenic inputs are unlikely to influence average concentrations significantly in the major global reservoirs. However, local accumulation of arsenic in the atmosphere, soils, rivers and coastal waters is of some concern, especially from the smelting of ores and the use of arsenical compounds.

REFERENCES

Ajtay, G. L., Ketner, P., and Duvigneaud, P. (1979). Terrestrial primary production and phytomass. In Bolin, B., Degens, E. T., Kempe, S., and Ketner, P. (Eds), *The Global Carbon Cycle. SCOPE 13*, pp. 129–181. John Wiley, Chichester.

Andreae, M. O. (1979). Arsenic speciation in sea water and interstitial waters: The influence of biological-chemical interactions on the chemistry of a trace element. *Limnol. Oceanogr.*, **24**, 440–452.

Andreae, M. O. (1980). Arsenic in rain and the atmospheric mass balance of arsenic. *J. Geophys. Res.*, **85**, 4512–4518.

Austin, L. S., and Millward, G. E. (1984). Modelling temporal variations in the global tropospheric arsenic burden. *Atmos. Environ.*, **18**, 1909–1919.

Bennett, B. G. (1981). Arsenic. In *Exposure Commitment Assessments of Environmental Pollutants*, Vol. 1, No. 1. MARC Report No. 23, pp. 44–59. Monitoring and Assessment Research Centre, Chelsea College, London.

Bhattacharyya, D., Jumawan, A. B., Jr., and Grieves, R. B. (1979). Charged membrane ultrafiltration of heavy metals from non-ferrous metal. *J. Water Pollut. Control Fed.*, **51**, 176–186.

Bowen, H. J. M. (1979). *Environmental Chemistry of the Elements.* Academic Press, London, 333 pages.

Boyle, R. W., and Jonasson, I. R. (1973). The geochemistry of arsenic and its use as an indication element in geochemical prospecting. *J. Geochem. Explor.*, **2**, 251–296.

Bruland, K. W., Bertine, K., Koide, M., and Goldberg, E. D. (1974). History of metal pollution in Southern California coastal zone. *Environ. Sci. Technol.*, **8**, 425–431.

Burton, J. D., and Statham, P. J. (1982). Occurrence, distribution and chemical speciation of some minor dissolved constituents in ocean waters. *Environmental Chemistry*, Vol. 2, pp. 234–265. A Specialist Periodical Report, Royal Society of Chemistry, London.

Cawse, P. A. (1974). *A Survey of Atmospheric Trace Elements in the U. K. (1972–1973)*. Report AERA R7660. UK Atomic Energy Authority, Harwell, 95 pages.

Chadwick. M. J., and Lindman, N. (1982). *Environmental Implications of Expanded Coal Utilization.* Pergamon Press, Oxford, England, 283 pages.

Chase Manhattan Bank (1981). World Coal Outlook to the Year 2000: Opportunities and Uncertainties. A Study Conducted by the Energy Economics Division. The Chase Manhattan Bank, USA.

Crecelius, E. A., Bothner, M. H., and Carpenter, R. (1975). Geochemistries of arsenic, antimony, mercury and related elements in sediments of Puget Sound. *Environ. Sci. Technol.*, **9**, 325–333.

DOE (1980). *Arsenic-bearing Wastes.* Waste Management Paper No. 20. HMSO, London, 66 pages.
Duce, R. A., Mohnen, V. A., Zimmerman, P. R., Grosjean, D., Cautreels, W., Chatfield, R., Jaenicke, R., Ogren, J. A., Pellizzari, E. D., and Wallace, G. T. (1983). Organic material in the global troposphere. *Rev. Geophys. Space Phys.*, **21**, 921–952.
Elseewi, A. A., Straughan, I. R., Page, A. L., Doyle, C. P., and Fox, C. A. (1984). Trace elements in flue-gas cleaning residues from the combustion of Western US coals. In *Environmental Contamination*, London, July 1984, pp. 633–640. CEP Consultants Ltd., Edinburgh.
Filby, R. H., Sandstrom, D. R., Lytle, F. W., Greegor, R. B., Khalil, S. R., Ekambaram, V., Weiss, C. S., and Grimm, C. A. (1981). Chemistry of trace element species in coal liquefaction processes. In Brinckman, F. F., and Fish, R. H. (Eds), *Environmental Speciation and Monitoring Needs for Trace Metal-containing Substances from Energy-related Processes*, pp. 21–38, NBS Special Publication No. 618.
Fowler, B. A. (1977). Toxicology of environmental arsenic. In Goyer, R. A. and Mehlman, M. A. (Eds), *Toxicology of Trace Elements*, pp. 79–122. John Wiley and Sons, New York.
Frisch, J. R. (1983). *Energy 2000–2020: World Prospects and Regional Stresses.* World Energy Conference Conservation Commission. Graham and Trotman Publishers.
Garrels, R. M., and Mackenzie, F. T. (1971). *Evolution of Sedimentary Rocks.* W. W Norton, New York, 397 pages.
Germani, M. S., Small, M., Zoller, W. H., and Moyers, J. L. (1981). Fractionation of elements during copper smelting. *Environ. Sci. Technol.*, **15**, 299–305.
Ghassemi, M., Iyer, R., Scofield, R., and McSorley, J. (1981). Effects of synfuel use. *Environ. Sci. Technol.*, **15**, 866–873.
Goetz, L., Springer, A., Pietra, R., and Sabbioni, E. (1981). *Mobilization of Heavy Metals from Fossil-fuelled Power Plants. Potential Ecological and Biological Implications: III. Heavy Metal Content in Coals Burnt in European Power Plants.* Report EUR 6998/III. CEC, Luxembourg, 56 pages.
IEA (1982). *Coal Prospects and Policies in IEA Countries: 1981 Review.* Organization for Economic Co-operation and Development (OECD)/International Energy Agency (IEA), Paris, 170 pages.
International Iron and Steel Institute (1984). *Steel Statistical Yearbook 1983.* Brussels, Belgium.
International Lead and Zinc Study Group (1983). *Trends in Production of Lead and Zinc.* London, England, 57 pages.
International Wrought Copper Council (1984). *Annual Report 1983–84.* International Wrought Copper Council, London, 17 pages.
Junge, C. (1979). The importance of mineral dust as an atmospheric constituent. In Morales, C. (Ed.), *Saharan Dust: Mobilization, Transport, Deposition, SCOPE 14*, pp. 49–60. John Wiley and Sons, Chichester.
Klein, D. H., Andren, A. W., Carter, J. A., Emery, J. F., Feldman, C., Fulkerson, W., Lyon, W. S., Ogle, J. C., Talmi, Y., Van Hook, R. I., and Bolton, N. (1975). Pathways of thirty-seven trace elements through coal-fired power plant. *Environ. Sci. Technol.*, **9**, 973–979.
Lantzy, R. J., and Mackenzie, F. T. (1979). Atmospheric trace elements: Global cycles and assessment of man's impact. *Geochim. Cosmochim. Acta*, **43**, 511–525.
Lepel, E. A., Stefansson, K. M., and Zoller, W. H. (1978). The enrichment of volatile

elements in the atmosphere by volcanic activity: Augustine Volcano 1976. *J. Geophys. Res.*, **83**, 6213–6220.

Lindau, L. (1977). Emissions of arsenic in Sweden and their reduction. *Environ. Health Perspect.*, **19**, 25–29.

Logan, J. A., Prather, M. J., Wofsy, S. C., and McElroy, M. B. (1981). Tropospheric chemistry: a global perspective. *J. Geophys. Res.*, **86**, 7210–7254.

Lowenbach, W. A., and Schlesinger, J. S. (1979). *Arsenic: A Preliminary Materials Balance.* EPA-560/6-79-005. US Environmental Protection Agency, Washington DC, 56 pages.

Mackenzie, F. T., Lantzy, R. J., and Paterson, V. (1979). Global trace metal cycles and predictions. *J. Int. Assoc. Math. Geol.*, **11**, 99–142.

Mance, G., and O'Donnell, A. R. (1984). The magnitude of land-based discharges of heavy metals entering the coastal waters of the United Kingdom and the North Sea. In *Environmental Contamination*, London, July 1984, pp. 542–548. CEP Consultants, Edinburgh.

Metal Bulletin Ltd. (1977). *Minor Metals Survey.* Metal Bulletin Ltd, London.

Milliman, J. D. (1981). Transfer of river-borne particulate material to the oceans. In *River Inputs to Ocean Systems*, pp. 5–12. Proceedings of a Review Workshop held at FAO Headquarters, Rome, Italy, March 1979. United Nations, New York.

MINIMET (1982). *Penarroya*, Paris, 50–80.

Morse, D. L., Harrington, J. M., Housworth, J., and Landrigan, P. J. (1979). Arsenic exposure in multiple environmental media in children near a smelter. *Clin. Toxicol.*, **14**, 389–399.

NAS (1977). *Arsenic.* National Academy of Sciences, Washington DC, 332 pages.

NRCC (1978). *Effects of Arsenic in the Canadian Environment.* Publication No. NRCC 15391. National Research Council Canada, Ottawa, 349 pages.

Nriagu, J. O., and Davidson, G. I. (1980). Zinc in the atmosphere. In Nriagu, J. O. (Ed.), *Zinc in the Environment. Part I: Ecological Cycling*, pp. 113–160. John Wiley and Sons, New York.

Ondov, J. M., Ragaini, R. C., and Biermann, A. H. (1979a). Elemental emission from a coal-fired power plant. Comparison of a Venturi wet scrubber system with a cold-side electrostatic precipitator. *Environ. Sci. Technol.*, **13**, 588–601.

Ondov, J. M., Ragaini, R. C., and Biermann, A. H. (1979b). Emission and particle-size distribution of minor and trace elements at two western coal-fired power plants equipped with cold-side electrostatic precipitators. *Environ. Sci. Technol.*, **13**, 946.

Pacyna, J. M. (1984). Estimation of the atmospheric emissions of trace elements from anthropogenic sources in Europe. *Atmos. Environ.*, **18**, 41–50.

Parker, J. G. (1978). *Occurrence and Recovery of Certain Minor Metals in the Smelting-refining of Copper.* Bureau of Mines Information Circular 8778. US Dept. Interior, Washington DC, 79 pages.

Peterson, J. T., and Junge, C. E. (1971). Sources of particulate matter in the atmosphere. In Mathews, W. H., Kellogg, W. W., and Robinson, G. D. (Eds), *Man's Impact on the Climate*, pp. 310–320. MIT Press, Cambridge, MA.

Piver, W. T. (1983). Mobilization of arsenic by natural and industrial processes. In Fowler, B. A. (Ed.), *Biological and Environmental Effects of Arsenic.* Topics in Environmental Health 6, pp. 1–50. Elsevier/North Holland, Amsterdam.

Predicasts (1982). *World Coal Supply and Demand.* Predicasts Inc., Cleveland, Ohio.

PSAPCA (1974). The concentration of lead, arsenic, mercury, cadmium, SO_2 and suspended sulphates downwind from the Tacoma smelter, the impact and control status and benefits from reduction. Interim Report for Presentation to the

ASARCO Impact Task Force. Puget Sound Air Pollution Control Agency, Seattle, Washington, 63 pages.

PSAPCA (1980). *Total Suspended Particulate Modelling Analysis in Seattle and Tacoma, Washington.* Puget Sound Air Pollution Control Agency, Seattle, Washington.

Sabadell, J. E., and Axtmann, R. C. (1975). Heavy metal contamination from geothermal sources. *Environ. Health Perspect.*, **12**, 1–7.

Sabbioni, E., and Goetz, L. (1983). *Mobilization of Heavy Metals from Fossil-fuelled Power Plants, Potential Ecological and Biochemical Implications. IV: Assessment Study of the European Situation.* Report EUR 6998/IV. CEC, Luxembourg, 165 pages.

Schwitzgebel, K., Meserole, F. B., Oldham, R. G., Magee, R. A., and Mesich, F. G. (1975). Trace element discharge from coal-fired power plants. In *International Conference on Heavy Metals in the Environment, Proceedings of the Toronto Conference*, pp. 533–551. Toronto, Ontario, Canada.

Schwitzgebel, K., Coleman, R. T., Collins, R. V., Mann, R. M., and Thompson, C. M. (1978). *Trace Element Study at a Primary Copper Smelter*, Vol. 1 and 2. EPA-600/2-78-065a. US Environmental Protection Agency, 50 pages.

USEPA (1976a). *Technical and Micro-economic Analysis of Arsenic and its Compounds.* EPA-560/6-76-016. US Environmental Protection Agency, Washington DC, USA.

USEPA (1976b). *Air Pollutant Assessment Report on Arsenic.* Office of Air Quality Planning and Standards, Research Triangle Park, North Carolina. US Environmental Protection Agency, Washington DC, USA.

Walsh, P. R., Duce, R. A., and Fasching, J. L. (1979a). Consideration of the enrichment, sources and flux of arsenic in the troposphere. *J. Geophys. Res.*, **84**, 1719–1726.

Walsh, P. R., Duce, R. A., and Fasching, J. L. (1979b). Tropospheric arsenic over marine and continental regions. *J. Geophys. Res.*, **84**, 1710–1718.

Whittaker, R. H., and Likens, G. E. (1973). Carbon in the biota. In Woodwell, G. M., and Pecan, E. V. (Eds), *Carbon and the Biosphere.* AEC Symposium Series 30, pp. 221–240, NTIS. US Dept. Commerce, Springfield, Virginia.

Williamson, D., and Fromson, S. (1982). *Ninth Annual Tin Review.* Metals Research Unit, London.

Woolson, E. A. (1977). Generation of alkylarsines from soil. *Weed Sci. Soc. Am. J.*, **25**, 412–416.

Woolson, E. A. (1983). Emissions, cycling and effects of arsenic in soil ecosystems. In Fowler, B. A. (Ed.), *Biological and Environmental Effects of Arsenic.* Topics in Environmental Effects of Arsenic. Topics in Environmental Health, 6, pp. 51–139. Elsevier, Amsterdam, 281 pages.

World Bureau of Metal Statistics (1984). *World Metal Statistics*, **37** (6).

Lead, Mercury, Cadmium and Arsenic in the Environment
Edited by T. C. Hutchinson and K. M. Meema

CHAPTER 18

Dispersion of Cadmium, Lead and Zinc in Soils and Sediments of a Humid Tropical Ecosystem in Nigeria

M. A. OLADE

Geology Department,
University of Ibadan,
Nigeria

ABSTRACT

Primary ores and naturally enriched rocks constitute one of the main sources of anomalous metal concentrations in the environment. In the search for economic mineral deposits, geochemists have delineated areas where excesses of such heavy metals may be hazardous to plants and/or animals.

Several studies on the cycling of cadmium, lead and zinc have focused largely on their pathways from industrial point sources and on their biological fluxes, while relatively little is known about the factors influencing their mobility, redistribution and 'early' pathways in rock–soil–sediment systems (geochemical cycle). A better understanding of such processes may aid in our assessment of the environmental impact of these heavy metal contaminants and the nature of metal uptake by plants. It can also lead to a better understanding of the potential influences of man on local and global cycles of these elements. Data are particularly limited for tropical regions where the high temperatures and rainfall may have specific influences.

Studies on the distribution of cadmium, lead and zinc in soils overlying background and mineralized bedrocks show that, during ferralitic weathering, these metals are released through oxidation and leaching. Cadmium and Pb are generally enriched in the topsoil while Zn is depleted. Within the subsoil, Cd and Zn are characterized by enhanced levels within the B_1 horizon in association with Fe-Mn sesquioxides whereas Pb is accumulated nearer the surface within the B_2 horizon. The dominant pathway for their dispersal is by aqueous transportation by downward percolating rainwater and readjustments of the groundwater table. In mineralized soils, the accumulations of Cd and Pb in the topsoil can reach toxic proportions.

Stream sediments are composites of rock and soil material eroded upstream of the catchment area. Because of the intensity of tropical weathering and metal solubility in water, Zn and Cd enter the streams mostly as solutes which are later precipitated in the sediments. In contrast, Pb shows a rather restricted dispersion train which may suggest transportation mainly as suspended sediment derived from the erosion of anomalous seepage-area soils.

The 'natural' geochemical cycles of Cd, Pb and Zn in tropical environments are generally similar, although their mobility characteristics and chemical forms in surficial materials differ. Considering the flux of these heavy metals from the soil to the streams, it is estimated that the rate of geochemical cycling is more rapid for Zn and Cd than for Pb. This longer residence time for Pb than for Cd or Zn is parallelled in studies done in temperate regions.

INTRODUCTION

In the search for mineral deposits geochemists have delineated areas where excesses of certain trace elements, such as cadmium, lead, zinc, molybdenum and selenium may be harmful to plants and/or animals (Warren *et al.*, 1971; Fletcher *et al.*, 1973; Thornton and Webb, 1981). Other workers, during similar investigations, have attempted to establish the factors influencing the mobility, redistribution and chemical forms of trace elements in rock–soil–plant systems, thereby contributing to our understanding of metal cycling in the environment (Jenne, 1968; Fletcher and Doyle, 1974; Olade *et al.*, 1978; Olorunfemi, 1984).

Cadmium, lead and zinc represent a coherent group of metals both from the metallogenic viewpoint, and as contaminants of the environment (Thornton and Webb, 1981; Peterson and Alloway, 1979; Davies, 1981). Previous studies on the biogeochemical cycling of these metals have focused largely on the pathways from industrial point sources, most commonly within either temperate or sub-tropical ecosystems (Wilkins, 1978; Warren, 1981; Martin and Coughtrey, 1982; Khan and Frankland, 1983). According to Peterson and Alloway (1979), an assessment of the environmental impact of heavy metal contaminants can only be better achieved if the main steps in their geochemical pathways within the various natural ecosystems are understood. However, a great majority of studies have been done in temperate regions, while both arid and humid tropical regions here received little attention.

The tropical rainforest is unique and highly activating due to a combination of extreme climatic conditions, luxuriant vegetative growth and attendant effects of intense weathering, leaching and soil erosion. Although there have been a number of synoptic studies on the nutrient status of tropical soils and the nature of plant communities, relatively little is known about the biogeochemical cycling of trace elements in tropical ecosystems.

This chapter describes the geochemical dispersion of cadmium, lead and zinc in surficial materials from background and anomalous areas within a humid tropical environment. The technique of regional geochemical reconnaissance based on stream sediment sampling is used as a tool in the study of the regional distribution of trace elements (Webb and Thornton, 1981).

DESCRIPTION OF STUDY AREAS

Over 60% of tropical West Africa is underlain by crystalline igneous and metamorphic rocks, mostly of Precambrian age. They comprise a complex assemblage of gneisses, migmatites, metasediments and granitic rocks which have been subjected to several episodes of deformation and metamorphism. Sedimentary rocks of Cretaceous–Tertiary age are largely confined to the coastal areas and interior basins (Figure 18.1), and they are mostly shales and sandstones with minor limestones. The only areas with known lead–zinc deposits, where mining activity has been intermittent, are along a 600 km long metallogenic belt within Cretaceous sediments of Nigeria's Benue Trough.

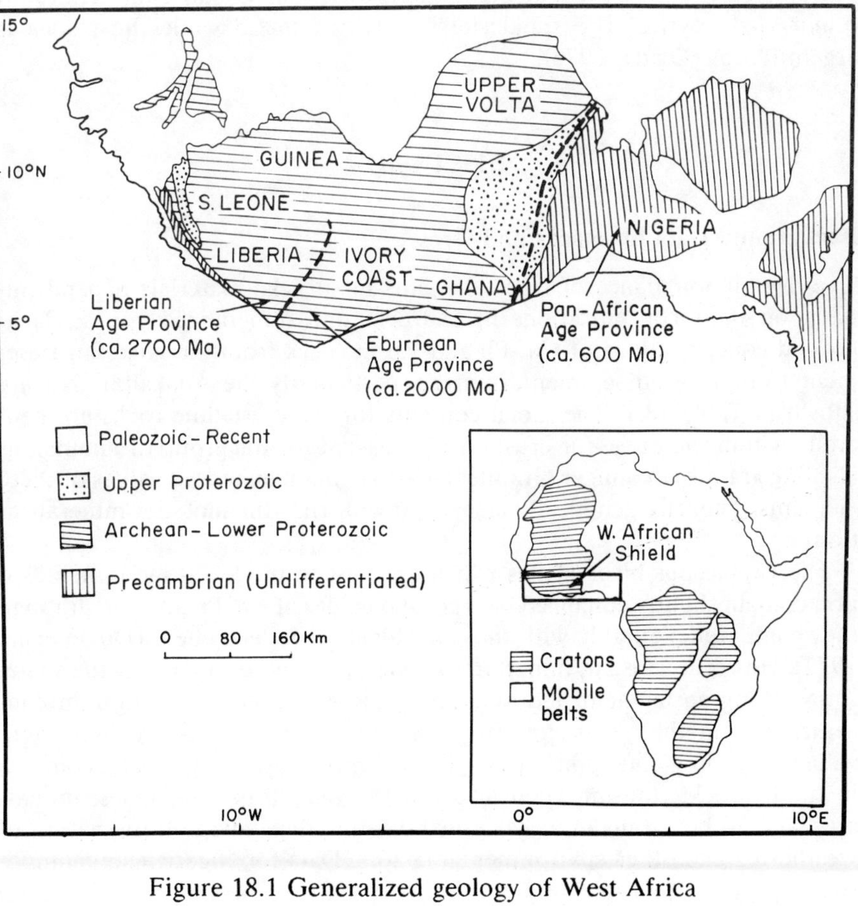

Figure 18.1 Generalized geology of West Africa

The present study on the dispersion of Cd, Pb and Zn is focused on areas around the lead–zinc mines where soils and stream sediments have developed from naturally enriched parent materials, including black shales, hydrothermally mineralized rocks and mine dumps. The heavy metal content of rocks, soils and stream sediments from uncontaminated areas overlying the crystalline basement and sedimentary rocks are presented for comparison.

The main study area around Abakaliki is one of the well known lead–zinc mineralized districts in Africa (Hawkes and Webb, 1954). The host rocks are black, carbonaceous shales with nodules of calcareous matter. Extensive weathering has generally converted the black shales to a bleached pale grey colour. Mineralization consists of several occurrences of hydrothermal vein deposits which have been mined intermittently for decades. However, because of the small-scale mining, contamination of the environment by mining activity is very limited and areas with undisturbed mineralization can still be found. Principal ore minerals are galena and sphalerite with minor chalcopyrite. The sphalerite is enriched in Cd while the galena is argentiferous (Olade, 1976).

DISCUSSION

Background Data for Rocks and Ores

The overall abundance of trace elements in surficial materials depend initially on their concentrations in bedrocks and mineralized zones. Background concentrations of Cd, Pb and Zn in rocks from the Nigerian Basement Complex and sedimentary areas, particularly the Abakaliki area are shown in Table 18.1. The metal contents for the crystalline rocks are generally within the range for crustal rocks, except for the group of anomalous alkaline granites (Younger Granites) which contain enhanced values of these elements. They are genetically associated with zinc–tin–niobium mineralization.

The Cretaceous black shales which are host to the lead–zinc mineralization contain slightly enhanced background values of Cd, Pb and Zn although they compare favourably with those for black shales elsewhere (Doyle *et al.*, 1973). However, close to mineralized zones, the metal contents of the shales show strong enrichment due to primary dispersion effects. Cadmium concentrations in the sphalerites from the mineralized zones are high, with values exceeding 2500 ppm (Table 18.1). On the basis of the metal contents in the bedrocks, the inherited levels of Cd and Pb in soils and sediments are likely to be anomalous near mineralization depending on the nature of weathering and other environmental factors. Data for Nigerian soil composition are given in Table 18.2.

Table 18.1 Trace element contents (ppm) of tropical rocks (Nigeria). Data from Olade (1976)

	Cd	Pb	Zn
Gneisses	0.25 (0.1–0.4)*	46 (28–65)	72 (45–92)
Granites	0.15 (0.1–0.3)	20 (12–30)	54 (45–64)
Anomalous granites	0.45 (0.2–1.5)	32 (10–54)	250 (70–800)
Shales	1.5 (0.5–4)	32 (20–36)	95 (80–115)
Black Shales	6.5 (5–12)	80 (60–115)	175 (160–220)
Sphalerite (Benue valley)	(2100–2600)	—	—
Average crustal rocks	0.2	13	70

* Range

Table 18.2 Trace element contents (ppm) of tropical soils (Nigeria)

	Cd	Pb	Zn
Background soils (crystalline rocks)	0.35 (0.2–0.8)*	42 (16–124)	74 (10–158)
Background soils (black shales)	4.5 (2–12)	78 (35–150)	118 (80–270)
Mineralized soils (black shales)	(12–30)	(2200–7500)	(500–1800)

* Range

Metal Dispersion in Sediments

Approximately 350 stream sediment samples were collected within the Abakaliki area with an overall sample density of about one per km^2 (Olade *et al.*, 1978). Samples were air dried, disaggregated and sieved to <100 mesh and analysed for Zn, Pb, Cd, Mn and Fe by atomic absorption spectrophotometry after hot nitric acid digestion. Data are given in Table 18.3. Organic matter content and pH of the sediments were also determined.

High concentrations of lead (more than 100 ppm and up to 1000 ppm) occur in sediments from areas contaminated by exposed mineralization and mining. The dispersion pattern of Pb is limited as the anomalies are confined to the mineralized zones. Anomalous Zn values (200 and up to 2000 ppm)

Table 18.3 Trace element contents (ppm) of stream sediments in the Abakaliki area of Nigeria. Values are based on nitric acid digestions (n = 350)

	Mean	Range
Cd	3.5	(1–20)
Pb	50	(10–99)
Zn	112	(20–2178)
Cu	21	(8–260)
Mn	1183	(20–10 000)
Fe	5.2	(0.8–16.0)
pH	5.9	(4.6–8.5)
Organic matter	6.4%	(1.3%–14.4%)

also outline the known mineralized areas with a dispersion pattern that is more extensive than that of Pb.

Several other localities contain isolated but anomalous values not related to mineralization. Cadmium levels are generally very low (<2 ppm) except close to mineralized zones where values up to 20 ppm are encountered (Table 18.3).

R-mode factor analysis was used to obtain inter-element relationships and factors controlling metal dispersion patterns. The following metal associations for a 3-factor model which accounts for 70% of total data variability were obtained.

Factor 1: Organic matter, Pb, Cu, Cd.

Factor 2: Zn, Fe, Mn

Factor 3: pH, Mn

Factor 1 which accounts for 60% of total variance is most strongly related to organic matter and may reflect metallo-organic complexing of ore elements. Zn–Fe–Mn which are closely associated in the second factor indicates Fe–Mn oxide co-precipitation with Zn in sediments, whereas the strong association of pH–Mn in Factor 3 may represent the effect of increased pH on Mn precipitation.

Metal Distribution in Soils

Distribution of heavy metals within soil profiles provides an insight into their behaviour during weathering and availability to plants. Due to high temperatures and heavy rainfall, the predominant soils of humid tropical regions are ferralsols (Young, 1976). They are usually characterized by very intense and complete weathering and leaching of the parent materials to kaolinite and Fe–Al oxides. The high ratio of precipitation to evaporation generally favours downward percolations, as it does in temperate region podsols. Due to very high rainfalls the humid tropical environment provides the optimum conditions for metal dispersion.

Soil differentiation with depth in ferralitic soils usually occurs by the development of four distinct horizons. The A horizon is usually restricted to a relatively thin layer of sandy to loamy soil with small to moderate amounts of organic matter at the top. The B_2 and B_1 'horizons' which may reach several metres in thickness consist of an upper reddish-brown ferruginous gravelly horizon and a lower yellowish kaolinitic horizon, respectively. Iron-rich layers or concretions (laterite) may occur anywhere within the B_1 or B_2 horizons. The C horizon is commonly made up of a mixture of clay and partially weathered parent material.

In order to establish the geochemical behaviour of Cd, Pb and Zn in ferralitic weathering, uncontaminated soil profiles overlying crystalline and sedimentary rocks, as well as those from mineralized but undisturbed areas, were sampled and analysed by AAS after nitric acid digestion. Results show that, although the absolute concentrations of these metals differ due to lithologic control, the dispersion patterns are generally similar (Figures 18.2 and 18.3).

In Profile 1, cadmium shows a slight enrichment in the topsoil, with values decreasing in the highly-leached B_2 horizon. Within the B_1 horizon with abundant clay, Cd increases in concentration and shows highest values within an Fe-rich layer (Figure 18.2). Zinc shows a similar pattern except for the relative depletion within the top soil. In contrast, lead is relatively enhanced within the top soil and B_2 horizon while it shows strong depletion within the B_1 horizon and slight enhancement near the base in association with Al-hydroxides. In Profile 2 with a fragmental soil overlying a weathered carbonaceous shale, Pb and to lesser extent Cd are enriched in the A horizon while Zn is impoverished (Figure 18.3a).

In the weathering profile overlying a mineralized zone which has not been disturbed by mining activity (Figure 18.3b), cadmium and zinc are characterized by fairly high levels reaching toxic proportions, but are depleted relative to parent material. On the other hand, Pb shows strong accumulation within the topsoil and B_2 horizon partly as a result of the formation of lead sulphates and carbonates. The slight enrichment of Zn and Cd within the B_1 horizon may also be attributed to the formation of secondary minerals (Figure 18.3b).

Factors Controlling Dispersion of Metals in Tropical Soils and Sediments

The concentrations of metals within tropical soils are strongly controlled

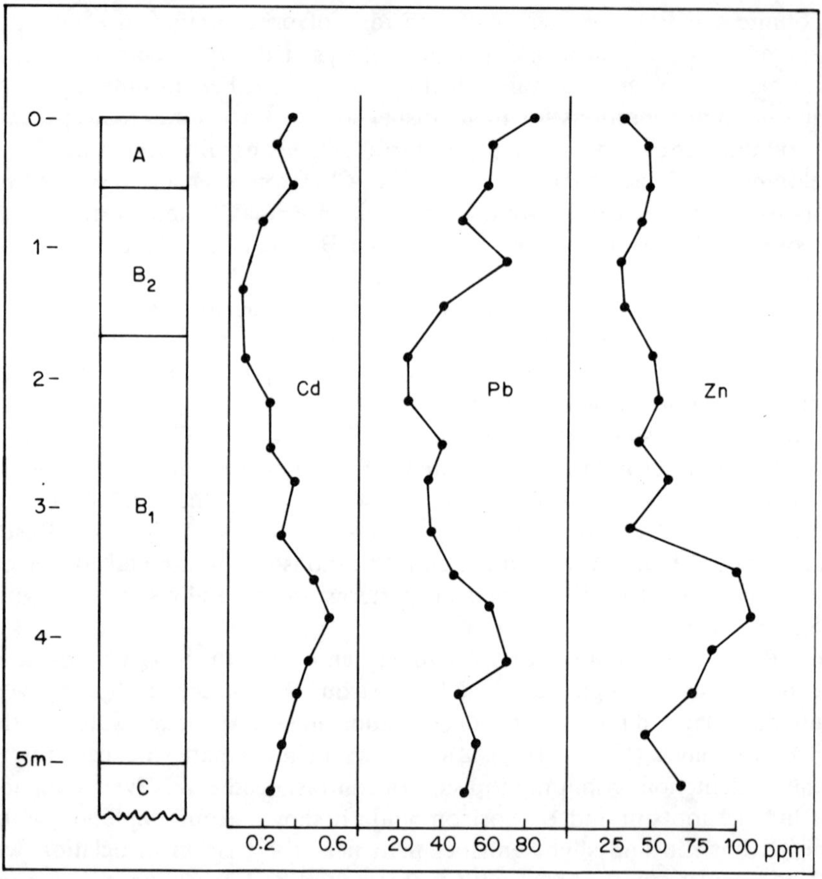

Figure 18.2 Metal distribution in weathering soil profile over crystalline rocks (profile 1)

by the geochemical characteristics of the underlying rock and the nature of weathering processes. Ferralitic weathering, which is prevalent in the humid tropics, results in oxidation and intense leaching of major and trace metals. The high ratio of precipitation to evaporation permits free drainage and transportation of metals down the profile. Among the major elements, silica is partially or totally leached while iron and alumina concentrate within the profile as oxides and hydroxides.

However, the degree of trace element re-distribution is usually dependent on the mobility characteristics and the effects of other environmental parameters such as organic matter content, Fe–Mn co-precipitation and formation of insoluble precipitates.

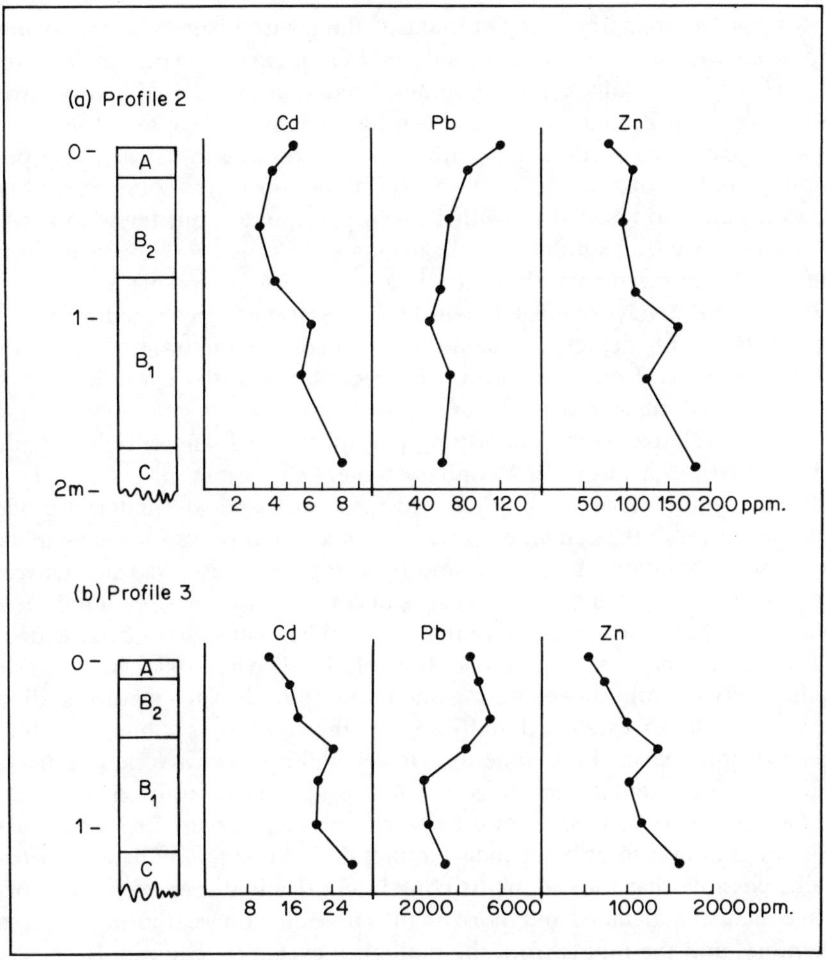

Figure 18.3 Metal dispersion in soil profile overlying (a) uncontaminated and (b) mineralized carbonaceous shale

Cadmium tends to be fairly mobile during tropical weathering and is redistributed within the soil profile. There is evidence from our results that, despite its mobility, it shows slight enrichment within the topsoil as well as in clayey zones rich in soluble Fe hydroxides, but is strongly leached within Al-rich zones. The most likely forms of Cd in tropical soils are as soluble or insoluble organo-metallic complexes and co-precipitates with Fe–Mn hydroxides. The relatively high Eh values in freely-drained humid tropical soils tend to promote formation and sorptive ability of Fe–Mn sesquioxides particularly with increasing pH conditions. Within stream sediments, results of R-mode factor analysis suggest that Cd, along with Pb and Cu, are strongly associated with finely divided organic matter in carbonaceous shale parent rock. However, in soils overlying mineralized zones Cd shows a very strong association with Zn and both are precipitated as secondary minerals.

Lead most often accumulates within the topsoil due to its relatively poor mobility and strong tendency to form metallo-organic complexes. Lead shows little or no association with Fe–Mn precipitates, but tends to readily form secondary lead sulphates and carbonates in soils overlying mineralization. In stream sediments Pb is closely associated with organic matter.

Zinc is a relatively mobile element both in soils and stream sediments and is most commonly depleted in the topsoil. The dispersion train of Zn is long, although there is strong evidence of co-precipitation with Fe–Mn hydroxides in the surficial environment. In mineralized soils, Zn is generally depleted in the soils relative to the underlying parent material, and enrichment due to precipitation of insoluble secondary minerals is common.

Stream sediments are largely derived from rock and soil material eroded from upstream in the catchment area. The major pathways for heavy metals are by sedimentation of particles and by sorption of dissolved metals from surface waters or groundwater seepages in contact with the stream sediments (Warren, 1981). The three main kinds of particles transported by stream are bed load, suspended sediment load and solutes (Davies, 1981). Considering the intensity of tropical weathering and metal solubility in aqueous medium, Zn and to a lesser extent Cd and Pb enter the streams as solutes which are later precipitated in the sediments. However, Pb shows a rather restricted dispersion train which may suggest that it is also partly transported as suspended sediment derived from erosion of anomalous bank or seepage-area soils containing insoluble secondary minerals (Ukpong and Olade, 1979). It is also possible that the sediments contain clastically dispersed mine debris.

In conclusion, detailed and more quantitative data on weathering rates, soil inventory, and the inputs from the biological cycle are required. Because of the rural setting of most communities within tropical Africa, the input of heavy metal contaminants into the environment from industrial activities is still minimal although growing steadily.

REFERENCES

Davies, B. F. (1981) Trace element pollution. In Davis, B. E. (Ed.), *Applied Soil Trace Elements*. pp. 287–351, John Wiley and Sons, Chichester.

Doyle, P., Fletcher, K., and Brink, V. C. (1973). Trace element content of soils and plants from the Selwyn Mountains, Yukon and Northwest Territories. *Can. J. Bot.*, **51**, 421–427.

Fletcher, K., Doyle, P., and Brink, V. C. (1973). Seleniferous vegetation and soils of the Eastern Yukon. *Can. J. Plant Sci.*, **53**, 701–703.

Fletcher, K., and Doyle, P. (1974). Regional geochemical reconnaissance and the molybdenum content of bedrock, soils and vegetation from the Eastern Yukon. In Hemphill, D. D. (Ed.) *Trace Substances in Environmental Health*, pp. 369–373, Univ. Missouri, Columbia.

Hawkes, H. E., and Webb, J. S. (1954). Geochemical prospecting in Nyeba lead-zinc district, Nigeria. *Bull. US Geol. Surv.*, 1000-B, 52 pages.

Jenne, E. A. (1968). Controls on Mn, Fe, Co, Ni, Cu and Zn concentrations in soils and water: the significant role of hydrous Mn and Fe oxides. In *Trace Elements in Water*, Adv. in Chemistry, Ser. 73, Amer. Chem. Society, Washington DC.

Khan, D. H., and Frankland, B. (1983). Chemical forms of cadmium and lead in some contaminated soils. *Env. Pollut.*, **6**, 15–32.

Martin, M. H., and Coughtrey, P. J. (1982). *Biological Monitoring of Heavy Metal Pollution.* Applied Science Publishers, London and New York, 475 pages.

Olade, M. A. (1976). On the genesis of lead-zinc deposits in Nigeria's Benue rift. *Nig. J. Min. Geol.*, **13**, 20-27.

Olade, M. A., Van de Kraats, A. H., and Ukpong, E. E. (1978). Effects of environmental parameters on metal dispersion patterns in stream sediments from the lead-zinc belt. Benue Trough, Nigeria. *Geologie en Mijnb.*, **58**, 341–351.

Olorunfemi, B. N. (1984). Factors influencing the geochemistry of lateritic soils in the green-schist-amphibolite meta-tectonic belt, S. W. Nigeria, *Chem. Geol.*, **42**, 261–275.

Peterson, P. J., and Alloway, B. J. (1979). Cadmium in soils and vegetation. In Webb, M. B. (Ed.), *The Chemistry, Biochemistry and Biology of Cadmium*, pp. 45–92, Elsevier/North-Holland and Biomedical Press, Amsterdam.

Thornton, I., and Webb, J. S. (1981). Regional distribution of trace element problems in Great Britain. In Davies, B. E. (Ed.), *Applied Soil Trace Elements*, pp. 381–439, John Wiley and Sons, Chichester.

Ukpong, E. E., and Olade, M. A. (1979). Geochemical surveys for lead-zinc mineralization, southern Benue Trough, Nigeria. *Trans. Inst. Min. Metall.*, **88**, B81–B92.

Warren, H. V., Delavault, R. E., and Fletcher, W. K. (1971). Metal pollution—a growing problem in industrial and urban areas. *Can. Min. Metall. Bull.*, July, 1–32.

Warren, L. J., (1981). Contamination of sediments by lead-zinc and cadmium—a review. *Env. Pollut.*, **2**, 401–436.

Wilkins, C. (1978). The distribution of lead in the soils and herbage of W. Pembrokshire. *Env. Pollut.*, **10**, 23–30.

Young, A. (1976). *Tropical Soils and Soil Survey.* Cambridge University Press, Cambridge, 468 pages.

Lead, Mercury, Cadmium and Arsenic in the Environment
Edited by T. C. Hutchinson and K. M. Meema

CHAPTER 19

The Cycling of Arsenic, Cadmium, Lead and Mercury in India

C. R. KRISHNA MURTI*

Pitambar Pant National Fellow and Principal Coordinator,
Integrated Environmental Programme on Heavy Metals,
Centre for Environmental Studies,
Anna University, Madras 600 025,
India

ABSTRACT

In this chapter the sources and levels of arsenic, lead, cadmium and mercury in India are examined, as well as the implications to human health and perspectives for research in the future.

Arsenic is found in vegetables at 2–30 $\mu g/100$ gm, in the soil at 118–212 $\mu g/100$ gm, and in drinking water at 3–15 $\mu g/100$ ml. The major sources of As pollution are from coal and from superphosphate.

The major contributing factor to high levels of cadmium in crops is sewage irrigation. The levels of both Cd and Pb in crops on soils irrigated by sewage were 2-3 times those on tubewell-irrigated soils. The high pH, organic carbon and cation exchange capacity appear to be factors in the retention of Cd in the sewage-irrigated soils. Soils of low CEC allowed crop accumulation of Cd more readily.

Lead intake via food is the major source of Pb to humans in India and can account for about 0.4 mg of daily intake. Other important sources are water heated in glazed pottery, tea, paints, pigments, pencil colours and auto-exhaust.

Mercury is not produced in India and the entire requirements of this metal are imported. Nevertheless, the industrial activity of Chembur is responsible for high concentrations of Hg in the air (up to 41.5 ng/m^3). Consumption of wheat and rice stored with mercury tablets, used as a preservative, can increase Hg intake to 0.1–0.4 mg Hg/day. In waters on the East Coast of India, the concentration of Hg in estuarine water is 5–15 times that of the backwater, 5–8 times that of mangroves, and 10–12 times that of normal sea water.

---

* Present address: Scientific Commission for Continuing Studies on Effects of Bhopal Gas Leakage, Cabinet Secretariat, Sadar Patel Bhavan, Parliament St., New Delhi 110001, India.

INTRODUCTION

This chapter addresses some of the problems related to the cycling of arsenic, cadmium, lead and mercury in India. Major developments in agriculture and industry have resulted in these elements showing substantial increases over their pristine levels in the environment. For want of reliable data, the task of making realistic estimates of the fluxes of these elements and consequently the degree of human exposure to them is beset with many difficulties. Much of the information that is likely to be useful in this context is locked up in reports prepared by government agencies for internal use and hence not easily accessible. Published information available to this author was confined to Indian sources. This overview is, therefore, restricted in its scope and indicates trends rather than absolute realities.

MINERAL PROCESSING AND COAL AND CHEMICAL INDUSTRIES IN THE INDIAN SUB-CONTINENT

India is known to have rich deposits of many ores and minerals, most of which are yet to be tapped for industrial use. Offshore prospecting has revealed commercially exploitable oil and gas wells in the Cambay basin, offshore Bombay, and in offshore basins of the Godavari, Krishna and Kaveri. The coal belt of Jharia and Raniganj in the Ganga basin has been mined for more than 150 years for power production and locomotion. The lignite deposits of Neyveli in Peninsular South India are currently being mined for power generation, cycling huge volumes of sub-soil water. In the absence of other viable options, India has decided to develop major super-thermal stations preferably after processing the locally available high ash coal.

As part of the constraints imposed by a developing economy in search of hard currency, considerable quantities of iron, manganese and other ores are exported, while at the same time large amounts are being used within the country. Statistics given in Tables 19.1–19.3 (Statistical Outline of India, 1982) illustrate the range of activities and products being produced in the country relevant to the cycling of metals and metalloids in general. Budgets need to be developed for their environmental impacts. The country does not produce mercury and consequently the entire requirements of this metal are imported. Substantial quantities of metallic cadmium and lead and arsenic salts are also imported. Total demand for zinc is between 1.2 and 1.3×10^5 tonnes. Indigenous production varies from 50 000 to 60 000 tonnes and the balance is made up by imports. For lead, the consumption of about 70 000 tonnes is met by indigenous production to the extent of 25 000 tonnes and an import of 45 000 tonnes. The requirement of copper is 90 000 to 100 000 tonnes, 60 000–70 000 being imported and 25 000–35 000 tonnes coming from local production. The annual consumption of 2.5×10^5 tonnes aluminum is substantially met by indigenous production. The estimated gap for zinc by the end of 1990 is 80 000–90 000 tonnes and for lead 50 000–60 000 tonnes (Raghavan, 1983). Data on rates of use are given in Table 19.3, and changes in production over the past 20 years in Table 19.4. From such information, inventories can eventually be constructed for the four elements under discussion, and these projected to scenarios of their cycling in air, water and soil and to interactions with biotic and abiotic pathways. What is of particular significance is the almost doubling, within a period of 20 years, of the rate of production of most of the materials of commerce and industry with a potential to diffuse lead, cadmium, mercury and arsenic in the main environmental compartments of the sub-continent.

Table 19.1 Industrial production of items with pollution potential. (Data from Statistical Outline of India, 1982)

		1980–81	1960–61
Pig iron	($\times 10^6$tonnes)	8.6	4.3
Saleable steel	($\times 10^6$ tonnes)	6.3	2.5
Nitrogenous fertilizers	($\times 10^3$ tonnes)	2164	98
Phosphatic fertilizers	($\times 10^3$ tonnes)	840	52
Sulphuric acid	($\times 10^6$ tonnes)	2.1	0.4
Caustic soda	($\times 10^3$ tonnes)	568.0	101.0
Soda ash	($\times 10^3$ tonnes)	540.0	152.0
Paper and board	($\times 10^3$ tonnes)	1145.0	350.0
Dry battery	(Million)	1291	—
Storage battery	(Million)	25	—
Hydrogenated oils	($\times 10^6$ tonnes)	748	340.0
Man-made fibre	($\times 10^6$ metres)	1328	550.0

ARSENIC LEVELS IN THE ENVIRONMENT

A survey was undertaken in Chandigarh, India to measure the arsenic content of the crops cabbage, pea, onion, radish, spinach and cooking banana, and of soil samples collected from the States of Panjab, Haryana, Himachal Pradesh and Uttar Pradesh (UP). The content of arsenic in these vegetables varied from 2–30 μg/100 gm. The mean value of arsenic in the soil in the study area varied from 118.0 to 212.7 μg/100 gm, although in some samples collected from the Meerut District of Uttar Pradesh values were found to be as high as 566 μg/100 gm. High arsenic in vegetables could mostly be related to the high arsenic content of soils. However, the correlation was not statistically significant (Datta and Kaul, 1978). Sample analyses of manure/fertilizers are given in Table 19.4, showing these fertilizers as

Table 19.2 Reserves of minerals and perspectives. (Data from Statistical Outline of India, 1982)

	Recoverable reserves 1979 (10^6 tonnes)	Likely depletion 1979–95 (10^6 tonnes)	Balance life after 1994–95 (years)
Non-coking coal	37 000	2 640	126
Prime coking coal	2 700	2 370	74
Iron ore (Haematite)	9 300	8 364	111
Bauxite	1 900	1 838	245
Limestone	53 000	51 980	519
Zinc	6.0	4.8	38
Lead	2.3	1.8	37
Manganese ore	75	41	14
Chromite	24	17.5	29
Copper	3.2	2.4	32

Depletion by 1994–95 has been officially estimated at 'contemplated rates of exploitation' and the balance life of the reserves, assuming 1994–95 consumption levels.

Table 19.3 Production of some important minerals. (Data from Statistical Outline of India, 1982)

		1980	1960
Coal	(× 10^6 tonnes)	107.0	52.6
Iron ore	(× 10^6 tonnes)	39.0	16.6
Limestone	(× 10^6 tonnes)	28.2	12.9
Bauxite	(× 10^3 tonnes)	1532	387
Copper ore	(× 10^3 tonnes)	1997	448
Manganese	(× 10^3 tonnes)	1740	1452
Gypsum	(× 10^3 tonnes)	817	997
Chromite	(× 10^3 tonnes)	305	107
Zinc concentrate	(× 10^3 tonnes)	60.3	9.6
Lead concentrate	(× 10^3 tonnes)	16.3	6.2
Gold	(kg)	2341	4995

Table 19.4 Arsenic content of manures/fertilizers used in India. (Data from Datta, 1976)

	mg/100 gm
Poultry manure I	13.3
Poultry manure II	23.7
Cow-dung	11.2
Superphosphate	187.8

potential factors for soil contamination by As and for subsequent plant uptake. The mean arsenic content of water collected from taps, wells, handpumps, irrigation pumps, ponds, canals, etc. in and around Chandigarh varied from 3–15 μg/100 ml. A value as high as 54.5 μg/100 ml was registered in one sample of water from a hand pump in an area where non-cirrhotic portal fibrosis was endemic (Datta, 1976, 1978a, 1978b; Datta and Kaul, 1976, 1978).

Range and mean concentration of As in samples of food, water and air collected from an independent study in metropolitan Bombay are shown in Table 19.5. Arsenic-containing coal is known to produce arsenious oxide, As_2O_3, on combustion. Coal mined in USA and Czechoslovakia has been reported to contain 1–10 and 1500 mg/kg, respectively, of arsenic. In a study of the captive fly ash collected from the Indra Prastha power station in New Delhi, the average arsenic content by EDXA was found to vary from 0.091 to 1.328% as elemental arsenic. The data for five samples of fly ash are summarized in Table 19.6. Since India uses annually 140 million tonnes of its coal reserves, one of the likely sources of arsenic pollution of water is coal. At the rate of about 1 mg/kg, the other important source of arsenic in the soil appears to be superphosphate. Other sources are limestone (0.1–20 mg/kg), smelting of copper and lead ores and the ceramic industries. A traditional use of arsenic as a mixture of arsenic oxide and copper sulphate in the seasoning of timber and as a fungicide, could also contribute to environmental contamination. No reliable data on these sources are available.

Table 19.5 Range and mean concentration of Hg, Cd and As in foods, drinking water and air at Bombay. (Data from Mishra and Sadasivan (1980) based on Anand, 1976a, 1978a)

	Mean concentration (range) (in ng/gm)		
Sample	Hg	Cd	As
Cereals	7.0 (0.75–31.1)	37.9 (15.2–79.8)	25.7 (1–88.2)
Vegetables	6.5 (0.75–30.3)	43.8 (16.5–98.1)	3.1 (1–8.2)
Fish	40.1 (4.6–282.6)	46.5 (15.9–175.6)	500 (20.5–1415)
Fruits	1.9 (1–2.8)	BDL*	BDL
Milk	(1.2–27.1) (no mean provided)	BDL	BDL
Drinking water (ng/ml)	1.9 (0.72–8.6)	BDL	BDL
Air	0.59 (0.19–0.91)	2.93 (0.97–6.0)	1.01 (0.59–1.42)
Calculated daily intake in μg	11.7	23.0	12.1

* BDL – below detection limit.

Table 19.6 Arsenic content of fly ash of the coal-fired Indra Prastha power station, New Delhi. (Data from VP Chest Institute, 1983)

	Arsenic as % of ash	Arsenic content in coal (ppm) 10% ash	20% ash
1	0.962	962	1924
2	1.328	1328	2656
3	0.539	1078	2156
4	0.851	851	1702
5	0.091	91	182

Arsenic content of the coastal waters and estuarine waters around Goa has been reported by Fondekar and Reddy (1979), as part of a general pollution study of the impact of fertilizer plants and other industrial activity on the coast of Goa on the West Coast of India. The values of the inshore waters were higher than those reported for sea water.

CADMIUM LEVELS IN THE ENVIRONMENT

Preliminary data summarized in Tables 19.7 and 19.8 on environmental pollution by cadmium were collected by a survey in Panjab using samples of fertilizers, water samples and common foodstuffs (Nath *et al.*, 1982). Wheat flour had the highest cadmium content. Water samples tested from various wells in Chandigarh had no detectable cadmium. The concentration of cadmium in crops grown with untreated municipal waste water around the cities of Panjab are given in Table 19.9. Levels in crops grown on land treated with municipal sewage wastes were two to three times higher than in those irrigated with well waters. The accumulation of DTPA-extractable Pb and Cd in soils receiving waste water from different sources is illustrated in Table 19.10. The increase was found to be substantially more in the industrial towns of Jullundur and Amritsar. The amount of Pb and Cd in

Table 19.7 Content of cadmium in some common foodstuffs in Panjab. (Data from Nath *et al.*, 1982)

	Cadmium (ppm)
Whole rice	0.079 ± 0.001
Whole maize	0.061 ± 0.001
Whole wheat	0.089 ± 0.004
Gram	0.074 ± 0.010
Gramdal	0.033 ± 0.025
Wheat flour	0.200 ± 0.040

Table 19.8 Content of cadmium (ppm) in fertilizers. (Data from Nath *et al.*, 1982)

Calcium nitrate	0.010–0.016
Superphosphate	1.190–1.350
Potash	0.030–0.060
Urea	0–0.020
Zinc sulphate	630–400

Table 19.9 Cadmium and lead content of crops grown in soils irrigated by tubewell and sewage water. (Data from Kansal and Singh, 1983)

	Cadmium (ppm dry weight)	Lead (ppm dry weight)
Tubewell-irrigated		
Maize	0.85 ± 0.10	1.98 ± 0.20
Berseem	0.69 ± 0.19	1.93 ± 0.43
Cauliflower heads	0.48 ± 0.14	1.27 ± 0.17
Cauliflower leaves	0.80 ± 0.11	2.69 ± 0.11
Spinach	0.50 ± 0.10	3.29 ± 0.20
Sewage-irrigated		
Maize	1.74 ± 0.24	3.82 ± 0.66
Berseem	1.67 ± 0.47	4.48 ± 1.00
Cauliflower heads	1.60 ± 0.53	1.80 ± 0.20
Cauliflower leaves	2.24 ± 0.74	5.23 ± 0.25
Spinach	2.59 ± 0.20	6.08 ± 0.35

Table 19.10 Average content of DTPA-extractable cadmium and lead in soils receiving tubewell and sewage irrigation. (Data from Kansal and Singh, 1983)

	Tubewell irrigation		Sewage irrigation	
	Cadmium	Lead	Cadmium	Lead
	(ppm in soils)			
Abohar	0.05	0.35	0.06	0.76
Bhatinda	0.02	0.20	0.04	0.69
Jalandhar I	0.05	0.85	0.17	0.89
Jalandhar II	0.05	0.822	0.17	1.09
Amritsar I	0.06	0.50	0.09	1.65
Amritsar II	0.06	0.50	0.14	1.58
Average	0.05	0.54	0.10	1.28

sewage-irrigated soils was twice as much as in tubewell-irrigated soils. The soils presumably pick up the elements from their direct discharge in soluble and suspended form or as part of organic matter. Path analysis revealed that pH, soil conductivity, organic carbon content and calcium carbonate content all exerted a direct influence on the availability of cadmium. The high value of the indirect effect of soil pH and extractable carbon indicated that interaction between these two factors alone was important in determining the cadmium content of the soils. Chelates of cadmium appeared to be more soluble than those of other metals. Due to the dissolution of the precipitates of cadmium carbonate, hydroxide and phosphate, the salts of cadmium could also contribute significant amounts to their available status. The relative influence of various soil properties on the content of heavy metals in plants grown on sewage-irrigated soils (n = 24) is given in Table 19.11.

Table 19.11 Relative influence of various soil properties on the content of lead and cadmium in plants on sewage-irrigated soils (n = 24)

Element	Multiple regression coefficient R^2	Simple correlation coefficient (r)				
		pH	EC	OC	$CaCO_3$	DTPA-extractable
Lead	0.45*	−0.17	−0.05	−0.06	0.39	−0.20
Cadmium	0.49*	−0.35	−0.30	−0.14	−0.03	0.32

* Significant at 0.01.

Studies on sorption of cadmium were carried out in some sewage-irrigated soils. The data was found to follow both Langmuir and Freundlich adsorption isotherms. Release of Cd by I.N. neutral ammonium acetate decreased with increases in pH, organic matter, $CaCO_3$, cation exchange capacity and clay content. Soils with high bonding energy constants and adsorption maxima released lesser amounts of Cd (Rana and Kansal, 1983). Total Fe, Cu, Zn and Mn and their DTPA-extractable functions also had a significant and positive correlation with Cd adsorption. The content of Cd in the surface layer (0–15 cm) of the soils analysed are given in Table 19.12. The sewage-irrigated soils with high pH, organic carbon and cation exchange capacity could retain higher amounts of Cd and retain it more tightly than 'normal' ones. These soils thus can act as a sink for the disposal of Cd. However, it may be that after a long disposal these soils become saturated with Cd and consequently its availability in soils will increase to an extent where plant growth would be adversely affected. This may also become hazardous to animals and human beings. Such reports have been made in Europe where sewage sludge amended soils have built up cadmium to the point of releasing some into milk of cows grazed on such pastures.

Table 19.12 Content of lead and cadmium in the surface (0–15 cm) layer of tubewell- and sewage-irrigated soils. (Data from Singh and Kansal, 1983)

	Tubewell-irrigated		Sewage-irrigated	
	Cadmium	Lead (ppm)	Cadmium	Lead
Abohar	3.9	14.0	9.3	20.0
Bhatinda	6.3	16.0	13.4	22.0
Jalandhar I	5.3	14.0	14.1	22.0
Jalandhar II	6.3	16.4	12.0	29.0
Amritsar I	5.3	13.0	11.0	25.0
Amritsar II	5.3	13.0	12.3	26.0

In alkaline soils of the Punjab, cadmium was retained by adsorption on the mineral interface and by interaction with organic matter and calcium carbonate. At high concentrations Cd precipitates as hydroxides. Sequential desorption with 1 M KCl and 0.05 M $Cu(CH_3COO)_2$ provided a measure of their exchangeable and chelated forms (Singh and Sekhon, 1977).

Limited data on the ambient levels of cadmium and lead in air, water, foodgrains, cooked food, vegetables, cigarettes and beedies (indigenous cigarettes) (Pandya, 1978; Pandya *et al.*, 1983) are summarized in Table 19.13. The content of cadmium in the fly ash collected from the coal-fired Indra Prastha thermal power station was fairly constant in samples collected

Table 19.13 Ambient levels of cadmium and lead in air, water and food samples collected in Ahmedabad. (Data from Pandya, 1978; Pandya *et al.*, 1983)

	Mean concentration (range) (ppm)	
	Cadmium	Lead
Air ($\mu g/m^3$) (n=19)	0.013 (0.006–0.037)	0.513 (0.091–1.99)
Food grain (n=9)	0.291 (0.074–0.34)	0.69 (0.55–1.05)
Vegetables (n=13)	0.32 (0.0–0.49)	0.74 (0.20–1.05)
Fruits (n=3)	0.38–0.88	1.8–2.5
Eggs, fish, meat (n=5)	—	—
Cooked food (n=7)	0.22 (0.12–0.30)	0.84 (0.40–1.25)
Cigarettes and beedies (n=16)	0.171 (0.077–0.48)	12.43 (7.37–20.41)
Water (n=7)	0.002 (0.001–0.004)	0.013 (0.008–0.23)

throughout 1983 (VP Chest Institute, 1983). The cadmium content of sea water estimated at ten monitoring stations located along 15 km of the Madras Coast over a period of one year showed a range of 0.001–0.047 mg/kg. The content of Cd in sediments collected from seven stations over this stretch showed levels ranging from 0.026–0.57 μg/gm dry matter. The content of the metal in samples of biota including 200 plankton, algae, molluscs, crabs and fish, showed Cd content of 0.02 to 7.39 μg/gm dry matter (ZSI, 1984). Seasonal fluctuations of cadmium in the Pichavaram mangrove forest of the East Coast of India have been reported to be 8.2–52.2 μg/gm as monitored at two stations (Subramanian, 1981). Animal feeds marketed commercially were found to contain 0.9–1.8 ppm Cd (ITRC, 1984).

LEAD LEVELS IN THE ENVIRONMENT

A study conducted in Ahmedabad with limited samples of air, food and water indicated that the ambient air lead concentration varied from 0.35 to 1.77 $\mu g/m^3$. Food grains contained 0.7 μg/gm whereas cooked food showed a mean of 0.8 μg/gm (Jathar *et al.*, 1981). Drinking water samples registered a level around 0.013 μg/gm. Food seemed to be a major source and can account for about 0.4 mg of daily intake. Intake of lead from air is only about 7.2% of the total daily intake. A selected bibliography compiled by Rawal *et al.*, (1982) on environmental exposure to lead and cadmium lists 122 publications from 1921–1982, covering not only ambient levels in air and water and the work-place of industries with a potential exposure to lead, but also some biological effects of lead in aquatic animals. The National Institute of Occupational Health, Ahmedabad, has reported the load of lead in ambient air due to emission of auto-exhaust in traffic-dense areas on Ahmedabad (Lahori *et al.*, 1972; Agarwal *et al.*, 1978, 1980).

Increased accumulation of Pb in soils receiving waste water from the industrial towns of Panjab has been reported (Rana and Kansal, 1983). The amount of Pb in sewage-irrigated soils was twice as much as in tubewell-irrigated soils. Data on DTPA extractable Pb as a function of various soil parameters did not show pH to have any direct effect but its indirect effect through organic carbon depressed the availability of Pb. Organic carbon exerted the maximum influence on the availability of Pb. Plants collected from waste water-irrigated soil had considerably higher concentrations of Pb than those from tubewell-irrigated soils. The increase occurring in the top soil was from 14.0 to 24.0 μg/gm.

Acidic media and higher temperatures facilitated leaching of Pb from glazed pottery sold in North India and 1–1.65 mg/litre could be recovered in tea and cooking waters (ITRC, 1984). One percent of hot decoctions of ten popular brands of tea sold in the North India market were found to contain 0.002–0.012 ppm Pb (ITRC, 1984). Other potent sources of lead identified by this study were paints, pigments and pencil colours. Commercial animal feed contained 9.66–5.94 ppm whereas over 8% of 650 samples of tap-water collected from Lucknow and Kanpur in Uttar Pradesh were found to contain 0.004–0.008 ppm Pb (ITRC, 1984). Content of Pb in samples of air, food grains, vegetables, fruits, cooked food, water, cigarettes and beedies have been reported by Pandya (1978). The content of Pb in fly ash collected in the Indra Prastha power station ranged from 110–260 g/gm in a twelve-month survey in 1983 (VP Chest Institute, 1983).

In a study of coastal water pollution of a 15 km stretch of the Coromandel coast in Madras, India, lead content of sea water was found to vary from traces to 0.4 mg/litre. The sediments examined registered a mean content of 15.0 μg/gm although the highest value noted was 41 μg/gm. The tissue content of Pb in samples of marine organisms collected over the region varied from traces to 25 μg/gm dry weight of samples (ZSI, 1984).

An important source of lead in the environment is auto-exhaust. The rapid expansion of road transport services in India is evident from data given in Table 19.14. According to a study done by the Environment Cell of the Bombay Municipal Corporation more than 60% of the 1730 tonnes of air pollutants in Bombay can be accounted for by automobiles alone (Despande *et al.*, 1972; Khandekar *et al.*, 1980).

Table 19.14 Number of vehicles (excluding rail track) using petroleum products. (Data from Statistical Outline of India, 1982)

	1978–79	1960–61
	($\times 10^3$)	
Buses	127	57
Trucks	412	160
Jeeps	119	32
Cars	723	256
Taxis	83	22
Two-wheelers	1678	88
Three-wheelers	139	6
Miscellaneous*	416	36

* Includes tractors, trailers, three-wheelers, tempos and pick-up vans.

MERCURY LEVELS IN THE ENVIRONMENT

The occurrence and distribution of mercury in water, particulate matter and sediment samples from three different biotopes, namely estuary, backwater and mangrove, have been studied on the East Coast of India (Kumaraguru *et al.*, 1979). The estuarine water contained 0.0015–0.0045 ppm mercury while the backwater had 0.002–0.0025 ppm and the mangrove water 0.003–0.0035 ppm. The mercury content of the particulate matter was 0.15–0.4 ppm, 0.28 ppm in the estuarine water, 0.13–0.30 ppm in the backwater and 0.18–0.35 ppm in the mangrove water. The range of concentrations of mercury in sediments was from 0.1–0.33 in the estuary, from 0.13–0.28 in the backwater and from 0.13–0.25 in the mangrove. Oysters of average wet body-weight had mercury contents ranging from 0.32 to 0.72 ppm. The levels of mercury in natural sea-water is reported to be 0.0003 ppm but in polluted waters the concentration can go as high as 0.03 ppm. On this basis, the concentration of mercury in the estuarine region was 5–15 times of the backwater, 5–8 times of the mangrove, and 10–12 times that of normal sea-water. The mercury content of the sediments in the biotopes was comparable with values reported for Lake Erie sediments (Annet *et al.*, 1972) and in the Calshot surface mud of Southampton (Raymont, 1972). The mercury content of the sediment, though less than that of particulate fraction, was 72–100, 50–100 and 33–75 times greater than that of the water of the three biotopes.

The total suspended particulate matter in the Bombay city air and the resulting dust vary over a wide range from 100 $\mu g/m^3$ to 900 $\mu g/m^3$ (the annual average US federal standard = 75 $\mu g/m^3$). The range and mean concentration of Hg in air (ng/m^3) over selected stations in the country are Ooty (Hill Station) BDL, Bangalore 0.02–1.70 (0.94), Calcutta BDL, Trombay 0.16–1.29 (0.52), and Chembur 6.92–41.1 (21.5). The industrial activity of Chembur is responsible for the high concentration of Hg observed. The measurement of elemental composition in size-separated aerosols helps in assessing contributions from different sources. Analysis of aerosols collected in the Trombay belt of Bombay showed that the highest concentration of Hg was in the small particulate range of <0.55 μm, deposited in the alveolar region (Mishra and Sadasivan, 1980).

Mercury content of stored grain using mercury tablets as a preservative increases with storage time and even a conservative estimate gives an average intake of Hg of 0.1–0.4 mg/day for those who consume wheat and rice stored with these tablets (Lalit and Ramachandran, 1977). The mercury content of foodstuffs, water and air at Bombay is given in Table 19.5. The calculated daily intake by the Bombay population of toxic elements is 11.7 μg Hg, 23.0 μg Cd and 12.1 μg As (for Hg and Cd the WHO fixed daily intake are 43 μg and 64 μg, respectively). On frying, the mercury burden of fish is reduced by 11–44% due to its volatility, while no change is observed for As and Cd (Anand, 1976a,b 1978a,b).

The bones and muscles of *Tilapia mozambique, Mugil dues umierie* and other varieties of fish from Thana and Mumbra creeks revealed concentrations greater than 0.5 ppm wet weight. In the Arabian sea, surface waters showed a range of 26–130 with an average of 78 ng/litre and at 10–100 metre depth the range was 0–204 with an average of 102 ng/litre. In the Bay of Bengal the surface showed 0–91 (average = 45) and at a depth of 10–100 m the range was 0–222 ng/litre. Hammer-headed shark and the white spot shark showed the highest concentrations of mercury (Qasim and Sen Gupta, 1980).

Analyses of samples of dust extracted from the lungs and lymph nodes of certain domestic animals from West Bengal, Orissa and Bihar provinces of India showed the presence of Hg and Pb in appreciable amounts. It was noted that the samples originating from West Bengal were generally richer in mercury and lead although the two elements were found in almost all samples derived from the coal belt and coal burning region of India (Dwivedi *et al.*, 1980; Krishna Murti, 1981). The Hg content of coal samples analysed in the USA ranged from 0.07 to 33 ppm (Bertine and Goldberg, 1971). The significance of mercury accumulating in the environment and entering the food chain of a variety of animals through burning of coal has been stressed (Falchuk *et al.*, 1977). Studies conducted with the fly ash captively collected from the Indra Prastha Estate in New Delhi have indicated 4–14 μg/gm of Hg in samples analysed over a year (VP Chest Institute, 1983).

There is no indigenous production of mercury in India and the annual requirement of 170–190 tonnes is entirely imported. There are no reliable data on the content of mercury in the ambient work environment associated with the use of mercury in the chloralkali industry and other heavy chemical and organic industries. The annual capacity for organomercurials permitted as fungicides is 91 tonnes but production in 1982–83 is reported to be 147 tonnes (Sarma, 1983). The caustic alkali industry, the main user of mercury in the electrolytic process, is well dispersed in India and big units are located on the banks of some major water bodies and the port towns of Dwaraka, Trombay, Greater Cochin, Madras and Tuticorinon. The difficulties of estimating the environmental diffusion into the major compartments are obvious. Coastal monitoring programmes have collected some useful information but much more remains to be done.

HUMAN EXPOSURE TO ARSENIC, CADMIUM, LEAD AND MERCURY IN INDIA

The available data on arsenic levels in water, food and air does not permit even tentative estimates to be made of the potential for human exposure. The figures suggested by the exploratory studies conducted in Bombay or Chandigarh on the daily intake of arsenic must be considered as tentative until more reliable data at a national level on the environmental levels of arsenic are generated. The use of arsenicals as therapeutic agents adds another degree of complexity to the problem. In the Chandigarh region, 29% of well-water samples, 50% of samples from handpumps and 37% of those from springs were found to be contaminated with arsenic (Datta and Kaul,

1976). There are reports of occurrence of non-cirrhotic portal hypertension in subjects consuming arsenic-containing medicine for several years (Datta, 1976). A number of clinical cases of idiopathic neuropathy reported by Chuttani *et al.* (1967) in Panjab have also been traced to the consumption of arsenic-contaminated opium (Datta, 1978a,b).

We can speculate on the outcome to human health of prolonged ingestion of arsenic through drinking water. Assuming 250 μg/litre as the mean content of water, a person who consumes 1500 ml of this water each day for 20 years will consume about 3 gm of arsenic only from water. It is reported in the literature that 3 to 25 grams of arsenic, when consumed over 1 to 22 years produces hepatic damage in the form of cirrhosis (Franklin *et al.*, 1956) and non-cirrhotic portal hypertension (Morris *et al.*, 1974). Arsenic ingestion is associated with cirrhosis in infants (Rosenberg, 1974) and adults (Luchtrath, 1972). From the data on the arsenic content of vegetables it is likely that the daily intake through this item of the diet may be about 70–100 μg. Most of the vegetables grown in Uttar Pradesh and Panjab were found to contain high amounts of arsenic, presumably derived from soil, water and manure (Datta and Kaul, 1978). One has to bear in mind the ability of humus to retain arsenic. Schroeder and Bolassa (1966) have reported that ash obtained on calcination of soil humus contains around 32 mg/100 gm of arsenic. Based on the data collected from Bombay on food, water and air, the daily intake of As in that city has been computed to be 12 μg (Anand, 1981).

Estimates of the daily intake of cadmium through food are given in Table 19.15 and the possible intake of cadmium through air at minimum and maximum ambient air pollution levels are given in Table 19.16. The intake through smoking cigarettes or beedies under Indian conditions has not been accounted for in exposure levels (Patel, 1980). From limited analytical data generated in Bombay the daily intake of cadmium from all sources works out to be 23 μg (Anand, 1981) as compared with the actual figure of 96 μg and calculated figure of 130 μg reported by Pandya (1978) from Ahmedabad.

Table 19.15 Daily intake (mg) of cadmium and lead through food. (Data from Pandya, 1978)

	India		USA
	Observed (based on cooked food)	Computed (based on calculation from content of raw food component)	
Cadmium	0.096	0.13	0.16
Lead	0.39	0.46	0.30

Table 19.16 Intake of cadmium and lead by man from air at minimum and maximum pollution levels. (Data from Pandya, 1978)

	Minimum pollution		Maximum pollution	
	Daily intake from air	%	Daily intake from air	%
Cadmium	0.10	0.10	0.6	0.62
Lead	1.40	0.36	29.9	7.15

Daily intakes of lead have been estimated by Pandya (1978) based on the lead levels in food, water and air. The daily intake is around 400 μg mostly through food and water, with only about 7.0% coming from air (Tables 19.15 and 19.16).

PERSPECTIVES

Increasing industrial development in India could lead to a significant increase in environmental arsenic, cadmium, lead and mercury. In order to assess the existing levels of ten metallic elements of environmental significance an Integrated Environmental Programme on Heavy Metals (IEPHM, 1980) has been initiated by the Department of Environment (Government of India). In Phase 1 of this programme six regional centres located at Chandigarh, Lucknow, Calcutta, Ahmedabad, Hyderabad and Madras have been identified to conduct a sample survey of Pb, Hg, Cd, Zn, Cu, Mn, Cr, Ni, As and Se in ambient air, water and food. The programme has been designed with due considerations for statistical variation and has an independent unit to monitor the quality of the analytical data generated at the six centres. The results of this first national coordinated study are to be available by early 1986.

As part of the Global Environmental Monitoring Services (GEMS) programme some limited experience has been gained to conduct biological monitoring of lead and cadmium in human blood and kidney cortex. Encouraged by this, a coordinated programme is currently being planned to conduct nationwide monitoring of lead and cadmium in the same six regional centres as indicated above. The per capita intake is very low compared with Japan, and, as such, this item is not likely to play a significant role in contributing to human exposure to toxic metals. However, with greater emphasis on the exploitation of the marine food resources with algae being considered as a human food source, there is need for greater vigilance in monitoring the quality of coastal waters and biota. The National Institute of Oceanography, Goa, and the newly created Department of Ocean Development (Government of India) have drawn up major plans for studies in this direction.Coastal pollution work which is being done with grants from the Department of Environment (Government of India) is expected to receive subtantial support to build up institutional infrastructure and expertise to keep a vigil on the 5000+ km coastline of India and the ecosystems associated with it.

Problems of bioaccumulation and biomagnification in the food webs have to be identified and investigated. It is proposed that an integrated programme on these aspects should be initiated as part of Phase II of the IEPHM funded by the Department of Environment likely to be commissioned during 1985–90.

India's contribution and participation in the biogeochemical cycles programmes sponsored by international agencies and scientific unions have been hitherto marginal. The national committee of SCOPE is currently addressing itself to the task of stimulating research activities in the major biogeochemical cycles including metals. Besides research, these envisage construction of inventories of toxic metals on a 'cradle to grave' basis, from their extraction and processing to diffusion in the environmental compartments and their impact on human health under the national Hazardous Wastes Control Programme. The Environmental Health Criteria documents on arsenic, lead, mercury and cadmium produced by international cooperation and released by WHO (WHO, 1976; 1977; 1981) do not reflect present situations appropriate to India and other countries with a high population density. Large numbers of people are likely to be exposed to toxic metals under situations totally different from those about which evidence has so far been used as the basis for the WHO documents. Hence, there is a need for more concerted studies on human exposure in the developing countries.

ACKNOWLEDGEMENTS

The author is grateful to the Department of Environment, Government of India, for financial support in coordinating the Integrated Environmental Programme on Heavy Metals, to all the principal investigators particularly Dr B.B. Chatterjee NOIH, Ahmedabad, Professor R. Nath, PGI, Chandigarh, Dr S.V. Chandra, ITRC, Lucknow, Dr J.C. Katyal, PAU Ludhiana, Dr U.C. Misra, V.P. Institute, Delhi, Dr A. Daniel, Joint Director, ZSI, Madras, for valuable information, reports and reprints. His thanks are due to the Indian National Science Academy, New Delhi for sponsoring his visit to Toronto, Canada to present this paper.

REFERENCES

Agarwal, A. L., Raiyani, C. V., Patel, T. S., Vittalmurthy, K. P. R., and Chatterjee, S. K. (1978). Role of traffic density and meteorology for air lead built up at crossroads of Ahmedabad. *Ind. J. Air. Poll. Cont.*, **1**, 86–94.

Agarwal, A. L., Raiyani, C. V., and Chatterjee, S. K. (1980). Auto-exhaust lead and its pollugenic effects. *Ind. J. Air Poll. Cont.*, **3**, 82–91.

Anand, S. J. S. (1976a). Determination of mercury in food by neutron activation. *Radiochem. Radioanal. Letters*, **27**, 313–320.

Anand, S. J. S. (1976b). Determination of mercury in drinking water. *Radiochem. Radioanal. Letters*, **27**, 43–50.

Anand, S. J. S. (1978a). Determination of mercury, arsenic and cadmium in fish by neutron activation. *J. Radioanal. Chem.*, **44**, 101–107.

Anand, S. J. S. (1978b). Mercury, arsenic and cadmium in the unfried and fried fish. *J. Radioanal. Chem.*, **47**, 93–97.

Anand, S. J. S. (1981). Studies on selected toxic metal contaminants in the environment. PhD Thesis, Bombay University, Bombay.

Annet, C. S., Fadow, M. P., Dikshi, F. M., and Stephenson, M. E. (1972). Mercury pollution and Lake Erie fishes. *Michigan Academician*, **4**, 325–337.

Bertine, K. K., and Goldberg, E. D. (1971). Fossil fuel consumption and major sedimentary cycle. *Science*, **173**, 233.

Chuttani, P. N., Chawla, L. S., and Sharma, T. D. (1967). Arsenic neuropathy. *Neurology*, **17**, 269.

Datta, D. V. (1976). Arsenic and non–cirrhotic portal hypertension. *Lancet*, **1**, 433.

Datta, D. V. (1978a). Arsenic adulteration in opium. *J. Assoc. Phys. India*, **26**, 1–5.

Datta, D. V. (1978b). Arsenic adulteration in opium. Arsenicosis—a real danger to health in developing countries. *Bulletin on Narcotics*, **29**, 41–44.

Datta, D. V., and Kaul, M. K. (1976). Arsenic content of drinking water in villages in northern India. A concept of arsenicosis. *J. Assoc. Phys. India*, **24**, 1–4.

Datta, D. V., and Kaul, M. K. (1978). Arsenic content of vegetables and soil in northern India. A concept of arsenicosis. *J. Assoc. Phys. India*, **26**, 395–398.

Despande, J. M., Dave, J. M., Jawanjal, R. B., and Yennawar, P. K. (1972). Lead contents of Bombay, Calcutta and Delhi air. *Ind. J. Occup. Health*, **15**, 12–19.

Dwivedi, K. K., Prasad, M. S., Rao, J. N., Dogra, R. K. S., Upreti, R. K., Ravi Shanker, Krishna Murti, C. R., Kapoor, S. S., Madan Lal, and Viswanathan, K. V. (1980). Trace elemental analysis of extracted dust from lungs and lymph nodes of domestic animals using X-ray fluorescence technique. *Intern. J. Environ. Anal. Chem.*, **7**, 205–221.

Falchuk, K. H., Goldwater, L. J., and Vallee, B. L. (1977). In McAuliffe, C. A. (Ed.), The biochemistry and toxicology of mercury. *The Chemistry of Mercury*, pp. 261–278, MacMillan Press Ltd, London.

Fondekar, S. P., and Reddy, C. V. G. (1979). Arsenic content in the coastal and estuarine waters around Goa. *Mahasagar*, **7**, 27–32.

Franklin, M., Bean, W. B., and Hardin, R. C. (1956). Fowler's solution as an actiologic agent in cirrhosis. *Am. J. Med. Sci.*, **219**, 589–596.

IEPHM (1980). *Integrated Environmental Programme on Heavy Metal Pollution*. Report of a Working Group. Department of Science and Technology. Government of India, New Delhi, Oct. 1980.

ITRC (1984). Industrial and Environmental Health. *A Bibliography of the Work done in India. Literature from 1945 through 1983*. Environmental Information System Industrial Toxicology Research Centre, Lucknow.

Jathar, V. S., Pandharkar, P. B., Pandya, V. K., Ravi, S. J., and Parameswarah, M. (1981). Intake of lead through food in India. *Bull. Rad. Protect*, **9**, 3–10.

Kansal, B. D., and Singh, J. (1983). Influence of the Municipal Waste Water and

Soil Properties on the Accumulation of Heavy Metals in Plants. *Int. Conf. Heavy Metals in the Environment*, Heidelberg—Sept. 1983. Vol. I, 413–416, CEP Consultants, Edinburgh.

Khandekar, R. N., Kelkar, D. N., and Vohra, K. G. (1980). Lead, cadmium, zinc, copper and iron in the atmosphere of Greater Bombay. *Atmos. Environ.*, **14**, 457–461.

Krishna Murti, C. R. (1981). *Biochemistry of Toxic Chemicals in Working and Living Environment. In Topics in Toxicology, 1981*, pp. 1–11, Keynote address Symposium on Toxicology Defense Science at Defense Research and Development Establishment, Gwalior, March 13, 1980.

Kumaraguru, A. K.; Selvi, D., and Venugopalan, V. K. (1979). Mercury in Vellar estuary and its toxicity to oysters. *Proc. Symp. Environ. Bull.*, 253–264.

Lahori, M. M. Ramprasad, G., Siamwala, R. S., and Sathwara, N. G. (1972). *Air Pollution by Lead Emission from Automobiles Exhaust in Ahmedabad City.* Seminar on Environmental Health Problems in Ahmedabad. March 11–12, 1972.

Lalit, B. Y., and Ramachandran, T. V. (1977). Concentration of mercury in wheat samples stored with mercury tablets as preservative. *J. Radioanal. Chem.*, **36**, 369–374.

Luchtrath, H. (1972). Cirrhosis of the liver in chronic arsenical poisoning of vinters. *German Med. Monthly*, **2**, 127.

Mishra, V. C., and Sadasivan, S. (1980). Gaseous and particulate air pollutant techniques, instruments and application. In Patel, B. (Ed.), *Management of Environment*, pp. 432–449, New Delhi Wiley Eastern Ltd.

Morris, J. S., Schmid, M., Newmen, S., Scheuer, P. J., Path, M. R. C., and Sherlock, S. (1974). Arsenic and non-cirrhotic portal hypertension. *Gastroenterology*, **66**, 86.

Nath, R. Lyall, V. Chopra, R., Prasad, R., Paliwal, V., Gulati, S., Sharma, S., and Chandan, R. (1982). Assessment of environmental pollution of cadmium in North India. *Bull. Post Grad. Inst.*, **16**, 202–208.

Pandya, C. B. (1978). Evaluation of occupational exposure to trace metals in some small and medium scale industries. PhD Thesis. Gujrat University, 1978.

Pandya, C. B., Patel, T. S., Parikh, D. J., Chatterjee, S. K., and Ramanathan, N. L. (1983). Environmental lead exposure as a health problem in India (An overview). *J. Environ. Biol.*, **4**, 127–148.

Patel, T. S. (1980). Studies on trace elements in the community environment. Ph. D. Thesis, Gujrat University, Ahmedabad.

Quasim, S. Z., and Sen Gupta, R. (1980). Present status of marine pollution in India. In Patel, B. (Ed.), *Management of Environment*, pp. 310–329, New Delhi, Wiley Eastern Ltd.

Raghavan, S. V. S. (1983). *MMT'C Crucial Role in Metallurgical Sector. The Hindu Survey of Indian Industry*, 1983, p. 59, Madras.

Rana, R. P., and Kansal, B. D. (1983). Sorption and release of cadmium in some sewage irrigated soils. *J. Ind. Soc. Soil Sci.*, **31**, 439–443.

Rawal, C. N., Pandya, C. B., Patel, T. S., and Parikh, D. J. (1982). *Environmental Exposure to Lead and Cadmium. A Selected Bibliography of Indian Publications from 1921 through 1982.* National Institute of Occupational Health, Ahmedabad and Indian Lead Zinc Information Centre, New Delhi.

Raymont, J. F. G. (1972). Some aspects of pollution in Southampton water. *Proc. Roy. Soc. London*, B 180, 451–468.

Rosenberg, H. G. (1974). Systematic arterial diseases and chronic arsenicism in infants. *Arch. Pathol.*, **97**, 360.

Sarma, P. R. (1983). *Imperatives in Pesticide Promotion. The Hindu Survey of Indian Industry*, p. 125, Madras.

Schroeder, H. A., and Bolassa, J. J. (1966). Abnormal trace metals in man: arsenic. *J. Chron. Dis.*, **85**, 106.

Singh, B., and Sekhon, G. S. (1977). Adsorption, desorption and solubility relationships of lead and cadmium in some alkaline soils. *J. Soil Sci.*, **28**, 271–275.

Singh, J., and Kansal, B. D. (1983). *The Accumulation of Cu, Zn, Cd and Pb in Sewage and Tubewell-irrigated Soils in Panjab.* National Conference on Lead, Zinc, Cadmium at Work Place: Environment and Health Care. New Delhi 14–15 Dec. 1982. Indian Lead Zinc Information Centre, New Delhi.

Statistical Outline of India (1982). Department of Economics and Statistics. Data Services Ltd, Bombay.

Subramanian, A. N. (1981). Some aspects of cycling of iron, manganese, copper, zinc and phosphorous in Pichavaram mangrove. PhD Thesis, Annamalai University, Chidambaram.

VP Chest Institute (1983). *Study of the Harmful Effects of Suspended Particulate Matter with Special Reference to Fly Ash.* Department of Biochemistry, VP Chest Institute, University of Delhi, Delhi.

WHO (1976). *Environmental Health Criteria. 1. Mercury.* Published under the Joint Sponsorship of the United Nations Environment Programme and the World Health Organization. World Health Organization, Geneva, 1976.

WHO (1977). Environmental Health Criteria. 3. Lead. Published under the Joint Sponsorship of the United Nations Environment Programme and the World Health Organization. World Health Organization, Geneva, 1977.

WHO (1981). Environmental Health Criteria. 18. Arsenic. International Programme on Chemical Safety. Published under the Joint Sponsorship of the United Nations Environment Programme, the International Labour Organization and the World Health Organization, World Health Organization, Geneva, 1981.

ZSI (1984). *The Effect of Pollution on Some Organisms in the Zooplankton, Benthos and Nekton contributing to the Food Chain in the Marine Environment.* Marine Biological Station, Zoological Survey of India, Madras, 1984.

Lead, Mercury, Cadmium and Arsenic in the Environment
Edited by T. C. Hutchinson and K. M. Meema

CHAPTER 20

Heavy Metal Pollution and the Need for Monitoring: Illustrated for Developing Countries in West Africa

M. A. OLADE
Geology Department,
University of Ibadan,
Ibadan, Nigeria

INTRODUCTION

Most of the countries within the West African sub-region are emergent nations which, for a long time, have been grouped among the Less Developed Countries of the world. Due to the low level of development, these countries have generally considered economic growth, social and educational development and industrialization as key development priorities, while protection of the environment has not been given the same importance.

However, in recent years, a few of these countries, such as Nigeria and Ivory Coast have achieved significant strides in their quest for rapid economic growth through industrialization. Thus, a number of factories, usually sited haphazardly, have developed, and with increased affluence generated by income from the export of natural resources, the problems of urbanization, population explosion and the increased use of automobiles have become very common.

It is well known that environmental pollution is a product of urbanization and technology, and other attendant factors of population density, industrialization and mechanization that serve to provide the necessities of the population. For example, in cities like Lagos (Nigeria) and Abidjan (Ivory Coast), the rural–urban migration activated by the search for increased incomes has resulted in the concentration of large populations in relatively small areas under poor conditions of sanitation. Traffic jams and the legendary 'go-slow' of automobiles are everyday occurrences in these cities.

The impact of pollution in the vicinity of overcrowded cities and from industrial effluents and automobile exhausts has reached a disturbing magnitude and is arousing public awareness. At present, relatively little data are available on the extent of environmental pollution because there are no agencies charged with the routine monitoring and protection of the environment.

This review chapter therefore focuses on the critical issues of heavy metal pollution in rapidly developing nations of West Africa and proposes avenues for effective monitoring.

SOURCES OF HEAVY METAL POLLUTION

The major sources of heavy metal pollution in urban areas of Africa are anthropogenic, while contamination from natural sources predominate in the rural areas. Anthropogenic sources of pollution include those associated with fossil fuel and coal combustion, industrial effluents, solid waste disposal, fertilizers and mining and metal processing. At present, the impact of these pollutants is confined mostly to the urban centres with large populations, high traffic density and consumer-oriented industries. Natural sources of pollution include weathering of mineral deposits, brush burning and windblown dusts.

Among the heavy metals, the most serious effect of pollution is presently associated with lead (Pb) emission.

Air Pollution

In West Africa, the emission of Pb from gasoline combustion by automobiles would account for about 80% of atmospheric pollution in urban areas, the remaining 20% being contributed from industrial point sources, refuse incineration and coal combustion by power plants. There has been a phenomenal growth in the use of automobiles within the last decade as a result of increased affluence. The level of Pb in supergrade gasoline in Nigeria is in the range of 0.7 gm/litre to 0.9 gm/litre which is higher than permissible levels in developed countries, where 0.4–0.5 gm/litre is common.

Preliminary studies show that the Pb emissions in air at ground level in Lagos (Nigeria) are far higher than for cities like London (UK) and New York, but similar to those of Brazil or the Caribbean due to perturbation by roadside traffic. Roadside soils and dust are also getting increasingly contaminated. In Lagos, Pb levels of 800 μg/gm have been obtained for dust in the densely populated areas of the city, while for roadside soils, values are in the range of 70–100 μg/gm. City roads are generally untarred, thus roadsides and adjacent houses are often very dusty. It is very common to find people eating raw fruits and food exposed at dusty roadsides. Moreover, children who play on roadsides and housefronts are often exposed to airborne emitted lead.

Other sources of atmospheric pollution are industrial emissions from fossil fuel combustion, and power plants, usually located within or at the precincts of the urban areas. Incineration of solid wastes is still the major mode of disposal in the urban areas, and this often releases pollutants into the air. Power plants using coal combustion are still common features in the cities.

Disposal of natural gas by flaring is widely practised in Nigeria. The city of Port Harcourt close to the oil-fields is lit brightly at nights by gas being flared in the surrounding areas. No data are available on the extent of metal pollution associated with this wasteful practice.

Water Pollution

The pollution of surface waters is significant, though on a local scale. However, the heavy runoff associated with perennial rainfall may mitigate the impact. Several small and heavy industries involved in activities such as battery and paint manufacturing, petroleum refining, cement and ceramic production, steel production, are now being located haphazardly, mostly near metropolitan centres. No centralized sewage systems exist, and the industrial effluents from the factories are usually discharged untreated into streams, lagoons, open drains and other water bodies. This has resulted in high, although not alarming, levels of lead, cadmium and copper in some localized areas.

Studies of heavy metal contents of surface waters at Ibadan show that levels of Pb may reach 50 μg/litre (Mombeshora *et al.*, 1983). In waters of the Lagos lagoon, Pb levels exceeding 120 μg/litre have been obtained (Olade unpublished data). It is a common practice to dump solid municipal waste into streams traversing urban areas. Because of the natural acidity (pH) of these waters and oxidation, heavy metal contaminants are being released into stream waters.

Although surface water pollution has often been assumed to be a minor problem in developing countries of tropical humid regions, it has become clear that our freshwater resources are limited and care should be taken to protect them. There is an urgent need to conduct environmental impact assessment studies before industries are located, so as to ameliorate environmental pollution from point sources.

Mining and smelting operations associated with base metals are very limited in occurrence. However, around the few lead–zinc and tin–zinc mines in Nigeria, acidic drainage from disused mine shafts and open pits and dumps have contributed to localized metal pollution (Olade, see this volume, Ch. 18).

As a result of the apparent shortage of potable water from municipal supply, many homes and industries have resorted to sinking wells and boreholes. Many of these sources of groundwater are sited close to domestic sewage tanks and solid waste dumps where they are likely to be contaminated. Water quality tests have shown that groundwaters, particularly those tapped from weathered regolith are acidic, which may promote oxidation and corrosion of steel casings and screens. There is a likelihood that heavy metals are being released into such waters although there is presently very little data available on groundwater quality.

Soil Pollution

The sources of soil pollution are quite similar to those for water pollution. However, in the quest for improved agricultural productivity and the need to rid crops of pests, fertilizers and pesticides are being applied indiscriminately in some countries of the sub-region. For example, Nigeria imported about one billion tons of commercial fertilizers in 1983, of which the heavy metal contents are unknown. It is well known that fertilizers may contain up to 400 ppm Pb which may constitute a major source of soil pollution, although phosphate lead is not often readily available to plants (Nriagu, 1978). Phosphate fertilizers also very often contain cadmium, which can be a contaminant of agricultural land and of crops.

It is not yet a common practice to apply sewage sludges as fertilizers primarily because of the lack of centralized municipal sewage systems. However, municipal composts and farmyard manure are being increasingly applied to croplands on a local scale. Heavy metal contents of municipal composts are similar to those of sewage sludges (up to 5000 ppm Pb) (Nriagu, 1978), and their applications may increase metal contents of farm soils.

An important problem associated with tropical weathering that has considerable implication for the susceptibility of soils to pollution, is the development of acid soils with low buffering capacities. Such soils lack one or more nutrients due to leaching, soil erosion and extractive farming practices. The soil acidity promotes sorption of toxic metals in both plants and soils and involves high levels of soluble aluminum. There is therefore considerable danger in the future if acid rains should develop.

Natural Sources

There are virtually no data available on the contributions of natural sources to environmental pollution. Critical issues include the amount of heavy metals released into the environment during brush burning which is a common feature of extractive farming practices. Also, there are questions of metal release into air and soils from the accumulation of litter in tropical rain forest regions. As a matter of fact, not much is known about the biogeochemical cycles of heavy metals in tropical ecosystems.

During the harmattan season in West Africa (December–March), several million tons of wind-blown dust from the Sahara are deposited in the coastal areas. Studies are still in progress on the heavy metal contents of the Saharan dust.

HUMAN HEALTH EFFECTS

Developing countries are generally preoccupied with immediate and more pressing problems of providing basic health care to their people, and with the eradication of major diseases amongst the large populations. They are not likely to worry about the health effects of metal pollution. Also, the medical manpower is still grossly inadequate to even maintain basic health services.

Nevertheless, a few minor incidences of metal pollution have occurred and public awareness of the apparent danger of metal poisoning is increasing. There was a recent incidence in Ghana of mercury poisoning from the consumption of dressed seeds preserved for planting but sold for consumption. Galena is widely used in some parts of Nigeria as mascara by women and may constitute a health hazard. Children run the greatest risk of lead poisoning in inner city areas where inhalation and ingestion of metals like Pb and Cd from dust and ashes of household items, such as paints and batteries, are common.

ENVIRONMENTAL MONITORING

Monitoring and systematic gathering of information on heavy metal levels in the environment are essential components of any pollution-control system. The establishment of such a system often presupposes the existence of minimum pollution standards and regulations. Most of the countries within the sub-region do not have such control standards, and environmental impact studies are often not required prior to the establishment of industrial projects.

Although some government agencies in these countries have some experience with monitoring on a limited scale, none have developed a comprehensive monitoring system. This may be attributed to the fact that economic growth and industrialization, without strings attached, are their main priorities of development. Moreover, there are no sufficient technical and manpower resources to maintain effective monitoring and pollution-control.

Despite these limitations, it will be disastrous if the developing countries make the same mistakes as the developed nations by sacrificing the environment at the expense of economic development. Because of the apparent lack of accurate information on the levels of heavy metals in the environment, it is most likely that pollution from point sources is more serious than documented. It is therefore essential to establish pollution standards and effective monitoring before environmental deterioration sets in.

Considering the fact that governments are the major sources of funding for most activities, the participation of government agencies is essential in the successful execution of any scheme of environmental monitoring. Thus, the first step in any such programme is to establish an agency to carry out these activities. Where the necessary expertise is lacking, assistance could be obtained from international agencies of the United Nations, such as UNEP and FAO, particularly in the establishment of laboratories and training of technical staff.

Research centres in universities and related institutions should also be encouraged to undertake systematic studies on specific problems related to environmental pollution. It is only by the coordination of efforts of government agencies and research-oriented institutions that the limited resources within these countries can be most efficiently utilized.

CONCLUSIONS

As developing countries of West Africa become industrialized and urbanized, heavy metal pollution is likely to reach disturbing levels. These countries should learn from the mistakes of the developed nations and recognize that rapid deterioration of the environment can occur. The critical issues however, are that preparations are not being made towards the protection of the environment.

There is a conspicuous lack of data on the nature and extent of metal pollution either at local or regional levels, particularly to assist in the understanding of metal cycling in the environment. No systematic studies are being carried out to examine the dynamics of tropical ecosystems.

Although most countries in the sub-region recognize the need to combat pollution, environmental controls are either non-existent or inadequate. Most industries discharge effluents into the environment without any prior treatment, and the manufacture of 'pollution-intensive' products are being shifted to the developing countries where strict controls do not exist. Even where controls exist on paper, there are difficulties and inadequate resources to enforce them.

The establishment of comprehensive monitoring systems and information gathering should be given priority by governments of the developing countries in the sub-region with support and encouragement from international agencies.

REFERENCES

Mombeshora, C., Osinbanjo, O., and Ajayi, S. O. (1983). Pollution studies on Nigerian rivers: the onset of lead pollution of surface waters in Ibadan. *Environ. Int.*, **9**, 81–84.

Nriagu, J. O. (1978). Lead in soils, sediments and major rock types. In Nriagu, J. O. (Ed.), *The Biogeochemistry of Lead in the Environment*, pp. 15–72, Elsevier/North-Holland.

Lead, Mercury, Cadmium and Arsenic in the Environment
Edited by T. C. Hutchinson and K. M. Meema

CHAPTER 21

Monitoring of Heavy Metals—A Proposed Strategy for Developing Countries

PATRICK DENNY

Aquatic Biology Research Unit,
School of Biological Sciences,
Queen Mary College, The University of London,
London, England E1 4NS.

INTRODUCTION

Emissions of toxic wastes from automobiles, mining, electrical power stations, heavy industry, large incinerators and from many other local point sources, is a worldwide problem. Protection of the environment is a global responsibility. However, at the Scientific Committee on Problems of the Environment (SCOPE) Workshop on Metal Cycling in the Environment (Toronto, September, 1984), it became clear that data came mainly from industrialized nations. Only limited amounts of information on the heavy metals of concern at the workshop, namely, lead, arsenic, cadmium and mercury, are yet available from the developing world.

There are several reasons for this lack of information. Firstly, in most developing countries, the degree of industrialization, and consequently the volumes of emissions from coal-fired power stations and heavy industry, are generally not at a level which give rise to obvious, widespread damaging effects upon human populations and natural environments. Nevertheless, localized and sometimes severe cases of pollution (e.g. from large cities and from mining and smelting areas) do occur. Secondly, countries of limited resources which have predominantly subsistence economies and which are ravaged by droughts, floods, disease and malnutrition, must consider the amelioration of immediate distress to their people as their first priority. For the present, food requirements, basic education, health care, and communication claim preference over environmental matters. Monitoring of, and protection from, pollution competes for these resources. Thirdly, within many Third World countries there is a lack of expertise and facilities to undertake the type of analysis and data-base assessment required.

THE PROPOSAL

Clearly the Third World is not yet in a position to participate fully in monitoring programmes. To overcome this I should like to propose the following strategy:

1. A three-tier system for monitoring be established, each tier catering for a particular need and incorporating an appropriate technique for sample collection, preparation and analysis.
2. Special books of techniques be prepared so that a uniformity of methods be established for purposes of international compatibility.
3. International collaboration and financial aid be sought to support this scheme. Such aid is essential for the establishment of functional laboratories, the preparation of appropriate books on methodology, the coordination of programmes of investigation and monitoring, and the training of personnel.
4. An overall coordinating body be established to optimise resources and alleviate wastage.

THE REASONING FOR THE ESTABLISHMENT OF A THREE-TIERED SYSTEM OF MONITORING

The monitoring system would consist of three levels of precision and accuracy.

First Level

Simple techniques sufficient to identify potentially detrimental concentrations of metals in the environment are required: Methods would be relatively cheap with a low degree of precision. Examples of target areas include large cities and industrial areas, and areas down-wind of mines, smelters and coal-fired power stations. The aquatic systems to be monitored would include rivers and lakes supporting major populations and/or industries, sewage outflows, and industrial and mining effluents.

Second Level

This is an intermediate level with a higher degree of precision required for the better understanding of metal cycling. For example, investigations may be needed to distinguish between different species of metals and their transfer in the food chain, or to know more accurately the concentrations of metals for environmental and human health purposes. The use of standard reference material would be introduced at this level.

Third Level

This employs the highest degree of precision and accuracy in the assay of heavy metals. Quality control would be most stringent. Both internal and external standard samples would be included in batch sample analyses. Cross-checking with other laboratories would be a matter of standard practice.

The relatively simple techniques proposed for the First Level would make it particularly suitable for use in undergraduate classes at universities and technical colleges. It would be an appropriate level for adoption initially by government laboratories (local and nation-wide) for their routine monitoring programmes. The advantage of this standard of analysis is threefold, namely, (i) it would educate a larger number of people in concepts of environmental protection and give them some practical knowledge, (ii) it would provide a pool of students with suitable qualifications for further training and, (iii) members of the public would become more familiar with their environment and become more aware of possible sources of pollution. It may also help local or central governments to pinpoint areas requiring more detailed study and populations or environments particularly at risk.

The Second Level would be more appropriate for use in government and university laboratories where fundamental research is carried out. Overall programmes of investigation made up of individual research projects would need to be identified carefully to provide the most suitable facilities.

The results from the Third Level analyses would provide the base-line data for monitoring programmes which could be used with confidence in global assessments. The establishment of environmental protection agencies and suitable laboratories in each country would be required in order to coordinate monitoring within and between regions.

PREPARATION AND DISTRIBUTION OF BOOKS

A basic necessity for the execution of this scheme is the preparation of handbooks of agreed techniques which provide details on methods of sample collection and analysis of heavy metals at each of the three levels. The books should be as comprehensive as possible, be underpinned by the necessary theory, and be distributed free of charge (or at subsidized costs) by international agencies. The organization of air monitoring networks, for example, in developing countries was addressed in a publication of the World Meteorological Organization (WMO, 1977).

INTERNATIONAL FUNDING

This proposal can only become a reality through international collaboration and financial aid. This would support educational and exchange schemes for training personnel, and establish functional laboratories in appropriate countries. The International Council of Scientific Unions (ICSU) and SCOPE, which is a sub-group and through which the current proceedings have evolved, are organizations which clearly should be involved. United Nations bodies (e.g. FAO, UNESCO, UNEP and UNDP) which are responsible for aid programmes to developing countries possess the infrastructure and potential to support training schemes and develop the necessary facilities, while the International Atomic Energy Agency, Vienna (IAEA) and the International Union for the Conservation of Nature (IUCN) are widely experienced in monitoring the environment. Scholarships and Training and Travel Awards from industries and organizations within the developed countries would provide another valuable contribution.

THE COORDINATING BODY

It is essential that an overall coordinating body for these programmes be set up. Such a body would help to optimize resources and alleviate overlap and wastage which all too frequently occurs when funds from a range of independent organizations are involved. They would establish priorities and could liaise with current monitoring centres, such as the Monitoring and Assessment Research Centre in London (MARC), the Institute for Environmental Studies at the University of Toronto, Canada or the Canadian Centre for Inland Waters (CCIW) at Burlington, Ontario, Canada, for a uniformity of procedure worldwide.

CONCLUSION

Specialist training for suitably qualified scientists, and the establishment of laboratories for metal analyses is desperately needed in many developing countries. In the long term, shortages of personnel can be ameliorated through education where projects using relatively simple methods of analysis are introduced into colleges and universities. Steadily, general knowledge would be increased, research would be encouraged and the core of local expertise would be strengthened. A reliable monitoring programme would become established and the public would become more aware of their environment. This offers the hope that some of the misfortunes of pollution perpetuated in the industrialized countries will not be repeated in the Third World.

REFERENCES

WHO GEMS (1977). *Air Monitoring Programme Design for Urban and Industrial Areas.* Pub. for Global Environmental Monitoring System by UNEP, WHO and WMO.

Index